Adhesion Aspects in
MEMS/NEMS

Edited by

S. H. Kim, M. T. Dugger and K. L. Mittal

CRC Press
Taylor & Francis Group
Boca Raton London New York

CRC Press is an imprint of the
Taylor & Francis Group, an **informa** business

First published 2010 by VSP

Published 2019 by CRC Press
Taylor & Francis Group
6000 Broken Sound Parkway NW, Suite 300
Boca Raton, FL 33487-2742

© 2010 by Taylor & Francis Group, LLC
CRC Press is an imprint of Taylor & Francis Group, an Informa business

First issued in paperback 2019

No claim to original U.S. Government works

ISBN 13: 978-0-367-44594-2 (pbk)
ISBN 13: 978-90-04-19094-8 (hbk)

Visit the Taylor & Francis Web site at
http://www.taylorandfrancis.com

and the CRC Press Web site at
http://www.crcpress.com

Contents

Part 3: Adhesion and Friction Measurements

Part 4: Adhesion in Practical Applications

Part 5: Adhesion Mitigation Strategies

Preface

Interfacial interactions between components play a crucial role in the manufacture and performance of MEMS/NEMS devices. This is ascribed to the very large surface-to-volume ratio in such components and, concomitantly, surface/interface phenomena become dominant in controlling the fate of such devices.

Adhesion is an interesting phenomenon. Depending on the situation/application, adhesion may be desideratum and in other situations, it can be an anathema. In the case of MEMS/NEMS devices, static adhesion between components (also known as 'stiction') is something to be avoided for the proper functioning of the device. Avoidance of adhesion constitutes the field of 'abhesion' which in a sense is 'negative adhesion'. In order to achieve abhesion, a plethora of techniques and materials have been developed to mitigate adhesion problems in a wide range of MEMS/NEMS products.

Even a cursory look at the literature will evince that currently there is tremendous R&D activity encompassing many facets (including adhesion) pertaining to MEMS/NEMS products and all signals indicate it will not only continue unabated, but will assume an accelerated pace. The current research emphasis is on the following topics: unraveling interfacial interactions and factors influencing such interactions; ramifications of these interactions in the functioning of devices/structures; developing novel or ameliorating the existing techniques for surface modification to attain desired surface characteristics; and development of various ways to mitigate adhesion problems.

In light of the tremendous relevance of interfacial interactions — and thus adhesion — and flurry of R&D activity in this burgeoning field of MEMS/NEMS products, we decided to make this book available as a single and easily accessible source of comprehensive information. This book is based on the Special Issue of the *Journal of Adhesion Science and Technology* (*JAST*) Vol. 24, Nos 15–16 (2010). The papers as published in the above-mentioned Issue have been grouped in a logical fashion in this book.

This book containing a total of 21 papers (reflecting overviews and original research) and covering many aspects of adhesion in MEMS/NEMS is divided into five parts as follows: Part 1: Understanding Through Continuum Theory; Part 2:

Computer Simulation of Interfaces; Part 3: Adhesion and Friction Measurements; Part 4: Adhesion in Practical Applications; and Part 5: Adhesion Mitigation Strategies. Topics covered include: numerical analysis of contact mechanics; equilibrium vapor adsorption and capillary force; contribution of fractal parameters to adhesion; effects of contacting surfaces on MEMS device reliability; adhesion model for micromanipulation; computer simulation of interfaces; vapor phase lubrication in MEMS; atomistic factors governing adhesion; adhesion and friction aspects at the nanoscale; friction of self-assembled monolayers (SAMs); interfacial adhesion and its implications in MEMS/NEMS technology; adhesion in MEMS/NEMS applications; molecular mobility and interface dynamics in organic NEMS; various adhesion mitigation techniques in MEMS/NEMS; superhydrophobic surfaces; plasma modification of polymer surfaces and its relevance in biomedical microdevices.

It is quite patent from the topics covered that many different aspects of adhesion in MEMS/NEMS are accorded due coverage in this book; concomitantly, this book represents a comprehensive treatise on this fascinating, mushrooming and technologically highly important field. Also we would like to point out that this book containing a wealth of information is the first book on the topic of adhesion in MEMS/NEMS. Moreover, we hope this book would serve as a fountainhead for new research ideas and new application vistas will emerge as the performance and durability/robustness of MEMS/NEMS are further enhanced.

This book should be of interest to both neophytes (as a gateway to this field) and veteran researchers as a commentary on the current research activity being carried out by luminaries in this field. An in-depth understanding of adhesion phenomena and development of more effective adhesion mitigation strategies would be a big step in the future of MEMS/NEMS technology.

Acknowledgements

Now it is our pleasure to acknowledge all those who helped in many and varied ways in materializing this book. First and foremost, we profusely thank the authors for their interest, enthusiasm and cooperation, without which this book could not be born. Second, we are most appreciative of the reviewers for their time and efforts in providing invaluable comments which certainly improved the quality of manuscripts. Next and last, our appreciation goes to the staff of Brill (publisher) for giving this book a body form.

SEONG H. KIM
The Pennsylvania State University
Department of Chemical Engineering
University Park, PA 16804-4400, USA

MICHAEL T. DUGGER
Materials Science and Engineering Center
Sandia National Laboratories
Albuquerque, NM 87185-0889, USA

K. L. MITTAL
P.O. Box 1280
Hopewell Jct., NY 12533, USA

Part 1

Understanding Through Continuum Theory

Numerical Analysis of Contact Mechanics between a Spherical Slider and a Flat Disk with Low Roughness Considering Lennard–Jones Surface Forces

Kyosuke Ono [*]

Mechanical Engineering Research Laboratory, Hitachi Ltd., Kirihara-cho 1, Fujisawa-shi, Kanagawa-ken 252-8588, Japan

Abstract

Although analytical and numerical analyses of the contact mechanics of a completely smooth sphere–flat contact have been done, the analysis of a realistic sphere–flat contact with a surface roughness whose mean height planes have a spacing greater than the atomic equilibrium distance has not been done thoroughly. This paper is a fundamental study of the elastic contact mechanics due to Lennard–Jones (LJ) intermolecular surface forces between a spherical slider and a flat disk with low roughness whose height is larger than equilibrium distance z_0. First, neglecting the effect of the attractive force at contacting asperities, adhesion contact characteristics of a 2-mm-radius glass slider with a magnetic disk are presented in relation to the asperity spacing σ between mean height planes. Results showed that the contact behavior at a small asperity spacing of ~0.5 nm cannot be predicted either by the Johnson–Kendall–Roberts or Derjaguin–Muller–Toporov theories. Second, contact characteristics of a 1-μm-radius sphere on a flat disk are presented to examine how LJ attractive force at contacting asperities can be evaluated. It was found that the adhesion force of contacting asperity is a function of separation in general, but it becomes almost constant when $\sigma = {\sim}z_0$. A simple equation to evaluate the LJ attractive pressure of contacting asperities is presented for the rough contact analysis. Third, numerical calculation methods for a sphere–flat contact including LJ attractive forces between the mating mean height planes and contacting asperities are presented. Then, adhesion characteristics of a 2-mm-radius glass slider and magnetic disk are calculated and compared with the previous experimental results of dynamic contact test. It is shown that the calculated LJ adhesion force is much smaller than the experimental adhesion force, justifying that the adhesion force observed at the separation of contact is caused by meniscus force rather than by vdW force.

Keywords

Nanotribology, sphere to flat contact mechanics, van der Waals forces, Lennard–Jones intermolecular forces, roughness effect, head–disk interface

---

[*] Tel.: +81-0466-98-4185; Fax: +81-0466-98-2119; e-mail: kyosuke.ono.jk@hitachi.com

Adhesion Aspects in MEMS/NEMS

1. Introduction

Adhesion contact characteristics between a perfectly smooth sphere and a flat due to intermolecular forces caused by Lennard–Jones potential have been elucidated well by analytical methods [1–4] and numerical calculation methods [5–7]. However, realistic engineering surfaces have roughness that is larger than the atomic equilibrium distance, even for magnetic head–disk interfaces. Adhesion and friction characteristics of nominally flat rough surfaces on the basis of the Greenwood–Williamson roughness contact model [8] and the Derjaguin–Muller–Toporov (DMT) adhesion model [2] were studied by Chang *et al.* [9] and Stanley *et al.* [10]. As an extension of these theories, Suh and Polycarpou [11] analyzed adhesion characteristics of rough surface contacts for nominally flat mating surfaces. In an asperity contact analysis, Lennard–Jones (LJ) surface forces at contacting and non-contacting asperities were taken into account, but LJ surface force due to mean height planes was not considered. However, it is natural to consider that LJ surface force due to mean height planes plays a dominant role when the spacing between the mean height planes is less than 0.5 nm. In addition, since a current magnetic head slider has an ellipsoidal surface, contact characteristics between a sphere and a flat disk should be analyzed. Even in an experimental measurement of contact characteristics using a simple spherical probe, the probe has a roughness with a height that is usually larger than the atomic equilibrium distance of $\sim$0.2 nm [12]. Thus, the adhesion characteristics between the spherical probe and flat disk would be influenced by the surface roughness. Therefore, contact characteristics between a sphere and a flat with a small roughness have not yet been studied in the history of contact mechanics.

In the field of head–disk interfaces, there has been a long-running argument about whether the attractive force between head and disk is caused by van der Waals (vdW) force [13–15] or meniscus force [16–18]. Ono and Nakagawa [19, 20] measured the dynamic adhesion force that was applied on the slider when 1- and 2-mm-radius glass sliders collided with a magnetic disk with a molecularly thin lubricant layer. They concluded that the adhesion force observed at the instant of separation after a short contact period of 15–30 μs was caused by the meniscus force rather than by vdW force. However, there were criticisms that the meniscus bridge could not be generated quickly enough to have this effect.

One of the motivations of this study was to calculate the vdW attractive force between a spherical glass slider and a magnetic disk and compare it with the experimentally measured value. The glass sphere and magnetic disk used for the experiment had a root-mean-square (rms) roughness heights of $\sim$0.33 and 0.52 nm, respectively. Since this is larger than the atomic equilibrium distance, development of an analytical method for asperity contact mechanics between a sphere and a disk is needed.

A numerical analysis method is presented for rough surface contact characteristics between a sphere and a flat, considering not only elastic deformation and vdW forces of mean height planes of the sphere and disk but also vdW forces of contacting asperities. In prior papers on adhesion of contacting rough surfaces [9–11], adhesion forces of contacting and noncontacting asperities were taken into account based on the DMT theory [2, 3]. In order to elucidate adhesion force of a sphere–flat contact with various scales of asperities more rigorously, fundamental contact characteristics of a 2-mm-radius glass sphere and 1-µm-radius asperity with a disk assuming a constant spacing between mean height planes due to contacting asperities are investigated using a numerical calculation method. In Section 3, an analytical model of asperity contact between a sphere and a disk and the assumptions used in this analysis are described. Then, basic equations and numerical calculation method are explained. In Section 4, contact characteristics between the 2-mm-radius sphere and disk are calculated assuming constant spacing between mean height planes in the contact area but ignoring vdW forces of the contacting asperities. In Section 5, to evaluate the adhesion force of contacting asperities of the rough surface, the contact characteristics between a 1-µm-radius spherical asperity and a disk are calculated assuming small-scale spacing due to a small-scale asperity height on the order of atomic equilibrium distance. From this analysis, it is found that $2\pi R \Delta \gamma$ approximates the attractive force at a contacting asperity when the small-scale asperity height is close to atomic equilibrium distance. In Section 6, using this adhesion force of a contacting asperity, an approximated numerical analysis method of rough surface contact between a sphere and a disk combining asperity vdW pressure with vdW forces between mean height planes is presented. The contact characteristics for the 2-mm-radius glass slider and magnetic disk are calculated using asperity parameter values measured in the experiment and compared with experimental ones. It is shown that the overestimated LJ attractive forces yield an adhesion force much smaller than the measured values.

2. Nomenclature

A: Hamaker constant (J)

E^*: Composite Young's modulus of two mating surfaces (Pa)

F_{el}: Elastic contact force of spherical slider (N)

F_{LJ}: Lennard–Jones attractive force (N)

F_{LJa}: Lennard–Jones attractive force due to contacting asperities (N)

F_{ex}: External force applied to disk surface by slider (N)

R: Radius of curvature of spherical slider (m)

R_a: Radius of curvature of asperity (m)

$O\text{-}r\theta z, O\text{-}s\varphi z$: Cylindrical coordinate system

P_{el}: Elastic contact pressure (Pa)

P_{ela}: Elastic contact pressure due to contacting asperities (Pa)

P_{LJ}: Lennard–Jones attractive pressure due to mean height planes (Pa)

P_{LJa}: Lennard–Jones attractive pressure due to contacting asperities (Pa)

a: Analytical radius of contact area (m)

$b(r)$: Surface contour of spherical slider (m)

d: Separation of spherical slider from mean height of undeformed disk surface (z-position of tip of spherical slider) (m)

$h(r)$: Spacing between spherical slider and mean height plane of disk surface (m)

h_0: Spacing between tip of spherical slider and mean height plane of disk surface (m)

r_{c}: Contacting radius between spherical slider and disk (m)

$w(r)$: Deformation distribution of disk (m)

$w_{\text{G}}(r)$: Deformation distribution of disk due to a unit force (Green function) (m/N)

z_0: Atomic equilibrium distance (m)

$\Delta\gamma$: Change in total surface energy due to contact (J/m^2)

σ: Spacing between mating mean height planes due to roughness asperities (m)

σ_{a}: rms asperity height (m)

ρ: Asperity density of rough surface (m^{-2})

3. Analytical Model and Numerical Calculation Method

3.1. Analytical Model and Basic Equations

Figure 1 shows the analytical model for asperity contact mechanics analysis of the spherical slider and flat disk treated in this paper. Since the geometries of the sphere and the disk are axisymmetrical, cylindrical coordinate system O-$r\theta z$, (O-$s\varphi z$) is fixed on the mean height plane of the original disk surface without considering deformation. The radius of the spherical slider is denoted by R. The z-position of the tip of the spherical slider is denoted by d, defined as separation. As is common in asperity contact mechanics, elastic and surface roughness properties of the spherical slider are included in those of the disk, and the slider is assumed as a smooth, rigid sphere. The disk has a composite Young's modulus and a composite roughness. It

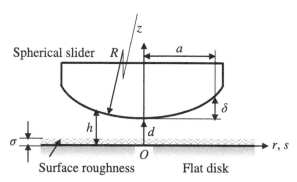

Figure 1. Analytical model of a spherical slider and a flat disk.

is assumed that the mating mean height planes are separated by σ in the contact region after asperity is compressed due to contacting force.

If we denote the observed position by (r, θ) and applied force position by (s, φ), disk deformation caused by a unit force is given by the following Green function [21]:

$$w_G(r, s) = \frac{1}{\pi E^* \sqrt{r^2 + s^2 - 2rs \cos(\theta - \varphi)}}. \tag{1}$$

Here, E^* is the composite Young's modulus given by

$$\frac{1}{E^*} = \frac{1 - v_1^2}{E_1} + \frac{1 - v_2^2}{E_2}, \tag{2}$$

where E_1 and E_2 are Young's moduli, and v_1 and v_2 are Poisson's ratios of the disk and the slider, respectively. Lennard–Jones (LJ) pressure acting on the mean height planes with spacing $h(r)$ is given by

$$P_{LJ}(h) = \frac{A}{6\pi h^3}\left\{1 - \left(\frac{z_0}{h}\right)^6\right\}. \tag{3}$$

Here, A is the Hamaker constant and z_0 is the atomic equilibrium distance [5]. Note that since z_0 is very small compared to the asperity height of the spherical glass slider and magnetic disk, P_{LJ} is almost equal to vdW pressure, ignoring the second term inside the braces. Since the z-position of the tip of the spherical slider is denoted by d, the contour $b(r)$ of the slider surface in the z-direction is written as

$$b(r) = d + \frac{r^2}{2R}. \tag{4}$$

Note that if $d < \sigma$, the slider penetrates into the disk through the asperity contact. If we denote the elastic deformation of the mean height plane of the disk surface by $w(r)$, the spacing $h(r)$ between the slider and disk is given by

$$h(r) = b(r) - w(r). \tag{5}$$

When the slider does not contact the disk, the disk deformation $w(r)$ is given by

$$w(r) = \iint P_{LJ}(h(s))w_G(r, s)s \, ds \, d(\varphi - \theta). \quad (6)$$

Under a non-contact condition, disk deformation can be determined by simultaneously solving the discretized equations from (1) to (6). Since the spacing $h(r)$ is altered by deformation $w(r)$, a convergent solution can be obtained through iteration process.

3.2. Numerical Analysis Method for Contact Characteristics

When the separation d decreases, the slider comes in contact with asperities on the disk surface. If we denote LJ pressure due to contacting asperities and elastic contact pressure due to contacting asperities by P_{LJa} and P_{ela}, respectively, the disk deformation is written as

$$w(r) = \iint (P_{LJ} + P_{LJa} - P_{ela})w_G(r, s)s \, ds \, d(\varphi - \theta). \quad (7)$$

In this paper, P_{LJa} and P_{ela} are not treated rigorously but are estimated by an approximation method.

As the separation d decreases further, asperity height is decreased due to the increased contact pressure. However, the asperity becomes too hard to be compressed due to the increased rigidity of the asperities and deformation of mean height plane becomes predominant [18]. This compressed asperity height is represented by the asperity spacing σ between the slider and the disk. This asperity spacing σ is still larger than z_0 in almost all actual cases. If we assume that σ is uniform in the contact area, the spacing $h(r)$ can satisfy the following inequality:

$$h(r) = b(r) - w(r) \geqslant \sigma. \quad (8)$$

When asperity density is not large and $\sigma < 1$ nm, P_{LJa} is smaller than P_{LJ}, as will be shown later. Therefore, at first we ignore P_{LJa}; the effect of P_{LJa} will be taken into account in Section 6.

When the mating surfaces come in contact with each other with an asperity spacing σ, P_{ela} is transmitted to the mean height surface resulting in mean surface deformation. Since the disk deformation is caused by the penetration of the rigid spherical surface, it will be reasonable to assume that P_{el} can be given by Hertzian contact pressure associated with mean plane deformation with contact radius of r_c, as follows:

$$P_{el}(r) = p_m\left\{1 - \left(\frac{r}{r_c}\right)^2\right\}^{1/2}, \quad p_m = \frac{2E^* r_c}{\pi R}. \quad (9)$$

Therefore, disk deformation is given by

$$w(r) = \iint (P_{LJ} - P_{el})w_G(r, s)s \, ds \, d(\varphi - \theta). \quad (10)$$

The reason for the minus sign on P_{el} is that elastic contact pressure acts in the $-z$ direction. Under the contact condition, a solution that simultaneously satisfies equations (1), (3), (4), (8), (9) and (10) must be obtained.

To perform the integration in equations (6) and (10), w_G was expanded into a Taylor series with respect to $k = 2rs/(r^2 + s^2)$, and then each expanded term was integrated by $d(\varphi - \theta)$ in the circumferential direction. The number of expanded terms sufficient to give a precise solution was examined by solving the Hertzian contact deformation caused by the Hertzian contact pressure, equation (9). Since a precise Hertzian contact deformation can be obtained using 1000 terms, 2000 terms of an expanded Taylor series were used for the calculation of the influence coefficient to save computing time. Integrations in equations (6) and (10) with respect to radial coordinate were performed numerically by expressing the continuous quantity by N representative points along the radial coordinate.

Figure 2 is a flowchart for numerical calculation. First, the initial separation value d is assumed. When $d > \sigma$, equation (6) is solved. Then, whether or not the disk deformation can satisfy the non-contact condition is examined. If the disk and slider do not contact each other, disk deformation and spacing are solved iteratively as a

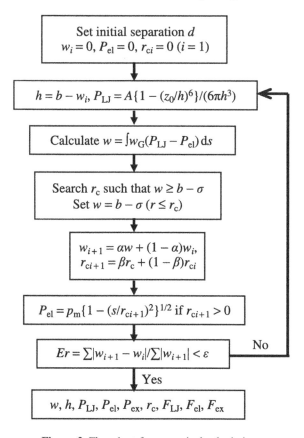

Figure 2. Flowchart for numerical calculation.

non-contact problem using equations (1), (3), (4), (5) and (6). When disk deforma-tion does not satisfy the inequality condition equation (8), the contact area radius r_c that satisfies $w(r_c) = b(r_c) - \sigma$ is numerically calculated, letting the disk defor-mation w be equal to $b - \sigma$ in the contact region of $r \leqq r_c$. Then, the new disk deformation $w(r)$ is calculated from equations (9) and (10) using the modified disk deformation $w(r)$, spacing $h(r)$, and contact radius r_c. This calculation procedure is repeated until a convergent solution is obtained. The convergence of the solution was determined when the Er value (the ratio of the sum of the absolute values of the differences of the successive values of w_{ij} to the sum of the absolute values of w_{ij}) was less than a small criterion value of ε. The discretized number of the radial coordinate is selected as $N = 400$ although $N = 100$ is enough for calculation of Hertzian contact deformation.

As shown in Fig. 2, in order to avoid the divergence of the solution due to the rapid changes of w and r_c, under-relaxation factors α and β are used for updat-ing deformation w and contact boundary r_c in the iterative calculation. Usually, the criterion of $\varepsilon = 10^{-6}$ was used, but the solution with $Er = \sim 10^{-2}$ was also considered to be an approximate solution because no clear difference between the solutions with $Er = \sim 10^{-2}$ and 10^{-6} was observed in the plots of w and r_c.

From the convergent solutions of LJ attractive pressure P_{LJ} and elastic contact pressure P_{el}, LJ attractive force F_{LJ}, elastic contact force F_{el}, and external force applied to disk F_{ex} were calculated by integration of the associated pressures over the analytical area as follows:

$$\text{LJ attractive force} \quad F_{LJ} = \sum_{1}^{N} P_{LJ}, \tag{11}$$

$$\text{Elastic contact force} \quad F_{el} = -\sum_{1}^{M} P_{el}, \tag{12}$$

$$\text{External force applied to the disk} \quad F_{ex} = \sum_{1}^{N} (P_{LJ} - P_{el}). \tag{13}$$

Since F_{LJ} and F_{ex} are external forces applied to the disk from the slider, the external forces applied to the slider from the disk are given by $-F_{LJ}$ and $-F_{ex}$.

4. Calculated Results for Contact Characteristics Ignoring Asperity LJ Forces

The parameter values used for the following calculations are listed in Table 1. These parameter values are the same as those of the 2-mm-radius glass slider and mag-netic disk used for the dynamic adhesion force measurements [19, 20]. Accurate values of the Hamaker constant and the atomic equilibrium distance are not known. A Hamaker constant of $A = 10^{-19}$ J was used as a standard value although this

Table 1.

Parameter values used for calculation

Physical parameter	Unit	Value
Disk Young's modulus E_1	GPa	163
Slider Young's modulus E_2	GPa	83
Disk Poisson ratio ν_1		0.3
Slider Poisson ratio ν_2		0.21
Slider radius R	mm	2
Hamaker constant A	J	10^{-19}
Atomic equilibrium distance z_0	nm	0.165

may overestimate the value for the head–disk interface. Equilibrium distance of $z_0 = 0.165$ nm was used since the same value is used as a standard value in the literature by Israelachvili [22] and Mate [23]. However, the effect of repulsive force is negligibly small when $\sigma \geqslant 0.3$ nm. The analytical area radius a was chosen to satisfy $\delta \ (= a^2/2R) = 10$ nm in Fig. 1.

Figure 3 shows the calculated contact characteristics when separation d is decreased from 2 nm to -5 nm in 0.2-nm steps: (a) disk deformation w, (b) LJ attractive pressure P_{LJ}, (c) elastic contact pressure P_{el}, (d) LJ attractive force $-F_{LJ}$, external force applied to the slider $-F_{ex}$, and Hertzian contact force F_{el} *versus* separation d, (e) external slider $-F_{ex}$ force *versus* contact radius r_c, and (f) iteration number IN and the value of $-10\log(Er)$. In this calculation the criterion of convergence was $\varepsilon = 10^{-6}$. We note from Fig. 3(a) that the disk first deforms upwards ($w > 0$) slightly due to the LJ attractive pressure while d decreases from 2.0 nm to 1.2 nm. But, at $d = 1.0$ nm, the slider comes in contact with disk asperities with a height $\sigma = 1.0$ nm. As d decreases further from 1 nm, the slider penetrates into the disk. The reason P_{LJ} is limited to a constant value at any separation (see Fig. 3(b)) is that spacing $h(r)$ in the contact region is equal to asperity spacing σ. When $\sigma = 1.0$ nm, we note from Fig. 3(b) and 3(c) that P_{LJ} is in the order of MPa, while P_{el} is one order larger than P_{LJ}. Therefore, disk deformation w is close to Hertzian contact deformation. Figure 3(d) shows that $F_{LJ} = 0.25$ mN at $d = -5$ nm and the minimum value of $-F_{ex}$ is ~ -0.03 mN at $d = 1.1$ nm. The relationship between contact radius r_c and external force F_{ex} is described by the Johnson–Kendall–Roberts (JKR) theory as $r_c^3 = 3R\{(-F_{ex}) + 3\pi\Delta\gamma R + [6\pi\Delta\gamma R(-F_{ex}) + (3\pi\Delta\gamma R)2]^{1/2}\})/4E^*$, whereas the relationship is described by the DMT theory as $r_c^3 = 3R\{(-F_{ex}) + 2\pi\Delta\gamma R\}/4E^*$ [24]. The contact radii given by the JKR and DMT theories are plotted in Fig. 3(e) for comparison using the relationship between $\Delta\gamma$ and A, which will be explained later. The difference between the JKR and DMT theories is small because the LJ attractive force is small and the present numerical results show values almost in between the external forces calculated from the JKR and DMT theories. It is seen

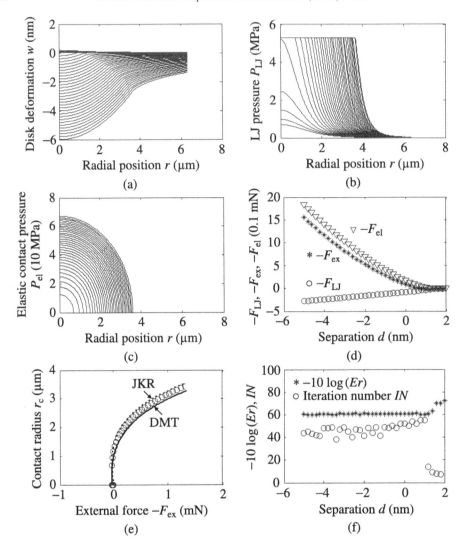

Figure 3. Calculated results for (a) disk deformation $w(r)$, (b) LJ pressure P_{LJ}, (c) elastic contact pressure P_{el}, (d) LJ force $-F_{LJ}$, external force $-F_{ex}$, and elastic contact force F_{el} *versus* separation d, (e) contact radius r_c *versus* external force $-F_{ex}$ and (f) iteration number (IN) and numerical error $-\log(Er)$ when d is varied from 2 nm to -5 nm by 0.2 nm ($R = 2$ mm, $\sigma = 1.0$ nm, $\varepsilon = 10^{-6}$).

from Fig. 3(f) that convergent solution with an error of less than 10^{-6} could be obtained with an iteration number of less than 100.

Figure 4 shows similar calculated contact characteristics when $\sigma = 0.5$ nm. The effect of σ on contact characteristics can be seen from a comparison between Figs 3 and 4. As seen in Fig. 4(f), convergence of the solution under contact conditions becomes worse and Er does not decrease to less than $\sim 10^{-2}$ even when the iteration number increases to 200. This is because the contact boundary does not converge to

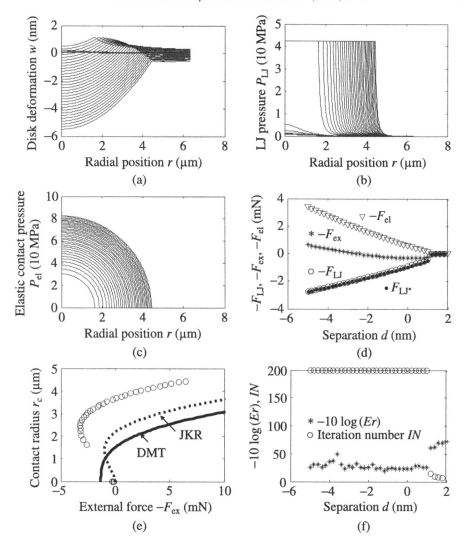

Figure 4. Calculated results for (a) disk deformation $w(r)$, (b) LJ pressure P_{LJ}, (c) elastic contact pressure P_{el}, (d) LJ force $-F_{LJ}$, external force $-F_{ex}$, and elastic contact force F_{el} *versus* separation d, (e) contact radius r_c *versus* external force $-F_{ex}$ and (f) iteration number (IN) and numerical error $-\log(Er)$ when d is varied from 2 nm to -5 nm in 0.2 nm steps ($R = 2$ mm, $\sigma = 0.5$ nm, $\varepsilon = 10^{-6}$).

a single value and shows a small cyclic variation. However, even when $Er = 10^{-2}$, variations of not only F_{LJ} and F_{el}, but also of $w(r)$, P_{LJ} and r_c are too small to be visible in the figure. Thus these results are considered to be reliable approximate solutions.

As seen from Fig. 4(a) and 4(d), disk deforms upwards slightly as in the same manner as in Fig. 3(a) while d decreases from 2.0 nm to 1.2 nm. Then, at $d = 1.0$ nm, the disk surface snaps into contact with the slider. It is noted that the disk deformation reaches 1 nm at the periphery of contacting area and that the contact radius abruptly increases as seen in Fig. 4(e). The reason why the contact radius increases appreciably compared to that from JKR theory is not clear, but is considered as follows. Since the spacing between mating mean height planes is equal to asperity height $\sigma = 0.5$ nm, repulsive pressure in equation (3) is almost zero and only elastic contact pressure opposes the attractive pressure in this model. Therefore, the total attractive pressure is stronger than that from the JKR model for a smooth spherical surface, where the effect of repulsive pressure inside the contact area is taken into account. Since P_{LJ} in Fig. 4(b) is 10 times larger than that in Fig. 3(b) and becomes comparable with P_{el} in Fig. 4(c) when $h = \sigma = 0.5$ nm, the disk adheres to the slider at the beginning of contact resulting in a large contact radius even at $d = 0.5–1.0$ nm. Note that the Tabor's parameter value of μ $(= (R\Delta\gamma^2/E^{*2}/z_0^3)^{1/3})$ is 1.64 in this case. The minimum value of $-F_{ex}$ is about -0.3 mN, and $r_c = 3.7$ μm at $F_{ex} = 0$, as seen from Fig. 4(e).

As seen in Fig. 4(d), as d is decreased, $-F_{ex}$ first decreases slightly and then decreases abruptly because the disk adheres to the slider at $d = \sim 1.0$ nm. After reaching the minimum value, $-F_{ex}$ increases due to the increase in elastic contact force, as seen in Fig. 4(d) and 4(e). Although Fig. 4 shows the calculated results in the approach process, almost the same characteristics without a visible hysteresis were obtained in the separation process when d was increased reversely.

Figure 5(a) and 5(b) shows the LJ attractive force F_{LJ} as a function of the minimum spacing $h_0 = b(0) - w(0)$ at the tip of the spherical slider for $\sigma = 1.0$ and 0.5 nm, respectively. For comparison, the vdW force of F_{vr} $(= AR/6h_0^2)$ between a rigid sphere and a flat disk and the meniscus force of $F_m = 4\pi R\gamma$ $(\gamma = 0.022$ J/m$^2)$ are shown in these figures. Noting that the separation d is decreased from 2 to -5 nm in 0.2-nm steps, h_0 is slightly smaller than d due to the elastic deformation under non-contact conditions. However, F_{LJ} is almost equal to F_{vr} at the same h_0 under the non-contact conditions. This indicates that the attractive pressure in a small area of the protruded tip of the sphere contributes to the total attractive

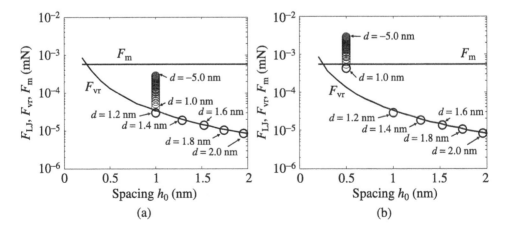

Figure 5. F_{LJ} (○), F_{vr} and F_m versus h_0 for (a) $\sigma = 1.0$ nm and (b) $\sigma = 0.5$ nm.

force. When h_0 decreases to less than 1.0 nm, the disk snaps into contact with the slider and h_0 jumps to σ. Therefore, no jump is observed when $\sigma = 1.0$ nm, but a clear jump of h_0 from 1.0 nm to 0.5 nm ($=\sigma$) is observed at $d = 1.0$ nm when $\sigma = 0.5$ nm. F_{LJ} jumps from the curve of F_{vr} to a value much larger than F_{vr} because of increase in contact area. F_{LJ} becomes larger than the meniscus force F_m when $d < 0.6$ nm.

The adhesion force at h_0 under contact conditions is approximately evaluated by summing the LJ force of rigid sphere contact F_{LJr} at h_0 ($=\sigma$) and the LJ attractive force within the contact area. F_{LJr}, including the repulsive term, is obtained by integrating P_{LJ} in equation (3) over the total surface and written as

$$F_{LJr} = \frac{AR}{6h_0^2}\left(1 - \frac{z_0^6}{4h_0^6}\right).$$

(14)

When $z_0/h_0 < 0.5$, the second term within braces can be neglected and $F_{LJr} = F_{vr}$. The LJ attractive force within the contact area can be obtained by multiplying the contact area πr_c^2 with $P_{LJ}(h_0)$. Thus the adhesion force under contact conditions is approximated by

$$F_{LJ*} = \frac{AR}{6h_0^2}\left(1 - \frac{z_0^6}{4h_0^6}\right) + \frac{Ar_c^2}{6h_0^3}\left(1 - \frac{z_0^6}{h_0^6}\right).$$

(15)

Here the repulsive term can be ignored when $z_0/h_0 < 0.5$. Using the calculated value of r_c and $h_0 = \sigma$, $-F_{LJ*}$ was calculated and plotted in Fig. 4(d) with a dot. It is clear that F_{LJ*} agrees well with F_{LJ}.

To take into account the LJ force at contacting asperities, contact characteristics between a small sphere and a flat are calculated next to make clear how the attractive force at contacting asperities is properly evaluated.

5. Evaluation of LJ Attractive Pressure at Contacting Asperities

5.1. LJ Attractive Force between a 1-μm-Radius Sphere and a Flat

In actual cases where asperity density is large and asperity spacing σ is more than 0.5 nm, adhesion characteristics cannot be evaluated well without taking into account the LJ attractive force of contacting asperities. Since an actual rough surface has fractal characteristics, it is natural to consider that an asperity has small-scale asperities whose asperity spacing is of the same order as z_0. If we focus on the smoothest engineering surfaces, such as a magnetic disk, head slider, and silicon substrate, the rms values of surface roughness are less than 0.5 nm and the mean asperity radius is near 1 μm. Therefore, the contact characteristics of a 1-μm-radius asperity with a flat were numerically calculated for various values of asperity spacing and the expression for the asperity attractive force was examined. The analytical model of asperity contact is similar to that shown in Fig. 1.

Figure 6 shows the calculated contact characteristics for $\sigma = 0.25, 0.20, 0.18,$ $0.165, 0.16$ and 0.15 nm when separation d decreases from 0.6 to -0.5 nm in steps

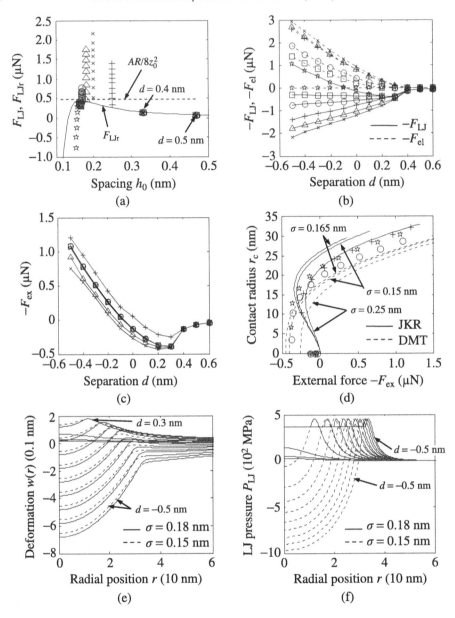

Figure 6. Contact characteristics of 1-μm-radius sphere and disk for $\sigma = 0.25, 0.20, 0.18, 0.165, 0.16$ and 0.15 nm (+: $\sigma = 0.25$ nm, ×: $\sigma = 0.20$ nm, △: $\sigma = 0.18$ nm, ○: $\sigma = 0.165$ nm, □: $\sigma = 0.16$ nm, ☆: $\sigma = 0.15$ nm). (a) F_{LJ} and F_{LJr} versus h_0. (b) LJ force $-F_{LJ}$ and elastic contact force F_{el} versus separation d. (c) External force $-F_{ex}$ versus separation d. (d) Contact radius r_c versus external force $-F_{ex}$. (e) Disk deformation $w(r)$. (f) LJ pressure P_{LJ}.

of 0.1 nm: (a) F_{LJ} versus h_0, (b) $-F_{LJ}$ and $-F_{el}$ versus d, (c) $-F_{ex}$ versus d, (d) r_c versus $-F_{ex}$, (e) disk deformation w, and (f) LJ pressure P_{LJ}. Figure 6(a) is similar to Fig. 5 but drawn in a linear scale diagram. In Fig. 6(a), the LJ force F_{LJr} for a rigid sphere–flat interface given by equation (14) is plotted with a solid line for comparison. The maximum attractive value, indicated with a dashed line, is given by $F_{\text{LJr max}} = AR/8z_0^2$. It is noted from Fig. 6(a) that F_{LJ} increases with decrease in h_0 along the curve of F_{LJr} in non-contact conditions. But after the contact at $h_0 = \sigma$, F_{LJ} increases from F_{LJr} with decrease in d because of the increase in the contact area when $\sigma > 0.16$ nm. In particular, when $\sigma = 0.2$ nm, the F_{LJ} increases to more than 2 μN at $d = -0.5$ nm, which is four times larger than the $F_{\text{LJr max}}$. The increase of F_{LJ} with a decrease in d is largest when $\sigma = 0.2$ nm. This is because P_{LJ} inside the contact area, given by equation (3), becomes maximum at $h = \sigma = 3^{1/6}z_0 = 0.198$ nm. Therefore, if the small-scale asperity on a roughness asperity is larger than z_0, attractive force at the contacting asperity should be considered as a function of separation d.

However, if σ is nearly equal to z_0, F_{LJ} has an almost constant value of $AR/8z_0^2$ at any separation as shown in Fig. 6(a) and 6(b). This is because vdW attractive pressure inside the contact area is cancelled by the repulsive pressure and the total attractive force is generated only from the attractive pressure in the surrounding area. This result can also be derived from $F_{\text{LJ}*}$ in equation (15), by substituting $h_0 = z_0$, the attractive force inside the contact area vanishes and the attractive force outside the contact area becomes $AR/8z_0^2$.

When $\sigma = 0.15$ nm (✳), however, F_{LJ} decreases rapidly with decrease in d after contact due to repulsive pressure effect as shown in Fig. 6(a) and 6(b). In this case, sphere and flat do not contact each other, Thus F_{el} is always zero, and $F_{\text{LJ}} = F_{\text{ex}}$, as seen from comparison between Fig. 6(b) and 6(c).

It is noted from Fig. 6(c) that the relationship of F_{ex} ($= F_{\text{H}} - F_{\text{LJ}}$) *versus* d is hardly influenced by σ. Particularly, $-F_{\text{ex}}$ *versus* d is almost identical when $\sigma \lesssim z_0$. This indicates that a variation of F_{LJ} due to σ is compensated with a reversal variation of F_{el}, as seen in Fig. 6(b).

In Fig. 6(d), the relationship between contact radius r_c and external force $-F_{\text{ex}}$ is compared with those from JKR and DMT theories for three cases of $\sigma = 0.25$ (+), 0.165 (○) and 0.15 (✳) nm. It is observed that the calculated results of r_c are close to JKR curve when $\sigma = 0.25$ nm, and are in between JKR and DMT curves when $\sigma = 0.15$ nm. However, the numerical results of r_c become closer to DMT curve when $\sigma = 0.165$ nm. This is because the total LJ force F_{LJ} is attributable to LJ pressure outside the contact area as in the DMT model when $\sigma = \sim z_0$ as stated above.

Typical disk deformation $w(r)$ and LJ pressure $P_{\text{LJ}}(r)$ are illustrated for $\sigma = 0.18$ and 0.15 nm in Fig. 6(e) and 6(f), respectively when d is changed from 0.6 nm to -0.5 nm in steps of 0.1 nm. Although the disk deformation is changed only slightly by the change of σ, P_{LJ} changes significantly with the change of σ after the disk surface comes in contact with slider ($d \leq 0.3$ nm). When $\sigma = 0.18$ nm, the disk surface contacts the spherical slider with a spacing of $h_0 = 0.18$ nm. Therefore, P_{LJ} inside the contact area becomes smaller than P_{LJ} in the surrounding area; P_{LJ} becomes maximum at $h = 0.198$ nm as explained above. When $\sigma = 0.15$ nm, the disk surface does not contact the slider because the spacing between the disk and the slider is larger than asperity height σ and the external force F_{ex} becomes equal to F_{LJ}.

Accordingly, it can be said that if and only if the small-scale asperity height is nearly equal to z_0, F_{LJ} at a contacting asperity can be expressed approximately as $AR/8z_0^2$ at any separation. If σ is a little larger than z_0, F_{LJ} at a contacting asperity would increase as separation decreases, as seen in Fig. 6(a) and 6(b). However, if we evaluate the largest adhesion force that can be observed in F_{ex} at the beginning and end of asperity contact, the LJ attractive force can be approximated by $AR/8z_0^2$ when $\sigma \leq 0.25$ nm. Moreover, as seen from Fig. 6(c), it is expected that the external force $-F_{ex}$ calculated from $F_{LJ} = AR/8z_0^2$ for the case of $\sigma = \sim z_0$ can be used for a wider range of σ.

On the other hand, using Derjaguin approximation, the attractive force of a sphere–flat contact has been derived from surface energy and given by

$$F_a = 2\pi R \Delta\gamma, \quad \Delta\gamma = \gamma_1 + \gamma_2 - \gamma_{12}, \tag{16}$$

where γ_1 and γ_2 are surface energies of the mating surfaces 1 and 2 before contact, and γ_{12} is that of the contacting surfaces after the contact. Since F_a should be equal to F_{LJr} in equation (14), we can obtain the general relationship between the Hamaker constant A and the surface energy difference $\Delta\gamma$ as follows:

$$\Delta\gamma = \frac{A}{12\pi\sigma^2}\left(1 - \frac{z_0^6}{4\sigma^6}\right), \tag{17}$$

where σ is the effective mean height of small-scale roughness.

When $\sigma \geqslant 1.16z_0$ ($= 0.19$ nm), the second term is less than 0.1. Thus, if the surface energy is determined by the pull-off force of sphere–disk contact that is affected by small-scale roughness, the Hamaker constant is approximately given by

$$A = 12\pi\sigma^2\Delta\gamma. \tag{18}$$

When $\sigma \leqslant 1.04z_0$ ($= 0.172$ nm), and the surface energy is determined from the pull-off force of sphere–flat contact, the Hamaker constant can be given by

$$A = 16\pi z_0^2\Delta\gamma. \tag{19}$$

When the surface energy is determined from the contact angles of liquids, the liquid molecules are considered to contact the solid surface at equilibrium distance. Thus, in this case, equation (19) can be used for the relationship between the surface energy difference and the Hamaker constant. The plotted curves of the JKR and DMT theories illustrated in Figs 3, 4 and 6 were calculated using equation (17).

5.2. Comparison between Asperity LJ Pressure P_{LJa} and Mean Height Plane LJ Pressure P_{LJ}

Next we consider the magnitude of averaged LJ pressure P_{LJa} due to contacting asperities and compare it with the LJ pressure P_{LJ} due to mean height planes. According to the calculated results described above, if the small-scale asperity spacing

σ is close to z_0, the LJ attractive force is given by $AR/8z_0^2$ at $h_0 = z_0$. This maximum attractive force is considered to be equal to F_a in equation (16) with surface energy determined from contact angles of test liquids. In this condition, P_{LJa} is given by

$$P_{LJa} = 2\pi R_a \Delta\gamma\rho p, \tag{20}$$

where R_a, ρ and p are mean asperity radius, asperity density, and probability of contacting asperity out of the entire asperity density ρ.

Since the probability of asperities contacting each other is usually less than 0.5 at separation d at which the minimum external force is observed, it can be said that P_{LJ} plays a dominant role if P_{LJa} at $p = 1.0$ is smaller than P_{LJ} given by equation (3), i.e.,

$$2\pi R_a \Delta\gamma\rho \leqslant \frac{A}{6\pi\sigma^3}\left\{1 - \left(\frac{z_0}{\sigma}\right)^6\right\}. \tag{21}$$

Since σ in inequality (21) is a large-scale asperity spacing, $(z_0/\sigma)^6 \ll 1$. In contrast, since the small-scale asperity spacing can be considered to be nearly equal to z_0, equation (19) holds. Therefore, by substituting equation (19) into inequality (21), we can obtain the condition where P_{LJ} plays a dominant role in attractive pressure compared to P_{LJa}, as follows:

$$\sigma \leqslant \left(\frac{4}{3}\frac{z_0^2}{\pi R_a\rho}\right)^{1/3}. \tag{22}$$

If we assume that $R_a = 1$ µm, then $\sigma \leqslant 1.05$ nm and $\sigma \leqslant 0.49$ nm for $\rho = 10$ and 100 µm^{-2}, respectively. Therefore, when the spacing σ of contacting asperities is less than 0.5 nm, the contact characteristics can be approximately analyzed by ignoring the LJ attractive force of contacting asperities and the contact mechanics analysis in Section 4 can be supported.

6. Calculation of Attractive Force of 2-mm Glass Slider Including Asperity LJ Pressure Effect

If asperity spacing σ is regarded as uniform in the contact area, contact characteristics including LJ pressure due to contacting asperities and mean height planes can be calculated from equation (7) in the same manner as equation (10) using P_{LJa} from equation (20). Ono and Nakagawa [20] measured the dynamic external force when 1- and 2-mm-radius glass spheres bounced on a magnetic disk with a thin lubricant layer. In experiments with a 2-mm smooth glass slider, a maximum attractive force of about 0.4–0.6 mN was detected at the end of the contact period of about 30 µs. To test our belief that the meniscus could be generated so rapidly and that the measured attractive force must, therefore, be a meniscus force, the maximum possible LJ attractive force was calculated by an approximated numerical analysis.

Before calculating eqaution (7), the statistical asperity contact characteristics of two flat rough surfaces, including the asperity LJ pressure P_{LJa} of equation (20), were first calculated following the conventional method similar to [9–11]. In this analysis, however, P_{LJa} was calculated from the attractive forces of contacting asperities given by equation (16). In contrast to the prior method, attractive pressure from noncontacting asperities was taken into account in the mean height plane LJ pressure P_{LJ}.

If we denote asperity height by z, mean asperity height by z_m, and asperity height distribution density function by $\phi(z - z_m)$ in the z-coordinate shown in Fig. 1, asperity elastic contact pressure P_{ela}, asperity LJ pressure P_{LJa} and real contact area ratio AR_a are, respectively, given by

$$P_{ela} = \frac{4}{3} E^* \rho R_a^{1/2} \int_d^\infty (z - d)^{3/2} \phi(z)\, dz, \tag{23}$$

$$P_{LJa} = 2\pi R_a \Delta \gamma \rho \int_d^\infty \phi(z)\, dz, \tag{24}$$

$$AR_a = \pi R_a \rho \int_d^\infty (z - d)\phi(z)\, dz. \tag{25}$$

Here, it is assumed that the asperity height has a Gaussian asperity height distribution with rms asperity height σ_a and mean asperity height z_m of the form:

$$\phi(z) = \frac{1}{\sqrt{2\pi}\sigma_a} \exp\left(-\frac{(z - z_m)^2}{2\sigma_a^2}\right). \tag{26}$$

Table 2 lists the surface roughness parameters for the 2-mm-radius glass slider and magnetic disk tested [20]. From Table 2, composite rms asperity height is $\sigma_a = (\sigma_{a1}^2 + \sigma_{a2}^2)^{0.5} = (0.15^2 + 0.31^2)^{0.5} = 0.344$ nm, and composite mean asperity height z_m is 0.56 nm + 0.52 nm = 1.08 nm. The composite asperity radius is given by $R_a = (R_{a1}^{-1} + R_{a2}^{-1})^{-1} = (1.36^{-1} + 3.06^{-1})^{-1} = 0.94$ µm. Since it is not easy to evaluate the equivalent asperity density and probability of asperity contact, various values of ρ were used for numerical calculation.

Table 2.
Surface roughness of the hemispherical R2 glass slider (scan area = 1 µm × 1 µm) and magnetic disk (scan area = 5 µm × 5 µm)

	Roughness			Asperity			
	R_a (nm)	R_q (rms) (nm)	R_p (nm)	Rms height (nm)	Mean height (nm)	Density (µm^{-2})	Mean radius (µm)
R2 slider	0.34	0.33	1.62	0.15	0.56	179	1.36
Disk	0.52	0.67	5.60	0.31	0.52	31.3	3.06

Figure 7 shows the calculated asperity contact characteristics between the flat glass slider and magnetic disk: (a) probability density function $\phi(z)$ and probability p of asperity contact as a function of normalized asperity height $(z - z_m)/\sigma_a$, (b) elastic contact pressure P_{ela} and LJ attractive pressure P_{LJa} versus normalized separation $(d - z_m)/\sigma_a$, (c) real contact area ratio versus $(d - z_m)/\sigma_a$, and (d) external pressure $P_{exa} (= P_{ela} - P_{LJa})$ versus $(d - z_m)/\sigma_a$. Material parameters of the two mating surfaces are as listed in Table 1. Surface energy $\Delta\gamma$ is assumed to be 0.06 J/m^2 ($A = 0.821 \times 10^{-20}$ J from equation (19)). The effect of asperity density ρ on contact characteristics is shown in Fig. 7.

As seen in Fig. 7(b) and 7(d), the external pressure P_{exa} has a minimum negative value at $(d - z_m)/\sigma_a = \sim 1$ and becomes positive when $(d - z_m)/\sigma_a < 0$. Figure 7(a) and 7(c) shows that the real contact area ratios are 0.013, 0.04 and 0.073 for $\rho = 31.3$, 100 and 179 μm^{-2}, respectively, at 50% probability of asperity contact. Since

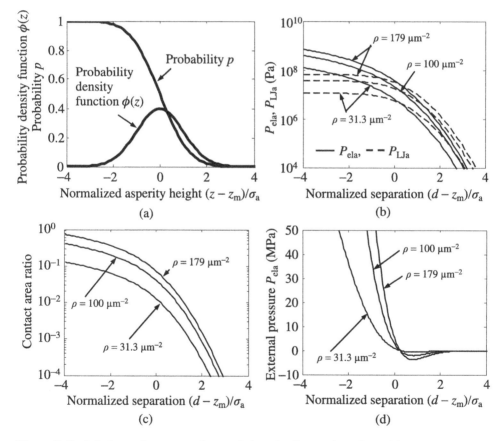

Figure 7. Statistical asperity contact characteristics of a flat rough surface with Gaussian asperity height distribution. (a) Probability density function $\phi(z)$ and probability p of contacting asperities versus normalized asperity height $(z - z_m)/\sigma_a$. (b) Asperity contact pressure P_{ela} (solid) and asperity LJ pressure P_{LJa} (dashed) versus normalized separation $(d - z_m)/\sigma_a$. (c) Real contact area ratio versus $(d - z_m)/\sigma_a$. (d) External pressure $P_{exa}(= P_{ela} - P_{LJa})$ versus $(d - z_m)/\sigma_a$.

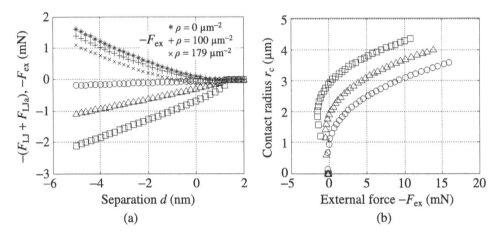

Figure 8. Contact characteristics of 2-mm-radius glass slider and magnetic disk including the effect of LJ pressures due to contacting asperities and mean height planes ($\bigcirc$: $\rho = 0\ \mu\mathrm{m}^{-2}$, $\triangle$: $\rho = 100\ \mu\mathrm{m}^{-2}$, $\square$: $\rho = 179\ \mu\mathrm{m}^{-2}$). (a) LJ force $-(F_{\mathrm{LJ}} + F_{\mathrm{LJa}})$ and external force $-F_{\mathrm{ex}}$ applied to the slider *versus* separation d. (b) Contact radius r_{c} *versus* external force $-F_{\mathrm{ex}}$.

the real contact area ratio at $\rho = 179\ \mu\mathrm{m}^{-2}$ increases to 0.55 at $(d - z_{\mathrm{m}})/\sigma_{\mathrm{a}} = -3$, it seems that an asperity density of 179 would be an overestimated value. Since we are interested in the region of negative external pressure, 50% probability of asperity contact would be sufficient for estimating the asperity attractive force.

Therefore, for the calculation of equation (7), it is reasonable to consider that the contact asperity spacing σ is $\sim z_{\mathrm{m}}$ ($= 1.08\ \mathrm{nm}$) and probability of contacting asperities p is ~ 0.5. When $p = 0.5$, P_{LJa} values in equation (20) become 5.56, 17.7 and 31.8 MPa for $\rho = 31.3, 100$ and $179\ \mu\mathrm{m}^{-2}$, respectively. These asperity LJ pressures were used to calculate asperity contact characteristics with equations (7), (8) and (9).

Figure 8 shows (a) $-(F_{\mathrm{LJ}} + F_{\mathrm{LJa}})$ and $-F_{\mathrm{ex}}$ ($= -F_{\mathrm{LJ}} - F_{\mathrm{LJa}} + F_{\mathrm{ela}}$) *versus* separation d and (b) r_{c} *versus* $-F_{\mathrm{ex}}$ for asperity densities of $\rho = 0, 100$ and $179\ \mu\mathrm{m}^{-2}$. These solutions have an accuracy of $\varepsilon = 10^{-6}$. As seen in Fig. 8(a), the total attractive force $-(F_{\mathrm{LJ}} + F_{\mathrm{LJa}})$ increases with an increase in asperity density and with a decrease in d, but the minimum negative value of $-F_{\mathrm{ex}}$ is not appreciable. In Fig. 8(b), the minimum values of $-F_{\mathrm{ex}}$ are 0.022, 0.05 and 0.151 mN for $\rho = 0, 100$ and $179\ \mu\mathrm{m}^{-2}$, respectively.

Figure 9 shows a comparison between the experimental dynamic indentation characteristics presented in a previous paper [20] and the calculated external force applied to the slider. The experimental relationship between force applied to the slider and displacement was calculated by differentiation and integration of the slider velocity in the bouncing process that was measured by a digital laser Doppler vibrometer. Experimental results show a small adhesion force at the beginning of the contact and a large adhesion force of -0.4 to -0.6 mN at the end of the contact. It should be noted that these force–displacement characteristics of the slider were

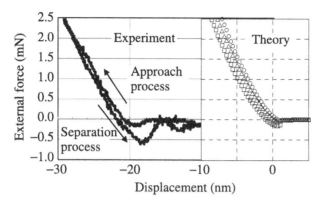

Figure 9. Comparison between experimental and theoretical indentation characteristics. Experimental results: lubricant thickness = 2 nm, mobile lubricant thickness = 0.98 nm, and impact velocity = 0.75 mm/s. Theoretical results: $\bigcirc \rho = 0\ \mu m^{-2}$ ($P_{LJa} = 0$ MPa), $\triangle \rho = 100\ \mu m^{-2}$ ($P_{LJa} = 17.7$ MPa), and $\square \rho = 179\ \mu m^{-2}$ ($P_{LJa} = 31.8$ MPa).

measured for only 30 µs contact time with magnetic disk with a mobile lubricant thickness of 0.98 nm. In the numerical calculation, the difference between approach and separation processes was too small to be visible in the figure. The calculated adhesion force is about 0.15 mN even when $\rho = 179\ \mu m^{-2}$. Therefore, the small adhesion force experimentally observed at the beginning of the contact might be caused by LJ adhesion force. However, the large adhesion force observed at the end of the contact is not caused by LJ attractive pressure. The meniscus force of lubricant with surface energy γ is given by $2\pi R\gamma(1 + \cos\theta_0)$ where θ_0 is the contact angle of lubricant on the glass slider and γ is the surface energy of the lubricant. Since $\gamma = 0.022$ J/m^2 in the experiment, the meniscus force $4\pi R\gamma$ is 0.55 mN when $\theta_0 = 0$ degrees. Therefore, it is natural to consider that the adhesion force observed at the end of the contact is caused by meniscus formation within a short contact period of 30 µs.

7. Conclusion

A numerical analysis method is presented for the rough surface contact characteristics between a sphere and a disk considering not only elastic deformation and LJ forces between mean height planes of the sphere and disk but also LJ forces of contacting asperities, and the intermolecular force between the glass slider and the magnetic disk is evaluated. The effects of LJ forces of contacting asperities and their elastic deformation and elastic contact force are taken into account. Fundamental contact characteristics of a 2-mm-radius glass sphere and a 1-µm-radius asperity model on the glass slider were calculated by assuming asperity spacing due to different scale asperity heights. The relationship between the Hamaker constant and the surface energy is discussed. This analysis suggests that LJ attractive force at a contacting asperity is given by $2\pi R\Delta\gamma$ at any separation if the spacing between the contacting asperity and the mating surface is close to atomic equilibrium distance. By using this adhesion force for contacting asperities, an approximate numerical analysis method for rough-surface contact between a sphere and a flat disk, including elastic deformation and LJ pressures of contacting asperities and mean height planes, is presented. The contact characteristics for the 2-mm-radius glass slider and magnetic disk are calculated using asperity parameters measured in the experiment and compared with experimental adhesion force values. It is shown that the calculated adhesion force is much smaller than the measured adhesion force, supporting the idea that the adhesion force measured at the end of the contact is generated from meniscus force rather than from vdW force.

References

1. K. L. Johnson, K. Kendall and A. D. Roberts, *Proc. R. Soc. Lond. Ser. A* **324**, 301–313 (1971).
2. B. V. Derjaguin, V. M. Muller and Y. P. Toporov, *J. Colloid Interface Sci.* **53**, 314–326 (1975).
3. M. D. Pashley, *Colloids Surfaces* **12**, 69–77 (1984).
4. D. Maugis, *J. Colloid Interface Sci.* **150**, 243–269 (1992).
5. P. Attard and J. L. Parker, *Phys. Rev. A* **46**, 7959–7971 (1992).
6. J. A. Greenwood, *Proc. R. Soc. Lond. Ser. A* **453**, 1277–1297 (1997).
7. J. Q. Feng, *Colloids Surfaces* **172**, 175–198 (2000).
8. J. A. Greenwood and J. B. P. Williamson, *Proc. Roy. Soc. (London)* **A295**, 300–319 (1966).
9. W. R. Chang, I. Etsion and D. B. Bogy, *ASME Trans. J. Tribology* **110**, 50–56 (1988).
10. H. M. Stanley, I. Etsion and D. B. Bogy, *ASME Trans. J. Tribology* **112**, 98–104 (1990).
11. A. Y. Suh and A. A. Polycarpou, *J. Appl. Phys.* **99**, 104328 (2005).
12. N. Yu and A. A. Polycarpou, *J. Colloid Interface Sci.* **278**, 428–435 (2004).
13. A. R. Ambekar, V. Gupta and D. B. Bogy, *ASME Trans. J. Tribology* **127**, 530–536 (2005).
14. S.-C. Lee and A. A. Polycarpou, *ASME Trans. J. Tribology* **126**, 334–341 (2004).
15. H. Matsuoka, S. Ohkubo and S. Fukui, *Microsys. Technol.* **11**, 824–829 (2005).
16. K. Ono and M. Yamane, *ASME Trans. J. Tribology* **129**, 65–74 (2007).
17. K. Ono and M. Yamane, *ASME Trans. J. Tribology* **129**, 246–255 (2007).
18. K. Ono and Y. Masami, *ASME Trans. J. Tribology* **129**, 453–460 (2007).
19. K. Ono and K. Nakagawa, *JSME International J. Ser. C* **49**, 1159–1170 (2006).
20. K. Ono and K. Nakagawa, *Tribology Lett.* **31**, 77–89 (2008).
21. K. Ono, *JSME Trans.* **51**, 2309–2316 (1985).
22. J. N. Israelachvili, *Intermolecular and Surface Forces*, 2nd edn, pp. 196–197. Academic Press (1992).
23. C. M. Mate, *Tribology on the Small Scale*, p. 153. Oxford University Press (2008).
24. X. Shi and Y.-P. Zhao, *J. Adhesion Sci. Technol.* **18**, 55–68 (2004).

Equilibrium Vapor Adsorption and Capillary Force: Exact Laplace–Young Equation Solution and Circular Approximation Approaches

D. B. Asay [a,*], **M. P. de Boer** [b,c] **and S. H. Kim** [a,†]

[a] Department of Chemical Engineering, Pennsylvania State University, University Park, PA 16802, USA
[b] Sandia National Laboratories, Albuquerque, NM 87185, USA
[c] Department of Mechanical Engineering, Carnegie Mellon University, Pittsburgh, PA 15237, USA

Abstract
The capillary adhesion force of an asperity of radius R as a function of vapor partial pressure is calculated using exact and approximate methods assuming a continuum model. The equilibrium between the capillary meniscus at the asperity and the adsorbate film on the surface is discussed through a disjoining pressure term. It is found that the two methods agree very well over a wide partial pressure range. Without taking into account the effect of the adsorbate film, the theoretical calculation results do not show the experimental partial pressure dependence of the capillary force except near the saturation vapor condition. The experimental capillary force trend with partial pressure can be explained when the presence of the adsorbate film is included in the calculation.

Keywords
Capillary forces, nanoscale, disjoining pressure, adsorption, equilibrium

1. Introduction

When a liquid meniscus is formed around the contact area of two neighboring surfaces, a force is exerted on the contacting surfaces due to the surface tension of the liquid and the curvature of the meniscus. This is called a capillary force. Capillary forces play important roles in studies of adhesion between particles and particles to flat or curved surfaces, adhesion of insects and small animals, particle processing, friction, etc. Compared to body forces, the relative strength of the capillary force becomes larger as the size of the object decreases, and, unless avoided, immediately leads to the "stiction" problem in microelectromechanical systems (MEMS).

---

[*] Current address: PPG, Allison Park, PA 15101, USA.
[†] To whom correspondence should be addressed. E-mail: shkim@engr.psu.edu

Adhesion Aspects in MEMS/NEMS
© Koninklijke Brill NV, Leiden, 2010

Even if the partial pressure of the vapor in the ambient is lower than its saturation pressure, a condensed phase can be formed in the narrow gap of two solid surfaces if the vapor molecule has a strong affinity toward the solid surface. This phenomenon is called capillary condensation and is explained well by the equilibrium relationship between the Laplace pressure due to the curvature of the condensed liquid meniscus and the vapor pressure (see Section 2.1). Alcohol vapors have been shown to be highly efficient to prevent MEMS failure compared to other coating based lubrication approaches [1–3], but are expected to form capillaries of the condensed liquid at the asperity contacts depending on the pressure of the vapor being adsorbed relative to its saturation vapor pressure (p/p_{sat}). Hence, it is critical to understand alcohol capillarity effects on the adhesion of nano-asperity silicon oxide surfaces.

Atomic force microscopy (AFM) has been used to quantitatively measure the capillary forces exerted by the liquid meniscus at the nano- and micro-scales, directly related to MEMS and other applications [4–10]. Figure 1 shows the pull-off force for silicon oxide surfaces in various alcohol vapor environments measured with AFM tips mounted on low spring constant cantilevers [10]. Of note in Fig. 1 is that starting at the saturation p/p_{sat}, the force increases substantially as p/p_{sat} decreases. There are a number of theoretical models that consider experimentally observed capillary force trends. In the following paragraphs, the assumptions used in these theoretical calculations and their validity will be briefly discussed.

One of the most widely used equations to describe the capillary force is $F_c = 4\pi R\gamma \cos\theta$ where R, γ and θ are the radius of the tip (modeled as a sphere), the liquid surface tension, and the contact angle of the liquid on the solid surface,

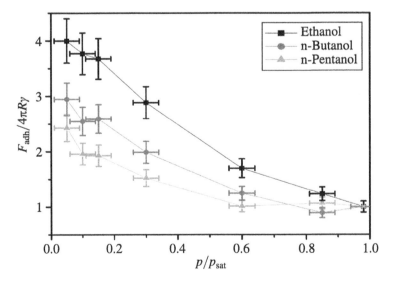

Figure 1. Pull-off force measured with atomic force microscopy for clean SiO_2 surfaces as a function of relative partial pressure (p/p_{sat}) of ethanol, n-butanol, and n-pentanol [10]. The force is normalized with $4\pi R\gamma$ (R = AFM tip radius and γ = surface tension of liquid). {Reprinted with permission from the American Chemical Society.}

respectively [11, 12]. This equation originates from the Young–Laplace equation which contains two principal radii of the meniscus (see Section 2.1) through several assumptions. One assumption is that the radius of the tip is much larger than the cross-sectional radius of the liquid meniscus which, in turn, is much larger than the meridional curvature (external curvature) of the meniscus surface [5, 11–13]. However, this assumption is not valid especially for the nano-scale relevant to AFM experiments. He *et al.* extended this model and derived an equation that does not require the large contact area assumption [7]. They found that the capillary force becomes larger than $4\pi R\gamma \cos\theta$ as the size of the tip decreases. In any case, these models predict no vapor pressure dependence of the capillary force, which is inconsistent with experimental results reported in the literature [4–10].

More elegant models use a circular approximation described by Orr *et al.* to calculate the capillary force due to the liquid meniscus without making the aforementioned assumption [14]. In this approximation, the meridional profile of the meniscus is modeled by an arc of a circle (cross section of a torus). Xiao and Qian assumed that the meniscus edges met with the tip and substrate surfaces with finite contact angles and estimated the size of the meniscus using the Kelvin equation at a given vapor partial pressure [5]. This model shows some vapor pressure dependence of the capillary force, but the exact shape deviates from the experimental data.

It should be noted that on clean silicon oxide surfaces, the contact angle of water and short-chain alcohols is near zero (completely wetting). Therefore, assuming a finite contact angle at the meniscus edge is inappropriate especially for water and alcohol condensation on hydrophilic surfaces. Bhushan, Butt and other groups have used the same or similar circular approximation for the case where the contact angles are zero at both tip and substrate surfaces [15, 16]. In this case, however, the theoretical predictions are that the capillary force is the same as $4\pi R\gamma$ (since $\cos\theta = 1$ when $\theta = 0°$) and does not change until the vapor pressure approaches the saturation pressure, where the capillary force diminishes. This prediction is in sharp contrast with Fig. 1 as well as with experimental observations previously reported for silicon oxide surfaces exposed to water and alcohol vapors [4–10, 17].

These theories neglect the role of the *adsorbate film* on the tip and substrate surfaces. The thickness trend of this film *versus* vapor pressure is known as an adsorption isotherm, and can be described by a Langmuir, BET or other isotherm model [18]. As we shall show, incorporating the effect of this film into the theory explains why the capillary force increases much larger than $4\pi R\gamma$ as p/p_{sat} decreases from saturation, i.e., below a value of 1. On most high surface energy solid surfaces (such as oxides and metals), adsorption of water or organic molecules readily occurs from the surrounding gas environments. The thickness of the adsorbate layer can vary from less than one monolayer to several molecular layers depending on the partial pressure of the molecule in the gas phase [19–21]. The adsorbed layer may have different molecular orientation or packing from the bulk liquid [21, 22]. Because both the liquid meniscus and the adsorbed film consist of the same molecules and are in equilibrium with the vapor, the adsorbed film should be considered a continuous film of the condensed phase composing the meniscus. This adsorbate film then acts as the *disjoining layer* in the mechanics terminology [23].

The effects of the disjoining layer in capillary force measurements have been discussed by Mate, White and their colleagues for lubricant films [24, 25]. However, the vapor pressure of lubricant molecules is extremely low. Therefore, the vapor-adsorbate equilibrium can be ignored, and the focus of that work was to calculate the disjoining pressure of the film from the capillary force measured during the stretching of the meniscus with tips attached to high spring constant cantilevers. In their experiments, a lubricant layer of a known thickness is applied to a substrate and the force *versus* displacement curve of an AFM tip of known radius is measured. The force measured during the stretching of the lubricant layer is analyzed to determine the effective meniscus curvature, which is directly related to the Laplace pressure and hence the disjoining pressure [25].

Wei and Zhao calculated the growth rate of a liquid meniscus in humid environments on the substrate and a finite contact angle on the tip surface [26]. They assumed a disjoining layer only on the substrate and calculated the capillary growth kinetics by considering vapor-phase diffusion. Their theory qualitatively supported the experimental results. Asay and Kim recently measured the effect of the vapor-adsorbate equilibrium on the AFM pull-off force measured in water and alcohol vapor environments [9, 10]. Their capillary force model uses the circular approximation with zero contact angles and includes the adsorbate films on both tip and substrate surfaces which are in equilibrium with the vapor. The good agreement between their model and experimental data obtained for alcohol vapors [10] motivated the present work. In the case of water, the agreement is poor and the discrepancy is attributed to the presence of solid-like structured water [9]. Therefore, only alcohol vapors are considered here.

In this paper, the effects of the adsorbate film (disjoining layer) in equilibrium with the vapor are considered by using the exact solution of the Young–Laplace equation for an axisymmetric meniscus (commonly known as a pendular-ring geometry) [14]. It is compared with the circular approximation using a simple trigonometric relation to estimate the meridional and axial radii of the meniscus [9, 10]. Both formulations use the adsorption isotherm and the Kelvin equation [10]. A polynomial fit describing the thickness of the adsorbate layer at a given vapor pressure relative to the saturation pressure, $h(p/p_{sat})$, has been generated to represent the experimentally determined isotherms of alcohols on clean SiO_2 surfaces [10]. The tip radius is varied in the models from 10 nm to 1 μm, mimicking typical AFM experiments. Both models show that as p/p_{sat} decreases from the saturation, the capillary force increases and reaches a maximum at $p/p_{sat} \sim 0.15$. The trend and magnitude calculated from these methods are in agreement with experimentally observed behavior for alcohol vapors. It is also confirmed that when the adsorption isotherm thickness is ignored, the capillary force is fairly insensitive to the alcohol vapor pressure except near the saturation vapor pressure region where the capillary force is predicted to diminish sharply.

2. Theoretical Calculation Details

In Section 2.1, we show that the capillary pressure of the meniscus and the disjoining pressure of the adsorbed layer are equal if both are in thermodynamic equilibrium with the vapor. In Sections 2.2 and 2.3, we derive exact and approximate models for the effect of disjoining pressure on capillary force as a function of p/p_{sat}. The assumptions are that the sphere and substrate are elastically rigid, and that the disjoining layer forms a continuous film. The disjoining layer is the adsorbate film which is in equilibrium with the vapor and the capillary meniscus. The calculation methods and results will be presented and discussed in Section 3, where they will also be compared to each other as well as to the theory which does not include the disjoining layer.

2.1. Equivalence of Capillary and Disjoining Pressures

The fundamental equation of capillarity, the Young–Laplace equation, is derived from a surface curvature argument [18]. Accordingly, a curved liquid–vapor interface supports a pressure difference

$$\Delta P = \gamma / r_e. \tag{1}$$

This quantity is equal to the capillary pressure. Here, the liquid surface tension is γ and the effective radius of curvature of the surface is r_e as defined by

$$\frac{1}{r_e} = \frac{1}{r_a} + \frac{1}{r_m}, \tag{2}$$

where r_a (the azimuthal radius) and r_m (the meridional radius) are the principal radii of curvature of the surface, as shown in Fig. 2(a). These quantities are positive when the center of the radius is inside the meniscus (r_a) and negative when outside (r_m). The liquid is assumed to be isobaric, meaning that r_e is constant along the meniscus surface, while r_a and r_m may vary along the surface.

The Kelvin equation is derived from thermodynamic equations and the Young–Laplace equation [18]. First, for a reversible process at constant temperature, the effect of the change in mechanical pressure on the free energy G_f of a substance is

$$\Delta G_f = \int V_m \, dP, \tag{3}$$

where V_m is the molar volume of the liquid. Assuming constant molar volume of the liquid substance V_m, and applying the Young–Laplace equation, equation (1), we obtain

$$\Delta G_f = \gamma V_m / r_e. \tag{4}$$

Thermodynamics relates the free energy of a substance to its vapor pressure. Assuming the vapor to be ideal,

$$\Delta G_f = R_u T \ln(p/p_{sat}), \tag{5}$$

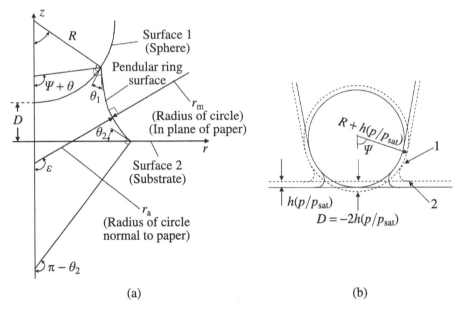

(a)   (b)

Figure 2. (a) Typical geometry considered for a pendular ring or liquid bridge between a sphere and a flat substrate [30]. {Reprinted with permission from Elsevier.} (b) Modified pendular ring geometry for a system where the equilibrium adsorbate layers are present on both the sphere and the substrate. In this modified geometry, $D = -2h(p/p_{sat})$. The interpenetration of the sphere into the surface 2 of (a) does not affect the force because the Laplace–Young solution treats only the capillary force and not the contact mechanics problem. In this paper, the force is calculated at point 1. Although individual components (axial surface tension and Laplace pressure) contributing to the capillary force vary with the position, the total capillary force does not vary with the position because the system is in mechanical equilibrium.

where R_u is the universal gas constant and T is temperature (K). Equating (4) and (5), and assuming thermodynamic equilibrium, the Kelvin equation [18] is found.

$$r_K = \frac{\gamma V_m}{[R_u T \ln(p/p_{sat})]}. \tag{6}$$

For example, $r_K = \frac{0.53}{\ln(p/p_{sat})}$ nm for water at 300 K. At thermodynamic equilibrium, r_e equals the Kelvin radius r_K, a negative number for $0 < p/p_{sat} < 1$.

Also, at thermodynamic equilibrium the disjoining pressure $P(h)$ of the adsorbed layer is related to the partial pressure of the vapor by [11],

$$\upsilon P(h) = k_B T \ln(p/p_{sat}). \tag{7}$$

Here, h is the equilibrium thickness of the film, υ is the liquid film molecular volume, k_B is Boltzmann's constant, T is the temperature and p/p_{sat} is the relative vapor pressure. According to Mate [27], the term $\upsilon P(h)$ represents the molecular interaction of the liquid film with the surface relative to that with the bulk liquid, i.e., the Gibbs free energy per molecule.

Changing (5) to its molecular form and equating it to (7), we have

$$P(h) = \Delta P. \tag{8}$$

For an undersaturated vapor, $\ln(p/p_{sat})$ is negative, so the pressures are negative and equal to each other, and both are in thermodynamic equilibrium with the vapor at partial pressure p/p_{sat}. Hence in thermodynamic equilibrium between the vapor, adsorbate and meniscus, the disjoining and capillary pressures are known from p/p_{sat}. Likewise, the adsorbate thickness can be measured experimentally. The theoretical models then calculate the capillary force and can be tested by comparing the calculated force *versus* p/p_{sat} trend with the measured data of Fig. 1.

2.2. Exact Solution of Laplace–Young Equation for an Axisymmetric Meniscus and Equilibrium Adsorbate Layers

The geometry of a liquid bridge spanning two solid surfaces separated by a distance D is represented in general as Fig. 2(a). This geometry is referred to as a pendular ring. The governing equation for the meniscus profile is [14]

$$R/r_K = -du/dx - u/x, \tag{9}$$

where $u = -\sin(\varepsilon)$ and the normalized meniscus coordinates are $x = r/R$ and $y = z/R$. The angle ε is formed between the negative z-axis and the normal to the meniscus. The two terms in (9) represent the meridional and the azimuthal curvatures, respectively. The boundary conditions at the points 2 and 1, respectively, are

$$u_2 = -\sin(\pi - \theta_2), \qquad y_2 = 0, \tag{10a}$$
$$u_1 = -\sin(\theta_1 + \Psi), \qquad y_1 = d + 1 - \cos(\Psi), \tag{10b}$$

where $d = D/R$ is the normalized separation between the sphere and the substrate and Ψ is often called the filling angle. θ_1 and θ_2 are contact angles of liquid with the sphere and substrate, respectively. Note that the contact angles are close to zero for water and alcohols on clean hydrophilic silicon oxide surfaces. The solution to the boundary value problem has been given by Orr *et al.* [14]. Corresponding to different meniscus shapes, $\pm$ symbols occur. The various cases must be carefully considered to implement the correct solution. When $r_K < 0$, as for an undersaturated vapor, the boundary value problem solution for the mean curvature simplifies to

$$R/r_K = \left(\frac{1}{y_1}\right) \int_{(\pi - \theta_2)}^{(\theta_1 + \Psi)} \left(\sin(\varepsilon) - \frac{\sin^2(\varepsilon)}{(\sin^2(\varepsilon) + c)^{1/2}}\right) d\varepsilon, \tag{11}$$

where the curvature-dependent parameter c is

$$c(R, r_K, \Psi, \theta_1) = (R/r_K)^2 \sin^2(\Psi) - 2(R/r_K)\sin(\Psi)\sin(\theta_1 + \Psi). \tag{12}$$

The normalized meniscus coordinates are given by

$$x = -\frac{r_K}{R}[-\sin(\varepsilon) + (\sin^2(\varepsilon) + c)^{1/2}], \tag{13}$$

$$y = \left(\frac{r_K}{R}\right) \int_{(\pi-\theta_2)}^{(\varepsilon)} \left(\sin(\varepsilon) - \frac{\sin^2(\varepsilon)}{(\sin^2(\varepsilon) + c)^{1/2}}\right) d\varepsilon. \qquad (14)$$

The axial components of the surface tension force F_s, the Laplace pressure force F_p and the total capillary force F_c are:

$$F_s = 2\pi\gamma R \sin(\Psi)\sin(\theta_1 + \Psi), \qquad (15a)$$

$$F_p = -\pi\gamma R^2 \sin^2(\Psi)/r_K, \qquad (15b)$$

$$F_c = F_S + F_p = 2\pi\gamma R[\sin(\Psi)\sin(\theta_1 + \Psi) - R\sin^2(\Psi)/(2r_K)]. \qquad (15c)$$

The sign convention for the force is that positive forces are attractive. For convenience, equation (15) is calculated at point 1, but the force is the same at any point along the meniscus. For example, F_c at point 2 can easily be calculated and is the same as at point 1.

Here we shall consider only the capillary force when the nano-scale tip is in contact with the substrate. This force can be measured in AFM experiments when the spring constant is small, because after pull off, the spring pulls away from the capillary [28]. Both tip and substrate are fully covered with the adsorbate film. This is the disjoining layer which is in equilibrium with the vapor and the capillary meniscus. The relationship between the thickness of the adsorbate film *versus* p/p_{sat}, $h(p/p_{sat})$, can be independently measured using various experimental techniques such as vibrational spectroscopy, ellipsometry, quartz crystal microbalance, etc. Consideration of the effect of the disjoining layer leads to the recognition that the pendular ring method exactly describes the problem at hand when the transformation

$$R \rightarrow R + h(p/p_{sat}) \qquad (16a)$$

is made and the normalized separation is equated with a negative displacement according to

$$d(p/p_{sat}) = -2h(p/p_{sat})/R, \qquad (16b)$$

as represented by the dashed lines in Fig. 2(b). The apparent interpenetration of the dashed sphere and the dashed substrate does not affect the force because the pendular ring solution treats only the capillary force and not the contact mechanics problem. In any case, the actual penetration between solid surfaces remains equal to zero, as represented by the solid lines in Fig. 2(b).

2.3. Circular Approximation with a Simple Geometric Relation

In the circular approximation, the meridional profile of varying radius r_m is replaced by a circle of fixed radius r_1 as illustrated in Fig. 3. Also, the azimuthal radius r_a is replaced by a radius r_2. We shall call r_1 and r_2 in the circular approximation the external and cross-sectional capillary radii, respectively. Equation (2) is then rewritten with r_1 and r_2 replacing r_m and r_a respectively.

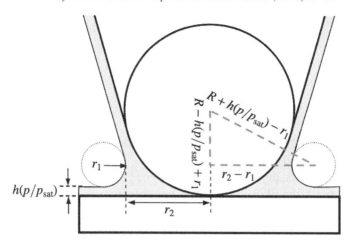

Figure 3. Circular (toroidal) approximation where the meridional curvature is replaced with a circle with a fixed radius, r_1 (external radius). Note that r_1 is negative because the center of the circle is outside the meniscus; so $-r_1$ is positive. At the position where r_1 and r_2 (cross-sectional radii) are parallel to the substrate, the Pythagorean theorem can be used to relate the tip radius, adsorbate thickness and two principal radii (see text).

When finite contact angles are considered at the liquid–solid–vapor three-phase line (in this case, the adsorption isotherm is ignored and only the equilibrium between the meniscus and the vapor phase is considered), the position of the circle (cross section of the torus) is adjusted such that the contact angles with the sphere and the substrate are consistent with the assumed contact angle values and the external and cross-sectional radii satisfy the mean curvature relationship at a given p/p_{sat}. However, when the adsorbate film is present on both silicon oxide tip and substrate surfaces, the contact angle of the meniscus edge to this disjoining layer is zero. This is because the molecules consisting of the meniscus and the adsorbate layer are the same. Thus, the adsorbate films on the sphere and substrate will meet with the circle at the tangential point as shown in Fig. 3. Notice that this tangential point is on the line connecting the center of the tip and the center of the external capillary radius. Hence, one can relate r_1 and r_2 with R and $h(p/p_{sat})$ using the Pythagorean relation:

$$[R - h(p/p_{sat}) + r_1]^2 + [r_2 - r_1]^2 = [R + h(p/p_{sat}) - r_1]^2. \qquad (17)$$

Note that r_1 is negative because the center of the circle is outside the meniscus. This equation can be solved for r_2:

$$r_2 = 2\sqrt{R \times (h(p/p_{sat}) - r_1)} + r_1. \qquad (18)$$

Once the adsorption isotherm, $h(p/p_{sat})$, is determined experimentally, then for a given $h(p/p_{sat})$ equations (18) and (2) expressed in terms of r_1 and r_2 (with $r_K = r_e$), can be solved simultaneously to determine r_1 and r_2. Once r_2 is found for a given R and $h(p/p_{sat})$, then the axial surface tension force F_s, the Laplace pressure force F_p, and the capillary force F_c (positive when attractive) are:

$$F_s = 2\pi r_2 \gamma, \tag{19a}$$

$$F_p = -\pi r_2^2 \frac{R_u T}{V_m} \ln \frac{p}{p_{sat}}, \tag{19b}$$

$$F_c = F_s + F_p. \tag{19c}$$

These are the forces calculated for the point where the r_1 vector is parallel to the substrate, i.e., ε defined in Fig. 2(a) is $90°$.

To calculate the force at point 1 of Fig. 2(b) (i.e., the same place as in the pendular ring exact solution), the filling angle Ψ is found from

$$\Psi = \sin^{-1}\left(\frac{r_2 - r_1}{R + h - r_1}\right) \quad (R - h > -r_1), \quad \text{or} \tag{20a}$$

$$\Psi = \frac{\pi}{2} + \sin^{-1}\left(\frac{-r_1 - (R - h)}{R + h - r_1}\right) \quad (-r_1 > R - h). \tag{20b}$$

The distance from the z-axis to point 1 is

$$r' = r_2 - r_1(1 - \sin(\Psi)). \tag{21}$$

The forces F_s, F_p and F_c can be calculated using the following equations:

$$F_s = 2\pi r' \gamma \sin(\Psi), \tag{22a}$$

$$F_p = -\pi (r')^2 \frac{R_u T}{V_m} \ln \frac{p}{p_{sat}}, \tag{22b}$$

$$F_c = F_p + F_s. \tag{22c}$$

Note that F_s and F_p calculated with equations (22a) and (22b) will be different from those calculated with equations (19a) and (19b) because the calculation position is changed. But, the changes in F_s are compensated by the changes in F_p and the total capillary force should be balanced along the meniscus surface. In the circular approximation, this compensation cannot be perfect since r_1 is fixed along the entire meniscus surface. However, the difference in F_c predicted between these two positions is very little (see Section 3.2) and negligible compared to typical experimental errors in AFM measurements.

3. Results and Discussion

In order to calculate the p/p_{sat} dependence of the capillary force accurately, one needs to know the exact adsorption isotherm, $h(p/p_{sat})$. In the previous study [10], it was found that the adsorption isotherms of ethanol, n-butanol, and n-pentanol were nearly the same. It is reasonable to use a best-fit average of all three isotherms to represent a general behavior of the capillary force exerted by alcohols. A fifth-order polynomial fit of the measured thickness of the disjoining layer *versus* p/p_{sat}

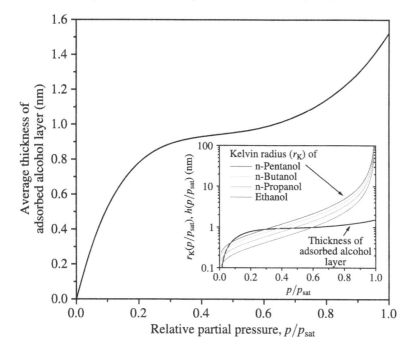

Figure 4. Model adsorption isotherm of alcohol on clean SiO$_2$ surface generated by averaging the adsorption isotherms of ethanol, n-butanol and n-pentanol at room temperature [10]. The inset compares the adsorption isotherm thickness with the Kelvin radius of various alcohols.

is [10]

$$h\left(\frac{p}{p_{sat}}\right) = 7.03\left(\frac{p}{p_{sat}}\right)^5 - 22.97\left(\frac{p}{p_{sat}}\right)^4 + 31.76\left(\frac{p}{p_{sat}}\right)^3$$
$$- 21.40\left(\frac{p}{p_{sat}}\right)^2 + 7.10\left(\frac{p}{p_{sat}}\right) - 0.003, \tag{23}$$

where the unit of h in equation (23) is nanometers. Figure 4 shows the empirical model isotherm curve of alcohols used in this study. For reference, the Kelvin radius, equation (6), is also plotted in the inset. Table 1 shows the γ_1 and V_m values used in this calculation. For comparison purposes, n-pentanol is assumed in Figs 5–7. As discussed in the Introduction, the simplest model for the capillary force is $F_c = 4\pi R\gamma \cos\theta$ independent of p/p_{sat} [11], and hence with $\theta_1 = \theta_2 = 0$, the force plots are normalized by $4\pi R\gamma$.

3.1. Exact Solution of Laplace–Young Equation

The calculation procedure for the pendular ring formulation is as follows. For each p/p_{sat} value, r_K is calculated from equation (6), and the value of $R' = R + h$ is determined from equation (23). Using equation (11), the value of Ψ is found iteratively using $y_1 = -2h/R + 1 - \cos(\Psi)$, and the value of c is then specified by equation (12). The capillary force can be found from equation (15c). The meniscus

Table 1.

Surface tension and molar volume of alcohols studied

	Surface tension (mN/m)	Molar volume (cm^3/mole)
Ethanol	21.8	58.4
n-Propanol	23.7	74.8
n-Butanol	24.6	91.5
n-Pentanol	24.9	108.7

profile is then established by equations (13) and (14). This procedure was carried out over the range of p/p_{sat} in a short loop using Mathcad™ software. As a check, plots were generated and overlaid with Fig. 2 of Bowles *et al.* [25], and agreement was excellent.

Results for the meniscus profiles for a range of p/p_{sat} values are given in Fig. 5(a). Near the substrate, each profile makes contact with the adsorbate layer of thickness according to equation (22). The sphere segments represent the transformation $R \rightarrow R + h(p/p_{sat})$.

The normalized components of the capillary force at point 1 and the normalized total capillary force $F_c/(4\pi R\gamma)$ are plotted *versus* p/p_{sat} in Fig. 6. As seen in Fig. 6(a) and (b), the Laplace pressure force is the main component of the force at point 1 except as p/p_{sat} approaches 1. $F_c/(4\pi R\gamma)$ is independent of where the force is calculated along the meniscus, while at point 2 the axial surface tension force is zero. As $p/p_{sat} \rightarrow 1$, the thickness of disjoining layer becomes negligible compared to the Kelvin radius. Here, $F_c/(4\pi R\gamma)$ in Fig. 6(c) decreases below one for all values of R. At point 1, this can be understood because as the Kelvin radius becomes very large, F_s is limited by the size of the sphere while F_p approaches zero. This effect is, therefore, more noticeable as R decreases.

As p/p_{sat} decreases below 1, $F_c/(4\pi R\gamma)$ increases above 1. In this realm, the disjoining layer thickness adds substantially to the meniscus area. Its relative importance continues to increase as p/p_{sat} decreases because its thickness does not fall off as quickly as does the Kelvin radius, as can be seen in Fig. 4.

For p/p_{sat} decreasing below 0.15, $F_c/(4\pi R\gamma)$ reduces to ~1. Here, the disjoining layer thickness again becomes negligible, and the expected value of 1 arises. At low p/p_{sat}, the data in Fig. 1 qualitatively do not agree with this trend. This is most likely due to inter-solid adhesion forces occurring when the surface coverage of the adsorbed alcohol is less than a monolayer [29], which are not accounted for in the pure capillary model. Their origin is discussed in more detail below.

3.2. Circular Approximation with Pythagorean Relation

By comparing the results from the exact solution to the results from the circular approximation, we can determine the degree to which the simpler latter method is valid. Figure 5(b) shows that the external radius (r_1) agrees well with r_m for

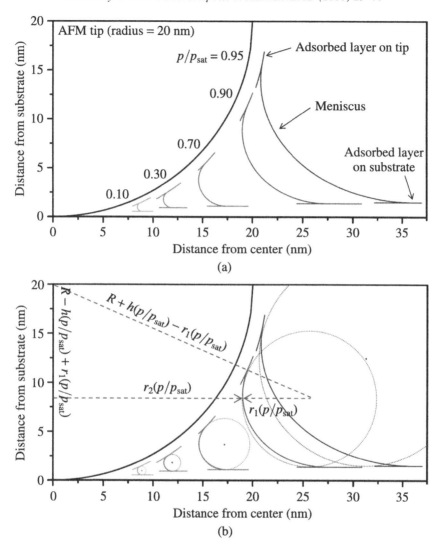

Figure 5. (a) Meniscus shape of n-pentanol obtained with the Laplace–Young exact solution for various p/p_{sat}. Also shown are the adsorbed n-pentanol layers on the tip and the substrate. (b) Comparison of the meniscus shapes calculated with the circular approximation (dotted lines) and the Laplace–Young exact solution (solid lines) for $p/p_{sat} = 0.1, 0.3, 0.7, 0.9$ and 0.95. The trigonometric relation among r_1, r_2, h and R is shown for the $p/p_{sat} = 0.90$ case.

$0 < p/p_{sat} \leqslant 0.90$ for $R = 20$ nm, and that the meniscus profiles are nearly circular. This agreement will extend to a wider p/p_{sat} range as R increases. Thus, the two methods agree well. The circular approximation tends to hold for r_m even for $D > 0$ for the constant pressure case since the Kelvin radius is fixed by p/p_{sat} (thermodynamic equilibrium). Note that the circular approximation does not work well for $D > 0$ in the constant volume case [30]. Furthermore, the cross-sectional radius

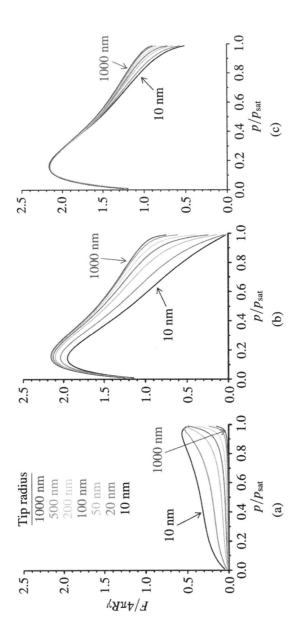

Figure 6. Capillary forces of n-pentanol calculated with (a)–(c) the Laplace–Young exact solution and (d)–(f) circular approximation as a function of p/p_{sat} (from 0.01 to 0.99) for different tip radii (10, 20, 50, 100, 200, 500, 1000 nm). (a), (d) Surface tension force and (b), (e) Laplace pressure force components calculated at point 1 as shown in Fig. 2(b). (c), (f) Total capillary force. The inset in (d) shows the filling angle calculated with the circular approximation, equation (20).

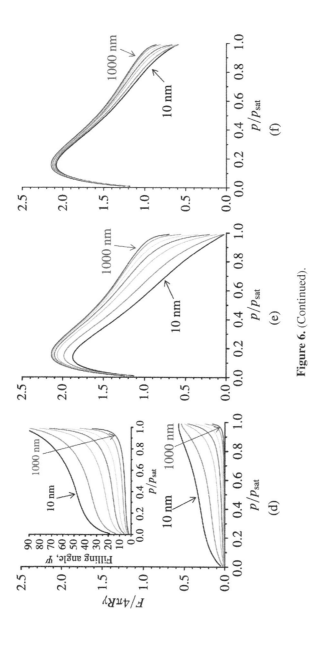

Figure 6. (Continued).

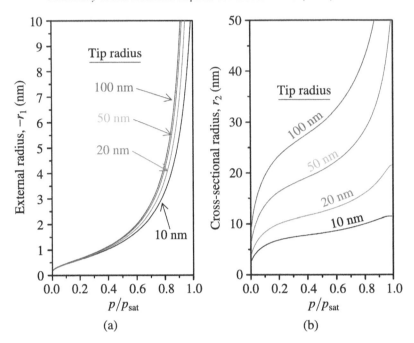

Figure 7. (a) External and (b) cross-sectional radii of the n-pentanol meniscus as a function of the alcohol partial pressure, $p/p_{sat} = 0.01–0.99$, for various tip radii calculated from the circular approximation, equations (2) and (18).

(r_2) agrees closely with the azimuthal radius r_a when parallel to the substrate. In this orientation r_1 and r_2 are nearly true principal radii.

Figure 7 plots how r_1 and r_2 vary as a function of p/p_{sat} of n-pentanol for R varying from 10 nm to 100 nm. In Fig. 7(a), the external radius (r_1) starts from ~0.3 nm at $p/p_{sat} = 0.03$ and increases slowly to 1.7–2.1 nm at $p/p_{sat} = 0.6$. Beyond this point, r_1 increases quickly with p/p_{sat}. Overall, the tip radius (R) dependence of r_1 is relatively weak until p/p_{sat} reaches ~0.8. The r_1 value at $p/p_{sat} = 0.03$ (which corresponds to a diameter of 0.6 nm) is very close to the effective size of n-pentanol molecules on the surface estimated from the adsorption isotherm measurement and the hard sphere diameter determined from self-diffusion coefficient measurements [21]. When p/p_{sat} reaches 0.15–0.18, the external diameter ($2 \times r_1$) of the meniscus becomes close to two molecular layers. It should be noted that the r_1 values shown here are just an estimate calculated from continuum models; the exact curvature or shape of the meniscus will be different especially when the meniscus height is only a few molecular layers tall.

Figure 7(b) plots r_2 as a function of p/p_{sat} for R varying from 10 nm to 100 nm. The cross-sectional radius (r_2) shows a strong dependence on the tip radius (R). For a given R, r_2 increases quickly as p/p_{sat} increases from 0.01 to ~0.2 and ~0.6–0.99. For $R = 10$ nm, r_2 increased to a value comparable to the size of the tip radius at $p/p_{sat} = 0.8$. For $R = 20$ nm, r_2 gets close to R at $p/p_{sat} \sim 0.95$.

The individual components from the surface tension and the Laplace pressure as well as the total capillary force at point 1 are calculated using equation (22) and plotted as a function of p/p_{sat} for varying tip radii in Fig. 6 for comparison with the exact solution results. For $R = 10$ nm, the circular approximation result deviates only 3%, on average, from the exact solution result when $p/p_{sat} < 0.97$. As R increases, the deviation is reduced. When $R = 100$ nm and 1000 nm, the average difference between the circular approximation and the exact solution is only 1.0% and 0.4%, respectively, for $p/p_{sat} < 0.90$. To the best of our knowledge, this is the first direct comparison between the exact and circular approximation methods for the nano-scale contact covered with an adsorbate film.

Regardless of the calculation point, the agreement between the circular approximation and the exact solution is remarkable. The capillary force calculated with equation (19c) is about 1% larger than the value calculated with equation (22c) at $p/p_{sat} \sim 0.15$ where F_c is the maximum. The difference between these two approximation results decreases to $\sim$0.5% as p/p_{sat} increases.

3.3. Comparison with the Experimental Data and the Case where the Adsorption Isotherm is Ignored

The predicted values of $F_c/(4\pi R\gamma)$ are plotted *versus* p/p_{sat} for short chain alcohols from ethanol through n-pentanol in Fig. 8(a) and (b) for the pendular ring exact solution and the circular approximation models, respectively. The surface tensions for these substances are approximately the same, while the molar volumes vary. The smallest molar volume gives rise to the smallest Kelvin radius at a fixed p/p_{sat}, according to equation (6). Thus, a larger value of $F_c/(4\pi R\gamma)$ should be observed for the smaller alcohol molecule when the $h(p/p_{sat})$ trend is nearly the same for all alcohols. The molar volume dependence of the capillary force is in good agreement with the experimental data shown in Fig. 1 [10].

There are some discrepancy between the theoretical calculation and the experimental data. The theory predicts that the AFM pull-off force measured for clean silicon oxide surfaces increases as the alcohol p/p_{sat} decreases from the saturation, reaches a maximum at $p/p_{sat} \sim 0.15$, then decreases as p/p_{sat} approaches zero. In the low partial pressure regime ($p/p_{sat} < 0.15$), the experimental data (Fig. 1) show that the measured force keeps increasing. This discrepancy must be mainly because the coverage of the adsorbed alcohol is less than a monolayer [21]. There may be direct interactions between silicon oxide surfaces and silanol groups of the two contacting surfaces, which are not considered in the capillary force calculation. In addition, the molecular orientations at sub-monolayer coverages are different from the full coverage film condition [21]. However, it is noteworthy that the experimental data show a small shoulder at $p/p_{sat} = 0.15$–0.2, which is the vapor pressure region where the capillary force is predicted to be the maximum.

Figure 9 compares the capillary forces calculated from the Young–Laplace exact solution (Fig. 9(a)) and circular approximation (Fig. 9(b)) methods for two cases: (1) the adsorption isotherm is taken into account, i.e., both the adsorbate film and

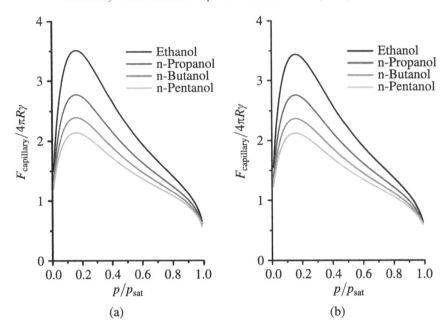

Figure 8. Capillary forces as a function of relative partial pressure (from 0.01 to 0.99) of ethanol, n-propanol, n-butanol and n-pentanol calculated with (a) the Laplace–Young exact solution method and (b) the circular approximation method at point 1 as shown in Fig. 2(b) for an AFM tip with $R = 20$ nm.

the meniscus are in equilibrium with the vapor phase (solid lines in Fig. 9) and (2) the adsorption isotherm is ignored, i.e., $h(p/p_{sat}) = 0$ nm, and only the meniscus is allowed to be equilibrated with the vapor phase (dotted lines in Fig. 9). The latter is the case considered in previous publications by others in the literature [15, 16]. The capillary force does not show any significant p/p_{sat} dependence. In the circular approximation, the cross-sectional radius of the meniscus (r_2) is found to be much smaller in the case where $h = 0$ is assumed compared to the case taking into account the adsorption isotherm. For $R = 20$ nm, for example, the r_2 value of the $h = 0$ case is 34% smaller than that of the $h(p/p_{sat})$ case at $p/p_{sat} = 0.1$, 33% smaller at $p/p_{sat} = 0.3$, 20% smaller at $p/p_{sat} = 0.7$, 13% smaller at $p/p_{sat} = 0.9$, and 10% smaller at $p/p_{sat} = 0.97$. Therefore, this reduction in the cross-sectional area results in a substantial lower F_p ($= \pi r_2^2 \Delta P$). One should note that this is the case where the liquid meniscus is formed due to equilibrium with the vapor phase, but the adsorbate layer is assumed not to be formed on the surface. This is physically possible only when the adsorbate molecule and the surface have unfavorable interactions. For example, water on low surface energy substrates such as hydrophobic surfaces will fall into this category. But in this situation, the contact angle will be close to 90°, instead of zero degrees. So, when the meniscus contact angle is zero, the effect of adsorbate layer on the surface must be taken into account.

Note that as p/p_{sat} decreases below 0.01, both models predict $F_c/(4\pi R\gamma)$ to be around unity. This is an artifact of the continuum model. As p/p_{sat} approaches

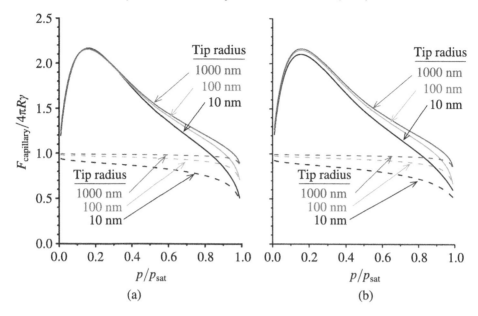

Figure 9. Comparison of (a) the Laplace–Young exact solution and (b) the circular approximation methods for cases with and without the adsorption isotherm. Solid lines are the results calculated with the adsorption isotherm (both the adsorbate film and the meniscus are in equilibrium with the vapor phase). Dashed lines are the results calculated without any adsorbed layer (only the meniscus is in equilibrium with the vapor phase). The plots are for the p/p_{sat} range of 0.01–0.99.

zero, the filling angle Ψ and hence the meniscus area decrease to zero. However, the Kelvin radius also approaches zero, and, therefore, the Laplace pressure approaches infinity. The product of the meniscus area and the Laplace pressure, equal to the capillary force, remains finite and decreases as the liquid contact angle increases. However, the validity of the continuum theory is questionable in this low pressure regime. In fact, we already mentioned that the Kelvin radius becomes smaller than the size of molecules when p/p_{sat} is lower than ~0.03. Therefore, one could presume the lower p/p_{sat} limit of the continuum theory to be ~0.03.

4. Conclusion

The importance of taking into account the adsorbate film is explained theoretically. Unless the partial pressure of the molecule in the gas phase (p/p_{sat}) is extremely low (as in ultra-high vacuum conditions), the adsorbate film will exist on the solid surface. Therefore, the adsorbate film, the capillary meniscus, and the vapor equilibrate with each other. Exact and approximate models for the capillary force *versus* p/p_{sat} have been developed for the sphere-on-flat geometry in which both sphere and flat surfaces are covered with adsorbate films. The agreement between the exact and approximate methods is excellent. The models show that without taking into account the adsorbate film, the theory cannot explain the large vapor pressure dependence of the capillary force. In the low vapor pressure regime where the solid surface is not fully covered with the adsorbate molecule, additional terms must be considered to explain the experimental data.

Acknowledgements

This work was supported by the National Science Foundation (Grant No. CMMI-0625493). Sandia is a multiprogram laboratory operated by Sandia Corporation, a Lockheed Martin Company, for the United States Department of Energy's National Nuclear Security Administration under contract DE-AC04-94AL85000.

References

1. S. H. Kim, D. B. Asay and M. T. Dugger, *Nano Today* **2**, 22 (2007).
2. D. B. Asay, M. T. Dugger, J. A. Ohlhausen and S. H. Kim, *Langmuir* **24**, 155 (2008).
3. D. B. Asay, M. T. Dugger and S. H. Kim, *Tribol. Lett.* **29**, 67 (2008).
4. M. Binggeli and C. M. Mate, *Appl. Phys. Lett.* **65**, 415 (1994).
5. X. Xiao and L. Qian, *Langmuir* **16**, 8153 (2000).
6. R. Jones, H. M. Pollock, J. A. S. Cleaver and C. S. Hodges, *Langmuir* **18**, 8045 (2002).
7. M. He, A. S. Blum, D. E. Aston, C. Buenviaje and R. M. Overney, *J. Chem. Phys.* **114**, 1355 (2001).
8. M. Nosonovsky and B. Bhushan, *Phys. Chem. Chem. Phys.* **10**, 2137 (2008).
9. D. B. Asay and S. H. Kim, *J. Chem. Phys.* **124**, 174712 (2006).
10. D. B. Asay and S. H. Kim, *Langmuir* **23**, 12174 (2007).
11. J. N. Israelachvili, *Intermolecular and Surface Forces*, 2nd edn. Academic Press, San Diego, CA (1992).
12. J. S. McFarlane and D. Tabor, *Proc. Royal Soc. London* **A202**, 224 (1950).
13. D. Xu, K. M. Liechti and K. Ravi-Chandar, *J. Colloid Interface Sci.* **315**, 772 (2007).
14. F. M. Orr, L. E. Scriven and A. P. Rivas, *J. Fluid Mech.* **67**, 723 (1975).
15. T. Stifter, O. Marti and B. Bhushan, *Phys. Rev. B* **62**, 13667 (2000).
16. H.-J. Butt and M. Kappl, *Adv. Colloid Interface Sci.* **146**, 48 (2009).
17. K. Strawhecker, D. B. Asay, J. McKinney and S. H. Kim, *Tribol. Lett.* **19**, 17 (2005).
18. A. W. Adamson, *Physical Chemistry of Surfaces*, 5th edn. Wiley, New York (1990).
19. D. B. Asay and S. H. Kim, *J. Phys. Chem. B* **109**, 16760 (2005).
20. A. L. Barnette, D. B. Asay and S. H. Kim, *Phys. Chem. Chem. Phys.* **10**, 4981 (2008).
21. A. L. Barnette, D. B. Asay, M. J. Janik and S. H. Kim, *J. Phys. Chem. C* **113**, 10632 (2009).
22. D. B. Asay, A. L. Barnette and S. H. Kim, *J. Phys. Chem. C* **113**, 2128 (2009).
23. B. V. Derjaguin and N. V. Churaev, *J. Colloid Interface Sci.* **49**, 249 (1974).
24. C. M. Mate, M. R. Lorenz and V. J. Novotny, *J. Chem. Phys.* **90**, 7550 (1989).
25. A. P. Bowles, Y.-T. Hsia, P. M. Jones, J. W. Schneider and L. R. White, *Langmuir* **22**, 11436 (2006).
26. Z. Wei and Y. P. Zhao, *J. Phys. D. Appl. Phys.* **40**, 4368 (2007).
27. C. M. Mate, *J. Appl. Phys.* **72**, 3084 (1992).
28. B. Cappella and G. Dietler, *Surf. Sci. Rep.* **34**, 1 (1999).
29. A. Fogden and L. R. White, *J. Colloid Interface Sci.* **138**, 414 (1990).
30. M. P. de Boer and P. C. T. de Boer, *J. Colloid Interface Sci.* **311**, 171 (2007).

Which Fractal Parameter Contributes Most to Adhesion?

D.-L. Liu [a], **J. Martin** [b] **and N. A. Burnham** [a]

[a] Department of Physics, Worcester Polytechnic Institute, Worcester, MA 01609-2280, USA
[b] Micromachined Products Division, Analog Devices Incorporated, Cambridge, MA 02139-3541, USA

Abstract
Surfaces can be characterized by three fractal parameters: root-mean-square (RMS) roughness, roughness exponent and lateral correlation length. Little work has been done on correlating these parameters with adhesion. In this study, we simulated the adhesion between an atomic force microscope (AFM) tip and sample surfaces with varying fractal parameters. And experimentally, we performed adhesion measurements on polycrystalline silicon sidewalls of varying topography using AFM. Both the simulations and the experimental data support the conclusion that surface roughness is a significant predictor of adhesion, with the adhesion dropping by more than an order of magnitude for a roughness change from 1 to 10 nm. For the roughness exponent, the simulations reveal a 20% decrease in adhesion as the roughness exponent varies from 0 to 1. The scatter in the experimental data was large since the range of the roughness exponent varied only from 0.85 to 0.99. For the lateral correlation length, the experiment showed a wide range of adhesion values for smaller correlation lengths and low adhesion for larger correlation lengths and we are still investigating the theoretical basis of this observation. Although much work is still needed, the work presented here should advance the fundamental understanding of the role of nanoscale fractal roughness in adhesion and might enable better design of future nanoscale devices.

Keywords
Adhesion, roughness, AFM, MEMS

1. Introduction

Industry is pushing towards the nanoscale, which provides the driving force for a better understanding of adhesion physics [1]. For example, adhesion ('stiction') is a significant cause of failure in microelectromechanical systems (MEMS). The microstructures in MEMS devices are susceptible to stiction because they typically have closely spaced, mechanically compliant elements with a high surface-to-volume ratio. To avoid stiction in MEMS devices, adhesion forces should be minimized [2–4].

It has been known for a long time that surface roughness is very important in determining the interaction force between contacting surfaces. Since the contacting surfaces are always rough (even though the roughness might be limited to the atomic scale), actual contact area occurs at the peaks of the inevitable surface irreg-

Adhesion Aspects in MEMS/NEMS

ularities, where the local contact pressure can be very high [5, 6]. Furthermore, the details of the distribution of asperities can also affect the adhesion [7–9]. However, how roughness affects adhesion is still not well understood at a fundamental level.

Many contact mechanics models, such as the Hertz model [10], the Johnson–Kendall–Roberts (JKR) model [11], and the Derjaguin–Muller–Toporov (DMT) model [12, 13], can be used to describe the surface forces experienced between a spherical tip on an AFM cantilever and a sample surface. With these models, surface forces such as adhesion and friction were also studied by experimentalists using AFM. However, all these adhesion models do not explicitly take into account the roughness of either the probe or the substrate. Surface roughness can significantly alter the true contact area between the probe and the substrate from that predicted by contact mechanics models, making analysis of the measured adhesion or 'pull-off' forces very difficult. Persson's more recent scaling approach [14, 15] does not consider individual asperities, but assumes the surface is smooth and differentiable at small scales.

Large-scale computational materials science now permits hierarchical studies of a few thousand atoms from first principles, or, of billions of atoms with less precise interatomic potentials. In 2005, Luan and Robbins [16, 17] studied contact between a rigid cylinder or sphere and a flat elastic substrate using molecular dynamics simulations, and the surfaces were composed of discrete atoms in their modeling. They predicted a breakdown of continuum models and suggested that contact areas and yield stresses would be underestimated by continuum theory and the friction and contact stiffness would be overestimated. It was found that atomic-scale roughness can have a large influence on adhesion, and that an atomistic approach is essential for predicting adhesion.

In comparison, recent experimental studies support a different interpretation. Many MEMS devices have a cantilever that bends or twists during operation, the size of which is comparable to the size of AFM tips. This makes AFM ideal for understanding the interaction between surfaces that have MEMS size. The tips were naturally oxidized silicon, a common MEMS material. To investigate the role of few-asperity contacts in adhesion, Thoreson *et al.* [18, 19] measured the pull-off force between different sized AFM tips (with different roughnesses) and sample surfaces that had well-controlled material properties with roughnesses less than 0.4 nm. It was observed that the uncorrected pull-off force was independent of the radius of the AFM tip, which was contrary to all continuum-mechanics model predictions. To explain this behavior, they corrected for the measured root-mean-square (RMS) surface roughness of the AFM tips. Although the magnitudes were higher than expected, the simple correction for the RMS surface roughness resulted in the expected dependence of the pull-off force on radius.

Liu *et al.* [20] continued Thoreson's work and used a single-asperity model to describe a smooth tip in contact with a rough surface. They predicted that there would be an optimal size of asperity that will yield a minimum of adhesion. Adhesion measurements were performed on silicon wafers with varying roughness using

AFM cantilevers with varying tip radii. It was found that the asperities contributed to the adhesion when they were large, while they diminished it when they were small. The adhesion force was observed to fall very quickly at low roughness, and it fell at higher roughness for larger feature size. Optimal roughness for minimal adhesion exists for each tip radius as expected by the simple single-asperity model. This work suggests a promising way to minimize the adhesion between two surfaces by tuning asperity height to feature size in MEMS devices. Although there was no direct molecular dynamics simulation result for how atomic-scale surface roughness affected adhesion, the consistency between the experimental results and the modeling in this work indicates that continuum theory is a useful predictor of the behavior of nanoscale adhesive contacts.

Despite these advances, problems arise due to the strong assumption that the statistics of the surface height fluctuation is uncorrelated Gaussian [21, 22]. As early as 1966 when Greenwood and Williamson [7] studied the contact of rough surfaces, they suggested that while many surfaces will display a Gaussian-type distribution for asperity heights, some will not. One way of progressing further might be through self-affine fractal analysis.

The use of fractals in the understanding of rough surfaces is becoming increasingly common [23–26]. Self-affine fractals are objects that are invariant under affine transformation — any set of translation, rotation and scaling operations in the two spatial directions of the plane [27]. Many physical processes, such as etching, fracture, erosion and thin-film growth, can produce rough-surface morphologies, which exhibit scale-invariant behaviors [26]. The microstructure of the fractal surface or the surface roughness is described by three parameters: the interface width σ, the lateral correlation length ξ and the roughness exponent α. The amplitude of surface fluctuations, σ, describes the property of the surface at large distances, which is also known as the root-mean-square (RMS) roughness. The lateral correlation length ξ describes the wavelength of surface fluctuations. The roughness exponent α, also known as the Hurst exponent, describes how 'wiggly' the local surface is. Its magnitude is less than or equal to one.

For a given surface imaged by AFM, as in Fig. 1(a), the three surface parameters can be extracted from its height–height correlation function, shown in Fig. 1(b). $h(\mathbf{r})$ is the surface height at position $\mathbf{r}$ [$=(x, y)$] in the plane of the surface. The height–height correlation function $H(\mathbf{r})$, which is defined as $\langle[h(\mathbf{r}) - h(0)]^2\rangle$, can be calculated directly from the surface height data $h(\mathbf{r})$ of the AFM topography image and then fitted to $2\sigma^2 f(r/\xi)$ to obtain the surface parameters σ, α and ξ, where $f(r/\xi) = (r/\xi)^{2\alpha}$ for $r \ll \xi$, and $f(r/\xi) = 1$ for $r \gg \xi$ [26]. In Fig. 1(b), the height–height correlation function approaches $2\sigma^2$ at large r and its slope at smaller r is equal to 2α. The correlation length ξ is the maximum distance between different surface height data that can be considered correlated.

For Gaussian surfaces, the corresponding roughness exponent α has the well-known value of $\frac{1}{2}$ [23, 24]. The microscopic details for this case are not important on large scales, and this is known as universality. However, experimental results have

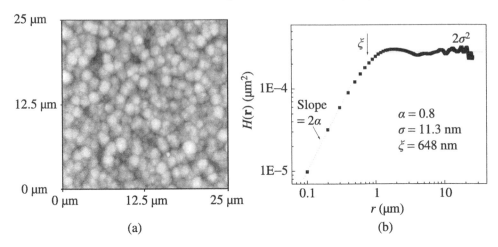

(a)   (b)

Figure 1. (a) AFM topographic image of a fractal rough surface. (b) Height–height correlation function $H(\mathbf{r})$ and its relation to the three fractal surface parameters RMS roughness σ, roughness exponent α and lateral correlation length ξ.

revealed that many rough surfaces behave anomalously with $\alpha \approx 0.6$–0.8, which exceeds $\frac{1}{2}$ [28–31]. Thus it is of crucial importance to the understanding of adhesion to go beyond Gaussian surfaces by investigating the influence of the microstructure of self-affine fractal surfaces.

In a theoretical study, Chow [32] developed a model that described solid–solid adhesion between deformable fractal surfaces. An integral equation was derived to predict the strength of adhesion as a function of the surface energy, the microstructure of fractal surfaces, and the effective elastic constant. Adhesion is maintained between a few asperities by van der Waals forces, depending on the distribution of asperity heights. Based on the general description of the microstructure of self-affine fractal surfaces, Chow obtained the analytical expressions for the asperity shape, for the nonlinear force–displacement relation, and for the non-Gaussian height probability function. Its solution predicts an orders-of-magnitude increase in adhesion as the roughness exponent decreases. Decreasing the ratio of RMS roughness to lateral correlation length will also increase adhesion. To date, there is no known experimental work related to this issue. Hence it is important to investigate the fundamental mechanisms on how the roughness exponent, the lateral correlation length, and the RMS roughness affect adhesion.

An assumption in Chow's method was that only the asperities on the surface contributed to the adhesion. In our work, numerical simulations of the tip-sample adhesion for surfaces with varying fractal parameters include both the asperities and the bulk of the sample. Experimentally, the data on MEMS sidewalls of varying topography — but with the same chemical treatment — were collected in order to unravel which of the three fractal parameters was the most important in determining adhesion. In the following, we describe our numerical and experimental approaches, present the results, discuss our conclusions, and state our remaining intriguing questions.

2. Method

2.1. Modeling the Adhesion between a Single Asperity and a Rough Fractal Surface

Chow's calculations were based on the adhesion between a fractal rough surface and a surface which was assumed to be rigid and flat. In order to have a better understanding of the experimental data, the analytical solution of the adhesion between a single asperity (AFM tip or MEMS beam) and a rough surface described by the three surface parameters must be found following the idea of Chow. Both the normalized force–displacement relationship f and the distribution of asperity heights will be calculated for a contact between a single asperity and a fractal rough surface and then used to derive the adhesion. The adhesion as a function of roughness exponent and RMS roughness will be plotted, and this should facilitate the later interpretation of the experimental data.

The interaction force between a molecule and a flat surface is [33]:

$$f_1(D) = -\frac{\pi C \rho}{D^4},\tag{1}$$

where D is the distance between the molecule and the surface, ρ is the number density of molecules and C is the coefficient of the molecule–molecule pair potential.

As shown in the diagram of Fig. 2, the interaction force between a molecule and a rough surface is:

$$f_2(D) = \int_{h_{\min}}^{h_{\max}} f_1(D + h_{\max} - h)\psi(h)\,dh$$

$$= \int_{h_{\min}}^{h_{\max}} -\frac{\pi C \rho}{(D + h_{\max} - h)^4}\psi(h)\,dh,\tag{2}$$

where $D + h_{\max} - h$ is the vertical distance between the molecule and the surface points which have the height of h. $\Psi(h)$ is probability of the surface height h, and we have $\int_{h_{\min}}^{h_{\max}} \psi(h)\,dh = 1$. Both the asperities and the substrate are considered since the interaction f_1 includes both.

The interaction force between a spherical AFM tip and a rough surface thus can be obtained by integrating the forces between all the molecules in the tip and the rough surface [32]:

$$F = \int_0^R f_2(z + h_c)\pi\rho(2R - z)z\,dz,\tag{3}$$

where R is the AFM tip radius and h_c is the separation at contact. Here $h_c \ll R$.

First, computer programs [34] were developed to simulate rough surfaces with differing fractal parameters. Then the probability function $\Psi(h)$ is calculated from

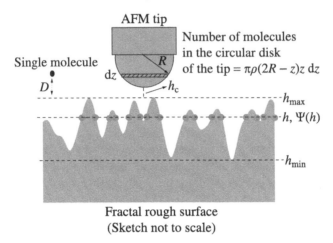

Fractal rough surface
(Sketch not to scale)

Figure 2. Modeling the adhesion between an AFM tip and a fractal rough surface. R is the AFM tip radius, h is the surface height, $\Psi(h)$ is probability of the surface data points which have height of h, and $\int_{h_{min}}^{h_{max}} \psi(h)\,dh = 1$. D is the distance between the molecule and the surface, and $D + h_{max} - h$ is the vertical distance between the molecule and the surface points which have height of h. Thus the adhesion can be integrated over all the surface heights and then the entire AFM tip and the result is equation (3). Here, ρ is the number density of molecules, C is the coefficient of the molecule–molecule pair potential and h_c is the separation at contact.

the generated surface heights of each surface. For given tip size R, the adhesion between the AFM tip and the surface will be found using equation (3).

2.2. Experimental

We performed adhesion measurements on a set of capped MEMS devices that had dry-etched trenches formed with processes that produced a range of surface topographies. The samples were treated with a few angstroms of vapor deposited diphenylsiloxane so they had low surface energy and could be stable for months [35]. Thus all surfaces were assumed to be chemically similar. Earlier studies indicated that the trench sidewalls might be surfaces with a large range of fractal parameters. Therefore, the sidewalls of the uncapped MEMS devices were broken off in order to perform the adhesion measurements. The devices were uncapped just prior to the adhesion measurements to reduce the possibility of contamination.

An AFM silicon cantilever with an integrated tip was employed in the experiment. Before the adhesion measurement, the spring constant of the cantilever was calibrated using a thermal method [36]. The spring constant was 3.6 N/m with a relative uncertainty of 10% and a precision of 5%. The radius of the tip was 196 nm and was measured by imaging a delta function grating (Mikromasch TGT01). The surface roughness of the AFM tip was less than 2 nm and was measured with the delta function grating by imaging a small portion of the tip.

A Veeco Autoprobe M5 AFM was used for the adhesion measurements in a laboratory. All measurements were made at a relative humidity of $39 \pm 12\%$ and a temperature of $20.2 \pm 0.8°C$. First, a 1 µm by 1 µm topographic image of the polysilicon MEMS sidewall was obtained before the force curve acquisition. After stopping the scanner, force curves were obtained at sixteen different locations and each of the average curves came from sixteen curves. Force curves were analyzed and the adhesion forces were measured directly from the force curves for each of the sixteen locations. Data points that were more than two standard deviations away from the average were rejected. If more than three data points were rejected in each data set, the experiment was repeated. As described in the Introduction, the height–height correlation function $H(\mathbf{r})$ was calculated from the surface height data $h(\mathbf{r})$ of the 1 µm by 1 µm topographic image (which had 256×256 data points), and then fitted by $2\sigma^2 f(r/\xi)$ to obtain the surface parameters σ, α and ξ. Next, the correlation between adhesion and RMS roughness, roughness exponent and lateral correlation length were studied and analyzed.

3. Results

3.1. Simulated Data

Adhesion between an AFM tip and two sets of two-dimensional fractal surfaces was simulated in order to establish the relationship between adhesion and the three surface parameters. The simulated AFM tip had a tip radius of 1 µm. The sizes of all the fractal surfaces were 24 µm $\times$ 24 µm, which is more than ten times larger than the diameter of the AFM tip so as to reduce the effect of limited size of the simulated surface. The surfaces had 256×256 data points. Adhesion between the AFM tip and each of these surfaces was calculated and the adhesion F was normalized by the Hamaker constant A, which is $\rho^2 \pi^2 C$. The simulation was done before the actual adhesion experiment was performed. Thus the dimensions of the simulation did not match very well with those of the experiment. Although the simulated surfaces were twenty-four times larger than the experimental surfaces and the AFM tip was five times larger than that in the experiment, a comparison of the trends in the simulated and experimental data should be valid since our interests are the magnitude and distribution of the surface heights.

The first set of simulated fractal surfaces has the same lateral correlation length of 3 µm and the same roughness exponent of 0.5, but the RMS roughness varies from 1 nm to 100 nm. Figure 3 shows the simulated data for adhesion as a function of RMS roughness. The adhesion drops quickly at low roughness, increases as RMS roughness increases at higher roughness, and there is optimal RMS roughness for minimal adhesion. This trend of the simulation result agrees well with the behavior of our previous simple analytical model and experimental data of Ref. [20].

The second set of simulated fractal surfaces has the same RMS roughness of 0.2 µm and the same lateral correlation length of 3 µm, but the roughness exponent varies from 0.1 to 1.0. The RMS roughness was chosen to be much larger than that in the experiment since the simulated surfaces had a much larger size. Figure 4 shows the simulated adhesion data as a function of roughness exponent. Adhesion

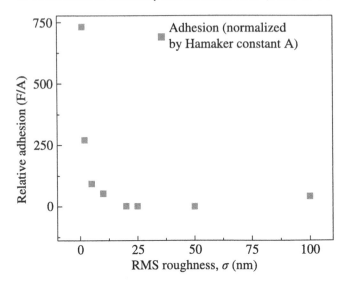

Figure 3. Plot of adhesion as a function of RMS roughness, based on the simulations. Adhesion was normalized by the Hamaker constant A.

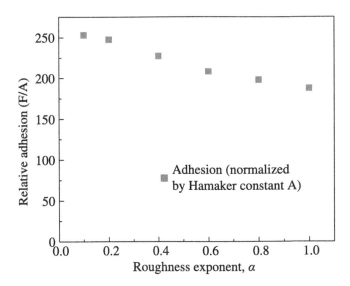

Figure 4. Plot of adhesion as a function of roughness exponent, based on the simulations. Adhesion was normalized by the Hamaker constant A.

decreases as the roughness exponent increases. A large roughness exponent means that the surface is locally flat. Smaller values stand for large deviations from flatness at shorter distances, and proportionally more of the material is closer to the inter- face. The importance of the closeness to the interface can be seen in equation (2) where the inverse-power-law dependence of the force on the vertical distance was

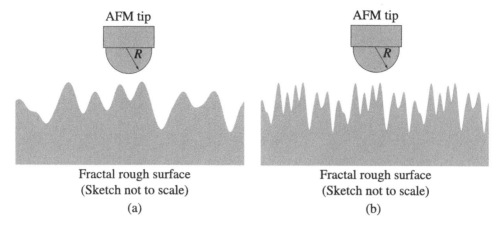

Figure 5. Schematics showing two surfaces with different lateral correlation lengths. (a) Fractal rough surface with a lateral correlation length of ξ; (b) Fractal rough surface with a lateral correlation length of $\xi/3$. The simulations would not predict a difference in adhesion between these surfaces because they have the same height distribution $\Psi(h)$. Thus there is no equivalent to Figs 3 and 4 for lateral correlation length.

shown. Thus, surfaces with lower roughness exponent are expected to have higher adhesion. The simulations follow the same trend as Chow's predictions, but rather than predicting orders of magnitude change, the simulations reveal only a 20% decrease as the roughness exponent varies from 0 to 1.

Figure 5 shows schematics for fractal surfaces with different lateral correlation lengths. The two surfaces have the same RMS roughness and the same roughness exponent but the lateral correlation length of the surface in Fig. 5(a) is three times of that of the surface in Fig. 5(b). Although the two surfaces have different lateral correlation lengths, they have the same height distribution, and hence share the same height probability function $\Psi(h)$. Thus when equation (3) is used to calculate the adhesion between the AFM tip and the two surfaces, the same result will be obtained. The simulation was not done for this case because it was not expected to reveal anything about how lateral correlation length affected adhesion. Thus there is no equivalent to Figs 3 and 4 for lateral correlation length.

3.2. Experimental Data

Adhesion data as a function of the individual surface parameters (RMS roughness, lateral correlation length and roughness exponent) are plotted in Fig. 6(a), 6(b) and 6(c), respectively. The error bars were obtained by averaging sixteen adhesion data points as described in the previous section and represent one standard deviation. In Fig. 6(a), the adhesion is over 150 nN only for RMS roughness of less than 10 nm. Similarly, in Fig. 6(b), adhesion is more than 150 nN for lateral correlation lengths of less than 75 nm. There are no apparent trends in the adhesion as a function of roughness exponent, Fig. 6(c).

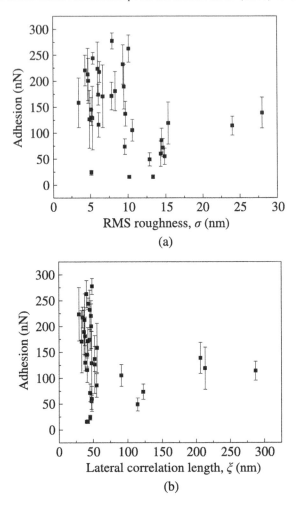

Figure 6. Experimental adhesion data between an AFM tip and MEMS sidewalls. (a) As a function of RMS roughness σ; (b) as a function of lateral correlation length ξ; (c) as a function of roughness exponent α. The error bars represent one standard deviation.

Because adhesion could be a function of more than one fractal parameter, it is plotted 'four-dimensionally' in Fig. 7(a); different sizes of the data points represent different adhesion levels from a few hundreds of nano-Newtons down to tens of nano-Newtons, as specified in the legend. To clarify viewing, the projected adhesion data on the $\alpha\sigma$, $\alpha\xi$ and $\xi\sigma$ planes are plotted in Fig. 7(b), 7(c) and 7(d), respectively. Error bars were omitted for simplicity. In Fig. 7(b) and (c), the scatter as a function of the roughness exponent α can be seen. Again, the higher adhesion values occur for lower values of RMS roughness and lateral correlation length ξ. In Fig. 7(d), the projection on the $\xi\sigma$ plane, the clustering of most of the higher adhesion data in the lower-left corner is striking.

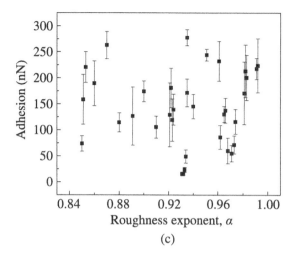

Figure 6. (Continued.)

4. Discussion

This discussion draws upon our earlier simple analytical model and experimental data set [20], the theoretical predictions of Chow [22], and the simulations and experimental data described in this article. The main interest is to establish which of the three fractal surface parameters — RMS roughness, roughness exponent, and lateral correlation length — or combinations of parameters contribute most to adhesion so that future devices could be designed to optimize adhesion and so that our fundamental knowledge of nanoscale adhesion advances.

Concerning the surface roughness, the simulation and experimental data in this work are in agreement with the simple analytical model and experimental data of Ref. [20]. Indeed, the shape of Fig. 3 is strikingly similar to Fig. 2(b) of [20]. In this work, there are relatively fewer experimental data at high surface roughness, thus the characteristic shape of Fig. 3 is less obvious, but the data are consistent with the earlier results. (For the previous study, the sample set was purposely designed to have a range of roughness values. Here, we chose samples that we hoped would have a large range of roughness exponents.) The fundamental principle is that the RMS roughness acts to separate the two main bodies — the tip and the sample — which causes a dramatic drop in the attraction between the bodies for a few nanometers of roughness, because of the inverse-square decay of the van der Waals force with distance. With increasing roughness, the asperities themselves make a significant contribution to the tip–sample adhesion [20, 37]. Surface roughness is a major determinant of adhesion.

As for the roughness exponent, Chow predicted an orders-of-magnitude decrease in adhesion for high roughness exponents (less 'wiggly' surfaces). An intuitive explanation for this prediction is that wigglier surfaces would form more contacts per unit area across an interface than a less jagged surface. Chow's approach took into

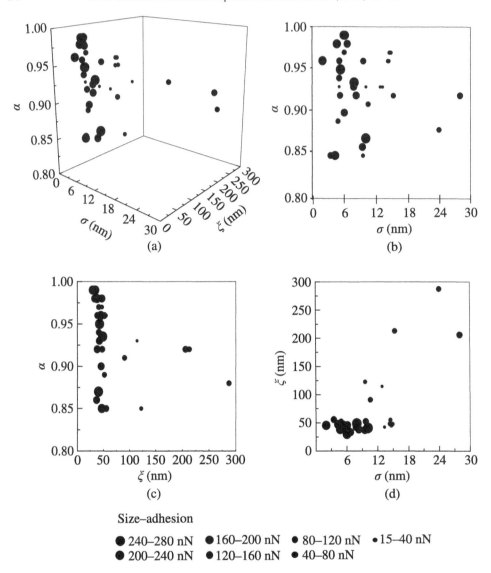

Figure 7. (a) 'Four-dimensional' experimental adhesion data, where the size of the data points is proportional to the adhesion. The three-dimensional position of any point is determined by the fractal parameters calculated from the topographic image. (b) Data projected onto the $\alpha\sigma$ plane; (c) the $\alpha\xi$ plane; (d) the $\sigma\xi$ plane. Note the clustering of most of the large adhesion points in the lower-left corner of (d).

account only the asperities at the surface; our simulation accounts for both asperities and the bulk. We predict a decrease in adhesion of only about 20% as the roughness exponent varies from 0 to 1, and a 10% change over the range corresponding to our experimental data, from 0.85 to 0.99. The scatter in the data was large, as seen in Figs 6(c), 7(b) and 7(c). A preliminary conclusion is that the roughness exponent contributes little to adhesion. Should experiments on samples with a larger range of roughness exponents be performed, this conclusion could be strengthened.

And for the lateral correlation length, Chow predicted that as the lateral correlation length increases, adhesion would increase. The methodology of our simulations leads to a prediction that the correlation length should have no influence on the adhesion. The observation from Figs 6(b), 7(c) and 7(d) is that adhesion is often high for small correlation lengths. In Fig. 7(d), most of the high adhesion points are clustered in the lower-left corner of the plot, corresponding to low surface roughness and low correlation length. A speculative interpretation is that asperities with short correlation lengths for their surface roughness have smaller effective radii. They, therefore, deform more easily, bringing the two larger bodies — the tip and sample — into closer proximity, and more force is required to pull them apart.

5. Summary

In summary, we investigated the role that three surface parameters (RMS roughness, roughness exponent and lateral correlation length) played in the adhesion between an AFM tip and a rough surface, both theoretically and experimentally. Adhesion was measured between an AFM tip and a set of sidewalls of MEMS devices. Together with the simulations, the previous and current experimental data support the conclusion that RMS roughness is a significant predictor of adhesion: adhesion falls very quickly at low RMS roughness. The simulation predicted that adhesion decreases as roughness exponent increases but not by orders of magnitude as stated by Chow. The experimental data are consistent with the simulation although the roughness exponent did not have a large range. The influence of lateral correlation length on adhesion needs to be further investigated.

References

1. J. A. D. Romig, M. T. Dugger and P. J. McWhorter, *Acta Materialia* **51**, 5837 (2003).
2. Y. P. Zhao, *Acta Mechanica Sinica* **19**, 1 (2003).
3. Y. X. Zhuang and A. Menon, *Tribology Lett.* **19**, 111 (2005).
4. N. Tas, T. Sonnenberg, H. Jansen, R. Legtenberg and M. Elwenspoek, *J. Micromech. Microeng.* **6**, 385 (1996).
5. L. Zhang and Y. P. Zhao, *J. Adhesion Sci. Technol.* **18**, 715 (2004).
6. Y. P. Zhao, L. S. Wang and T. X. Yu, *J. Adhesion Sci. Technol.* **17**, 519 (2003).
7. J. A. Greenwood and J. B. P. Williamson, *Proc. R. Soc. Lond. A* **295**, 300 (1966).
8. K. N. G. Fuller and D. Tabor, *Proc. R. Soc. Lond. A* **345**, 327 (1975).
9. L. Suresh and J. Y. Walz, *J. Colloid Interface Sci.* **183**, 199 (1996).
10. H. Hertz, *J. Reine Angewandte Mathematik* **92**, 156 (1882).
11. K. L. Johnson, K. Kendall and A. D. Roberts, *Proc. R. Soc. London A* **324**, 301 (1971).
12. B. V. Derjaguin, V. M. Muller and Y. P. Toporov, *J. Colloid Interface Sci.* **53**, 314 (1975).
13. B. V. Derjaguin, V. M. Muller and Y. P. Toporov, *J. Colloid Interface Sci.* **67**, 378 (1978).
14. B. N. J. Persson, *Phys. Rev. Lett.* **87**, 116101 (2001).
15. B. N. J. Persson, *Phys. Rev. Lett.* **88**, 129601 (2002).

16. B. Luan and M. O. Robbins, *Nature* **435**, 929 (2005).

17. B. Luan and M. O. Robbins, *Phys. Rev. E* **74**, 261111 (2006).

18. E. J. Thoreson, J. Martin and N. A. Burnham, *J. Colloid Interface Sci.* **298**, 94 (2006).

19. E. J. Thoreson, J. Martin and N. A. Burnham, *J. Microelectromech. Syst.* **16**, 694 (2007).

20. D.-L. Liu, J. Martin and N. A. Burnham, *Appl. Phys. Lett.* **91**, 043107 (2007).

21. M. Marsili, A. Maritan, F. Toigo and J. R. Banavar, *Rev. Mod. Phys.* **68**, 963 (1996).

22. T. S. Chow, *Phys. Rev. Lett.* **79**, 1086 (1997).

23. T. Vicsek, *Fractal Growth Phenomena*. World Scientific, Singapore (1992).

24. A.-L. Barabasi and H. E. Stanley, *Fractal Concepts in Surface Growth*. Cambridge University Press, New York (1995).

25. T.-M. Lu, H.-N. Yang and G.-C. Wang, *Mater. Res. Soc. Symp. Proc.* **367**, 283 (1995).

26. Y.-P. Zhao, G.-C. Wang and T.-M. Lu, *Characterization of Amorphous and Crystalline Rough Surface: Principles and Applications*. Academic Press (2001).

27. F. Family and T. Vicsek, *Dynamics of Fractal Surfaces*. World Scientific, Singapore (1991).

28. T. Halpin-Healy and Y.-C. Zhang, *Phys. Rep.* **254**, 215 (1995).

29. S. He, G. Kahanda and P.-Z. Wong, *Phys. Rev. Lett.* **69**, 3731 (1992).

30. M. A. Rubio, C. A. Edwards, A. Dougherty and J. P. Gollub, *Phys. Rev. Lett.* **63**, 1685 (1989).

31. S. V. Buldyrev, A.-L. Barabasi, F. Caserta, S. Havlin, H. E. Stanley and T. Vicsek, *Phys. Rev. A* **45**, R8313 (1992).

32. T. S. Chow, *Phys. Rev. Lett.* **86**, 4592 (2001).

33. J. Israelachvili, *Intermolecular and Surface Forces*. Academic Press, London (1991).

34. D.-L. Liu, D. Qi, S. Teng and C.-F. Cheng, *Chinese Phys.* **9**, 353 (2000).

35. J. Martin, in: *Nanotribology: Critical Assessment and Research Needs*, S. M. Hsu and Z. C. Ying (Eds), 1 st edn, p. 177, Chap. 14. Kluwer, Dordrecht (2003).

36. G. A. Matei, E. J. Thoreson, J. R. Pratt, D. B. Newell and N. A. Burnham, *Rev. Sci. Instrum.* **77**, 083703 (2006).

37. F. W. Delrio, M. P. De Boer, J. A. Knapp, E. D. Reedy Jr, P. J. Clews and M. L. Dunn, *Nature Mater.* **4**, 629 (2005).

Effects of Contacting Surfaces on MEMS Device Reliability

Y. Du *, W. A. de Groot, L. Kogut **, Y. J. Tung and E. P. Gusev

MEMS Research and Innovation Center (MRIC), QUALCOMM MEMS Technologies,
2581 Junction Avenue, San Jose, CA 95134, USA

Abstract
A multi-asperity contact model with adhesion is presented to analyze the real contact of microdevices, including contact force, contact area, and pull-off force. Since the macroscopic contact of two contacting surfaces is the statistical average of many individual micro/nano contacts, the key point is to include an accurate analysis of the discrete contacts. A finite element model which is applicable to a wide range of material properties was developed to study adhesion and plasticity of a single contact during both loading and unloading cycles in a previous investigation. The best fits of the results by comparing with the existing models are introduced in this work for both the loading and unloading (pull-off force) analyses. Pearson Type distributions which can independently adjust the skewness and kurtosis of surface heights are used to study the multi-asperity contact. The significance of this work is not only to predict adhesion of the existing devices, but to provide an optimal design of surface characteristics and material properties to improve device reliability.

An overview of the status of surface roughening indicates that various techniques exist to tailor surface morphological properties, but that these techniques mainly target the mean and root-mean-square (rms) values of the roughness. This approach insufficiently utilizes all available techniques to minimize adhesion force between surfaces. New approaches that include surface roughness process development should include a range of skewness and kurtosis values of the resulting surface as a design consideration in addition to the mean and rms values of the roughness.

Keywords
Adhesion, pull-off force, elastic–plastic contact, asymmetric rough surface, surface engineering

1. Introduction

A growing number of commercial applications and disruptive technologies focus on MEMS devices. This includes inertia sensors, gyroscopes, pressure sensors, micro-transducers, RF varactor diodes, filters, switches and so on. The MEMS field is not new by any means — it has a history of more than 40 years. However, most of these years it was in early phases of research and development of ideas and conceptual

---
* To whom correspondence should be addressed. Tel.: +1 408 546 2073; Fax: +1 408 546 1225; e-mail: yand@qualcomm.com
** Present address: RAFAEL — Advanced Defense Systems Ltd., P.O.B. 2250, Haifa 31021, Israel.

Adhesion Aspects in MEMS/NEMS

prototype devices. It is only recently we started to witness significant advances and a shift to commercialization of several applications. To a large extent, this transition is due to significant progress in understanding and improving reliability of MEMS devices.

Reliability of microdevices is governed by many factors. Some of them are defined by device design, others by materials and processes or operating conditions. MEMS package environment also plays an important role. For devices experiencing contacts between functional parts during operation, contacting surfaces play a major role, in particular their materials and morphology. A prime example of such complex technologies could be found in modern hard drives. This is also true for MEMS switches, whether Ohmic or capacitive. An overview of MEMS reliability can be found elsewhere (e.g., by Maboudian and Howe [1] and de Groot *et al.* [2]).

It is the purpose of this paper to focus only on one important aspect of contacting surfaces in MEMS devices, i.e., their morphology and its effect on device reliability. First, we describe how single contacts during loading and unloading can be modeled. The model is then generalized to a realistic case of multi-asperity contact. This will be followed by a discussion on how to correctly measure surface roughness and which parameters are important for MEMS reliability. Finally, we will review different techniques for surface morphology modifications and their practical aspects.

2. Single-Asperity Contact Model

An accurate characterization of the contact between rough surfaces is important for understanding device physics in various applications. These include sealing [3], friction [4], performance and life of machine elements [5], and electrical conductivity [6], to name a few. The high level of interest in this subject is evident from the number of published articles so far (see, e.g., reviews [7–9]).

In all the aforementioned cases the effect of adhesion on contact deformation and stress fields can be significant if the size of the contacting bodies is on the micrometer scale or smaller. The sources of the adhesion force can be capillary condensation, molecular van der Waals forces, chemical bonds, charge accumulation, etc. [1, 10, 11]. It is important to accurately model the contact, including adhesion between rough surfaces and study its dependence on design parameters and environmental conditions. The first step is to define the physical model of the rough surface. Surfaces are not perfectly smooth on the nanoscale, and can be defined by the multiple asperities that make up the roughness. There is considerable evidence that asperities on real surfaces can be modeled by any shape and any distribution of heights. However, the most used model is known as the GW (Greenwood and Williamson) model which is schematically described in Fig. 1 and assumes that the asperities are spherical near their summits with a uniform radius of curvature R, but different heights, and of Gaussian distribution [12, 13]. In Fig. 1, the symbols z, d, and ω denote the asperity height, separation of the surfaces, and interference, respectively, measured from the reference plane defined by the mean of the original asperity heights. The

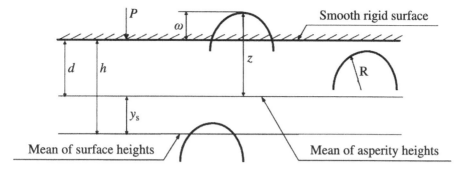

Figure 1. Contact between a deformable rough surface and a smooth rigid surface.

separation h is measured from the reference plane defined by the mean of the original surface heights. P is the contact force. It assumes asperities are far apart and there is no interaction between them, and finally no bulk deformation occurs during contact. Its roughness can be represented by standard deviation σ. Therefore, the macroscopic contact of two surfaces is the statistical average of many individual micro/nano contacts. Depending on the penetration of the asperity sphere into the flat, more and more asperities are in the contact. It is necessary to understand the contact information of each nano-asperity first. Subsequently, the result can be used to build a multi-asperity model which will be discussed in the next section. In addition to the GW model, many other approaches have been proposed to study the effect of adhesion on contact such as in [14] and [15].

 It is worth mentioning that for small contacts at the nanoscale, the effect of adhesion should dominate the contact due to large surface-to-volume ratio and lead to some novel phenomena compared with contacts at the microscale, especially for unloading. An effective method to study a nanocontact is to use molecular dynamics model (e.g., [16]). But this could be very time consuming and does not provide practical/simple equations to use. Several analytical and numerical works exist for microcontacts, although they are not expected to be used for nanocontacts directly. For microcontacts with contact radii from 0.1 µm to 10 µm, the material yield strength is much greater than that of the macroscopic material. Although strain gradient plasticity (a strain gradient leads to enhanced hardening due to the generation of geometrically necessary dislocations [17]) is not included in the microcontacts analysis, a higher hardness than that of bulk material is used to substitute for this phenomenon. In [18], a general finite element model (FEM) with ANSYS was developed to include the effect of adhesion on contact deformation and stresses. It is, therefore, applicable to a wide range of material properties. The results were compared with the existing theoretical models and another FEM method developed by Kogut and Etsion (KE) [19] who gave simple curve-fit equations for the relations among force, displacement and contact area. The comparisons demonstrate that the KE model agrees well with the results reported in [18] for the large-loading case, although it shows a discrepancy during the small-loading for the Johnson, Kendall, and Roberts (JKR) region [20], which typically works for a combination of soft solids, large radius of curvature, and large adhesion energy. The KE model is not expected to be accurate in the small-loading JKR region due to the fact that it neglects the change of contact profile due to adhesion.

In this work, the KE model is used for single-asperity contact due to its simplicity and relative accuracy. The KE model found that the entire elastic–plastic regime extends over interference values in the range $1 \leqslant \omega/\omega_c < 110$ with a distinct transition in the mean contact pressure at $\omega/\omega_c = 6$. The critical interference, ω_c, that marks the transition from elastic to elastic–plastic deformation is given by [21]:

$$\omega_c = \left(\frac{\pi K H}{2E}\right)^2 R, \tag{1}$$

where H is the hardness of the softer material which is related to the yield strength σ_Y by $H = 2.8\sigma_Y$ and K, the hardness coefficient, is related to the Poisson ratio ν of the softer material by $K = 0.454 + 0.41\nu$. E is the Hertz elastic modulus defined as:

$$\frac{1}{E} = \frac{1 - \nu_1^2}{E_1} + \frac{1 - \nu_2^2}{E_2}, \tag{2}$$

where E_1, E_2 and ν_1, ν_2 are Young's moduli and Poisson's ratios of the contacting surfaces 1 and 2, respectively. Up to the transition interference of $\omega/\omega_c = 6$ a plastic region develops below the contact interface while the entire contact area itself is elastic. Above $\omega/\omega_c = 6$ the contact area contains an inner elastic circular core that is surrounded by an external plastic annulus. This elastic core shrinks with increasing interference and finally disappears completely at $\omega/\omega_c = 68$. From here on the entire contact area is plastic but the mean contact pressure continues to grow until it becomes constant and equal to the hardness at $\omega/\omega_c = 110$, marking the beginning of fully plastic contact.

The dependence of contact area A and contact force P on ω in the elastic–plastic regime was presented in a dimensionless form through normalizing the relevant parameters by their critical values at the onset of plastic deformation, i.e., A_c, P_c and ω_c, respectively.

The dimensionless expressions can take on the general forms:

$$A/A_c = b(\omega/\omega_c)^m, \tag{3}$$
$$P/P_c = c(\omega/\omega_c)^n, \tag{4}$$

where

$$A_c = \pi R\omega_c, \tag{5}$$
$$P_c = \frac{2}{3}KH\pi R\omega_c. \tag{6}$$

The values of the constants b, m, c and n are determined by the contact mode in four different deformation regimes and are summarized in Table 1.

Table 1.
The values of b, n, c and m in the various deformation regimes (K, the hardness coefficient, is related to the Poisson ratio ν of the material by $K = 0.454 + 0.41\nu$)

Deformation regime	Constant			
	b	m	c	n
Fully elastic, $\omega/\omega_c < 1$	1	1	1	1.5
1st elastic–plastic regime, $1 \leqslant \omega/\omega_c \leqslant 6$	0.93	1.136	1.03	1.425
2nd elastic–plastic regime, $6 \leqslant \omega/\omega_c \leqslant 110$	0.94	1.146	1.40	1.263
Fully plastic, $\omega/\omega_c > 110$	2	1	$3/K$	1

Unloading as a follow-up stage of loading plays an important role in the reliability of microdevices because its separation modes, which are determined by material properties and maximum loading parameter, may affect the magnitude of the pull-off force (a mechanical force needed to separate two parts). When the pull-off force reaches the available restoring force the system can provide, the contacting parts cannot separate and the surfaces remain in contact. In microdevices, this phenomenon is referred to as stiction.

An interesting result in [18] was the identification of two distinct separation modes, i.e., brittle and ductile separation. In this FEM model, the unloading is allowed to be elastic and/or plastic as dictated by the results of the simulation. Ductile separation is defined as occurring when considerable stretching occurs during unloading. In reality, it may be associated with plasticity during unloading or material transfer. A subsequent study [22] used this model to conduct a series of simulations to determine the influence of four non-dimensional parameters (including the maximum load parameter) on the contact and separation modes. Other studies support the finding that unloading plasticity may be caused by large loading plasticity [23] or high adhesion energy [16]. Although unloading plasticity is widely observed in experiments and numerical simulations, it is hard to theoretically predict the unloading plasticity in detail because of the complex combinations of many phenomena, such as elasticity, plastic deformation, chemical reactions, adhesion, material transfer, etc.

In [22] pull-off force *vs.* interference for unloading was given and compared with other existing models. As a basis for comparison, the curve-fit to the finite element simulations for R_{res} with plastic deformation given by Etsion, Kligerman and Kadin (EKK) [24] was considered:

$$R_{res}^{EKK} = R \cdot \left[1 + 1.275 \left(\frac{E}{\sigma_Y} \right)^{-0.216} \left(\frac{\omega}{\omega_c} - 1 \right) \right]. \tag{7}$$

Note that here R_{res}^{EKK} is the residual radius of curvature at the summit of the deformed and fully unloaded sphere. This value is found to be greater at the summit than at other radial positions indicating that the sphere has been flattened more at its apex than elsewhere. The resulting pull-off force is then $F_{adh}^{EKK} = 2\pi\Delta\gamma R_{res}^{EKK}$ where $\Delta\gamma$ is the adhesion energy. Although the EKK model was built without adhesion, a comparison between the pull-off force predicted by [22] and the F_{adh}^{EKK} of the EKK model shows the ratio is close to unity for a wide range of cases. So equation (7) is used to calculate pull-off force.

3. Multi-Asperity Contact Model

It is very important to build a multi-asperity contact model based on the results for individual nanocontacts for device reliability studies. With the invention of the atomic force microscope (AFM), a variety of surfaces can be imaged and characterized at atomic level. For example, Figs 2(a) and 3(a) are 2×2 µm^2 AFM images of two contacting sample surfaces of the same commonly used material but at different process conditions. They both contain 512×512 data points using AFM tapping mode. Any rough surface can be statistically represented by the following roughness parameters:

$$R_a = \frac{1}{N}\sum_{i=1}^{N}|y_i - \bar{y}|, \tag{8}$$

$$R_q = rms = \sigma = \sqrt{\frac{1}{N-1}\sum_{i=1}^{N}(y_i - \bar{y})^2}, \tag{9}$$

$$R_{max} = \max(y_i) - \min(y_i), \tag{10}$$

$$Skewness = \frac{1}{N}\sum_{i=1}^{N}\left(\frac{y_i - \bar{y}}{\sigma}\right)^3, \tag{11}$$

$$Kurtosis = \frac{1}{N}\sum_{i=1}^{N}\left(\frac{y_i - \bar{y}}{\sigma}\right)^4, \tag{12}$$

where y_i is the surface height of each data point measured from the mean surface height $\bar{y}$. N is the total number of data points on the scanned images. R_a and R_q (also called rms or σ) represent how widely values are dispersed from the mean. Skewness characterizes the degree of asymmetry of a distribution around its mean. Positive skewness indicates a distribution with an extended right tail (many tall asperity peaks). Kurtosis characterizes the relative peakedness or flatness of a distribution compared with the normal distribution. Positive kurtosis indicates a relatively peaked distribution as compared with the normal distribution. Note that kurtosis defined here is called traditional kurtosis. Somewhere we may see another treatment of kurtosis called excess kurtosis which equals traditional kurtosis minus 3. Therefore, for a normal distribution traditional kurtosis equals 3 and excess kurtosis equals 0.

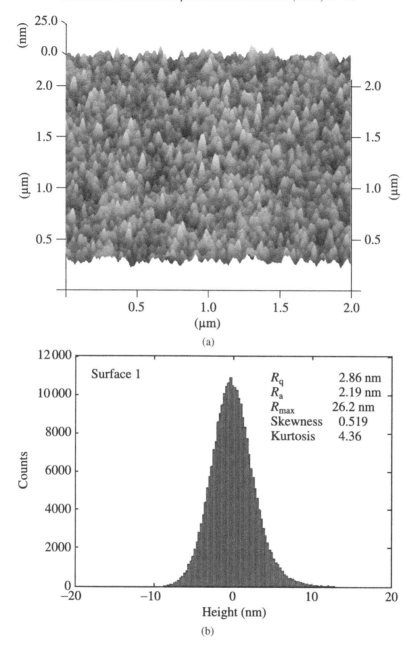

Figure 2. (a) AFM image of contacting surface 1. (b) Surface height distribution with roughness data.

A macroscopic contact is actually the integration of many individual asperity contacts. More asperities involved into contact cause more adhesion force and more pull-off force to break the contact, which indicates that for such a case, surfaces are more prone for stiction to occur. The most intuitive solution to decrease the number

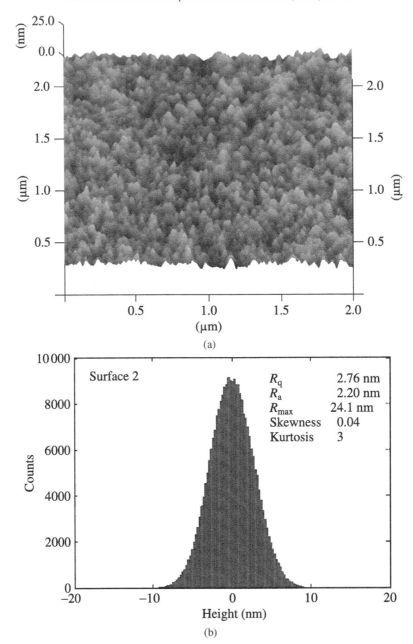

Figure 3. (a) AFM image of contacting surface 2. (b) Surface height distribution with roughness data.

of asperities into contact is to roughen the surface to have greater R_q. But surfaces with similar R_q may exhibit significantly different adhesion characteristics depending on the magnitudes of skewness and kurtosis such as discussed by Komvopoulos [25]. Figures 2(b) and 3(b) show the surface height distributions and their roughness data from their corresponding AFM data. Two surfaces have comparable R_q, but surface 1 has higher skewness and kurtosis than surface 2.

It is important to note that probabilistic approaches are limited by the dependence of statistical roughness parameters on the sampling length and resolution of the measuring instrument. Hence, they cannot provide unbiased information for the surface topography [26]. This shortcoming can be circumvented by using scale-invariant parameters to characterize the surface topography. Fractal geometry has been widely used to study engineering surfaces exhibiting random, multi-scale topographies due to its intrinsic advantages of scale invariance (i.e., independence of measurements on instrument resolution and sample length) and self-affinity (generalized self-similarity) [26]. Komvopoulos and Ye [27] developed a comprehensive contact analysis that incorporates the successive occurrence of elastic, elastic–plastic, and fully plastic deformations of contacting asperities on rough surfaces with fractal topography description.

If one contacting surface is much smoother and harder than the other one, the contact can be simplified as a deformable rough surface in contact with a smooth rigid surface. Image and data analyses make it possible to create tools to find each asperity on the surface with its height and radius of curvature based on the input of AFM data. Computations can then be done by integrating the contact of each asperity using formulas mentioned in the previous section. The randomly picked material properties for the calculation are $E = 70$ GPa, $H = 10$ GPa, $\Delta \gamma = 0.3$ J/m^2. Note that for a real material, a higher hardness than the material bulk hardness should be used to compensate for the grain strain hardening. Figure 4 demonstrates the pull-off force *vs.* contact force for two surfaces shown in Figs 2(a) and 3(a). Arbitrary units are used due to the random selection of hardness and adhesion energy. It is shown that for the same contact force, more pull-off force is needed to separate surface 2 from its counterpart. Thus from the perspective of preventing permanent adhesion, surface 1 is better than surface 2 which could be the combined effect of rougher R_q, higher skewness and kurtosis. At the same time, statistical methods can replace the physical representation of the asperity distribution.

The standard deviations σ_s and σ correspond to the asperity and surface heights, respectively and are related by [28]:

$$\frac{\sigma_s}{\sigma} = \sqrt{1 - \frac{3.717E - 4}{\beta^2}}, \tag{13}$$

where β is a surface roughness parameter defined as:

$$\beta = \eta R \sigma \tag{14}$$

and η is the area density of the asperities.

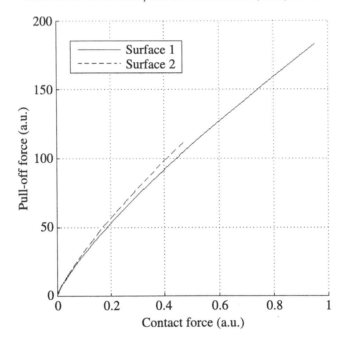

Figure 4. Pull-off force *vs.* contact force for surfaces 1 and 2 in Fig. 2(a) and Fig. 3(a), respectively.

All length dimensions are normalized by σ and the dimensionless values are denoted by superscript *. Hence, y_{s}^{*} is the difference between h^{*} and d^{*}, where h and d are as defined in the previous section, and is given by [29]:

$$y_{\mathrm{s}}^{*} = h^{*} - d^{*} = \frac{1.5}{\sqrt{108\pi\beta}}. \tag{15}$$

The total number of asperities is:

$$N = \eta A_{\mathrm{n}}, \tag{16}$$

where A_{n} is the nominal contact area. The number of asperities in contact is:

$$N_{\mathrm{c}} = \eta A_{\mathrm{n}} \int_{d^{*}}^{\infty} \phi^{*}(z^{*}) \, \mathrm{d}z^{*}, \tag{17}$$

$\phi^{*}(z^{*})$ is the dimensionless asperity heights probability density function, assumed to be Gaussian

$$\phi^{*}(z^{*}) = \frac{1}{\sqrt{2\pi}} \frac{\sigma}{\sigma_{\mathrm{s}}} \exp\left[-0.5\left(\frac{\sigma}{\sigma_{\mathrm{s}}}\right)^{2}(z^{*})^{2}\right]. \tag{18}$$

Once these expressions are known, the total contact area, A, and contact force, P, are obtained by summing the individual asperity contributions using a statistical model:

$$\bar{A} = \eta A_n \int_d^\infty A(z - d)\phi(z)\,dz, \tag{19}$$

$$\bar{P} = \eta A_n \int_d^\infty P(z - d)\phi(z)\,dz. \tag{20}$$

The integrals in equations (19) and (20) use equations (3) to (6) with relevant parameter values provided in Table 1 for the different deformation regimes of the contacting asperities.

If the second surface roughness or Young's modulus is not neglected, then the problem is more complicated. One existing approach for including both surfaces in a model is to consider the contact between an "equivalent" rough surface and a smooth surface. The equivalent surface parameters can be obtained by combining the properties of both surfaces [29].

The two methods introduced above are able to predict the real contact area and pull-off force based on AFM data for contacting surfaces of real devices. To help device design considerations, it is desired to have a statistical distribution which can separately adjust the four roughness parameters, mean height, R_q, skewness, and kurtosis to represent any surface topography. If it is assumed that contact happens only at the highest asperities, and the contact cut-off plane is determined by loading, the most important parameter for calculating the contact is really the shape of the high end tail of the distribution instead of roughness alone. To minimize adhesion force in microdevices, heavy-tailed and skewed distributions are desired. The normal Gaussian distribution as in equation (18) has a symmetrical shape, so skewness is always zero. To allow for non-zero skewness, McCool introduced the Weibull distribution [30], but its kurtosis is always dependent on skewness, so it is not applicable for the design flexibility that roughness tailoring calls for.

To allow for an independent adjustment of skewness and kurtosis, the Pearson system of frequency curves was first introduced into contact mechanics modeling by Kotwal and Bhushan in 1996 [31]. Pearson curves were published by Pearson first in 1893, with subsequent modifications and extensions. Depending on the values of skewness and kurtosis, seven types of Pearson distributions can be used. They have been widely used in many fields, such as financial markets, flood frequency analysis, etc. A complete description of this theory can be found in the book *System of Frequency Curves* [32]. It also shows special promise in contact mechanics of microdevices to help surface design and improve device reliability. It was found in [31] that a range of positive skewness (between 0.3–0.7) and a high kurtosis (greater than 5) significantly lowers the real area of contact and meniscus contribution, implying low friction and wear.

Tayebi and Polycarpou [33] used Pearson distributions to independently model the effects of skewness and kurtosis on the static friction. In [33], they substituted a Gaussian distribution into equations (19) and (20) with the probability density function of the Pearson distribution to calculate the contact area and contact force.

The Chang–Etsion–Bogy (CEB) [4] elastic–plastic contact model was used for the single contact case. Positive skewness and high kurtosis were proved again as desirable parameters to decrease the static friction coefficient. In a follow-up study [34], the CEB model were replaced by the KE model previously mentioned in this paper, and it was found that the CEB model underestimated the contact force, especially if a large plastic deformation occurs. The curve-fit results of the adhesion force in the KE model were used to calculate the pull-off force by subtracting adhesion force from contact force. However, the equations for adhesion force provided by the KE model are obtained by integrating the adhesion pressure over the profile outside of the contact area during loading. To make it applicable as a pull-off force prediction during unloading, it is assumed that the loading–unloading process is purely reversible and thus the same equations can be used during unloading. Such is not the case for most contacts involving plastic deformation during loading or unloading. This could be the reason that the simulation using Pearson distributions underestimated the experimental data.

In this current study, the nanocontact model discussed in the previous section is integrated using a Pearson distribution, especially using the residual radius of curvature in equation (7) to predict the pull-off force. A Pearson Type IV distribution is used as an example to study the effect of skewness on the pull-off pressure *vs.* contact pressure (the force normalized by the surface area). Three surfaces with the same mean height (=0 nm), R_q (=4 nm), kurtosis (=10), but different skewness varying from 0 to 1.5 were studied. As shown in Fig. 5 even with the same R_q, higher skewness can lower the pull-off pressure effectively. A complete study of

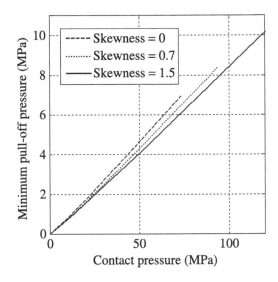

Figure 5. Minimum pull-off pressure *vs.* contact pressure using Pearson Type IV distribution for three surfaces with the same mean height (=0 nm), R_q (=4 nm), kurtosis (=10), but different skewnesses varying from 0 to 1.5.

the effects of various roughness parameters on contact mechanics is expected to lead to a design of surfaces that have the ability to mitigate stiction.

4. Engineering Surface Roughness

Advanced modeling of surface roughness can guide the device design for maximum contact life. But physically achieving the desired roughness and asperity distribution is not a trivial accomplishment. Most approaches to determining the adhesion–surface roughness correlation involve taking the existing roughness distribution as characterized by AFM or other technique and modeling the projected adhesion force. Not much effort has been expended into modeling adhesion force as a function of roughness distribution, determining the optimal roughness distribution needed to minimize the adhesion, and then trying to create this roughness by means of process design and control. To create such a desirable "designer" surface topography puts the burden squarely in the field of processing, either in the frontend or the backend. Engineering surface topography on the 1–20 nm characteristic scale is still relatively immature. An effort will be made in this section to summarize the status of "texturing" techniques. Detailed tailoring of specific roughening parameters using such techniques can be done by a combination of process equipment control settings and AFM roughness characterization.

For simplicity, different roughening techniques will be categorized by the type of process that is suggested, such as deposition, additive, subtractive, transfer, or imprint. Although some of the techniques overlap, each of these will be described and some examples given with advantages and disadvantages where applicable or known.

4.1. Deposition Techniques

These are the most straightforward techniques for tailoring surface roughness. They can be used when the deposited materials and process allows this flexibility. For example, the properties and thus the surface roughness of polysilicon films deposited using low-pressure chemical vapor deposition (LPCVD) depends on the nucleation and growth of the silicon grains. These can be controlled by deposition rate, partial pressure of hydrogen, total pressure, and the presence of impurities. Tayebi and Polycarpou [34] use polysilicon surfaces with various roughness and adhesion behavior to develop models that predict adhesion forces.

For many materials used in MEMS device architecture, such surface roughness control by means of controlling process conditions is not possible. And even if roughness control through process control is possible, variations in mechanical, electrical, or optical properties as the result of grain size variations can put limitations on the range of roughness control allowed for a given specific device operation. Stress in polysilicon films can vary over a wide range from compressive to tensile with deposition temperature and silane pressure. This would affect device operation if polysilicon is used as structural element. Alternative means for surface texturing may, therefore, be beneficial.

4.2. Additive Techniques

These techniques comprise the processes that add surface features on top of existing surface morphology. This could be in the frontend, whether interfacial inclusions or on a gas–solid surface, or as a backend solution, typically as a gas–solid surface additive. The prime mechanism by which to create such roughness is particle formation in the vapor or liquid phase. Although particle formation on the nanoscale is still difficult to control, efforts are ongoing to create "designer" particle surface roughness and to estimate the roughness effect on adhesion force.

Recent work on reducing real contact area by means of deposited particles has been done by Ashurst and coworkers [35, 36]. In this work it was shown that ligand-stabilized gold nanoparticles could be conformally deposited onto all surfaces using a CO_2-expanded liquid/supercritical drying process. In order to permanently attach these particles to the surfaces, a Self-Assembled Monolayer (SAM) was deposited as an adhesive overcoat. In another example of particle based surface topography modification, Patton *et al.* [37] successfully deposited particles in the 10-nm range on contacting surfaces of MEMS contact switches, and showed extended lifetime over non-coated switches. In addition to providing modified real contact area, these added conductive particles provided beneficial functions such as lubrication and current distribution.

The above referenced work used conductive particles. For applications that involve low resistance contact operation or device operation with contacting surfaces that do not include an applied electric field, conductive particles are an option. For applications that involve an applied electric field, buried or surface exposed conductive particles could pose reliability issues by functioning as charge traps, causing voltage offsets, or by facilitating device dielectric breakdown. In those applications, particles of a dielectric nature are preferred.

Additive techniques can be applied during frontend processing where typically a single surface is treated or backend processing where both contacting surfaces are roughened by particle deposition. After deposition, an overcoat with, for example, a SAM will help attach the particles more rigorously to the surface creating yet different surface properties through a change in surface energy.

4.3. Subtractive Techniques

These techniques mainly cover processes by which a material is selectively etched away on a nanoscale, leaving small residual islands that form the asperity tops. Such processes include deep reactive ion etching (DRIE) of SiO_2 or Si [38], where reductions in the inhibitor-film-forming species O_2 causes polymer residue buildup at distributed surface locations. These non-continuous (typically fluorocarbon) films act as nano-masks and can be leveraged for surface roughness creation. This roughness can be controlled by gas concentration, pressure, RF power, and temperature control.

Other techniques involving nanomasking include those where residue islands are artificially introduced and selected areas of the surface are etched. Wei *et al.* used E-beam deposition of non-continuous Ag film on silicon dioxide, followed by vacuum annealing [39]. This annealing caused Ag molecules to migrate and agglomerate, resulting in nano-scale Ag islands. The density and size of these islands can be varied by varying initial Ag film thickness and annealing rate. Varying process type (chemical, sputter), chemistry, pressure, RF power, and temperature of subsequent SiO_2 etch processes all contribute to the overall roughness distribution.

Additional subtractive techniques include chemical etch processes or oxidation followed by oxide removal processes that preferentially etch and/or oxidize along grain boundaries. These techniques can be applied to films with distinct grain structure and can be applied to functional layers as well as sacrificial layers. Subtractive techniques are usually limited to a single surface and in the frontend process only.

4.4. Transfer Techniques

These techniques rely on creating a roughness pattern on a sacrificial layer which is subsequently covered with one or more layers of a functional material. Upon removal of the sacrificial layers, the remaining film that was deposited over the sacrificial layer exhibits the roughness pattern of the sacrificial layer but with inverse topography. This technique is also limited to frontend processing and on a single surface only.

For example, for processes using amorphous silicon as sacrificial material, roughness can be tailored by fiber laser crystallization, see Dao *et al.* [40]. Exposure of a-Si to a pulsed Nd-YAG laser causes crystallization of the top layer. This forms polycrystalline silicon films. The growth of these polycrystalline films is accompanied by a roughening of the surface. Dao *et al.* have reported rms surface roughness from 9.9 to 40 nm, well within the range of useful roughness to reduce adhesion. If subsequent thin films are deposited and the sacrificial layer is removed, the thin films will be left with a roughness that complements the initial poly-silicon roughness. The roughness of the Nd-YAG exposed polycrystalline film depends on the energy per shot, laser pulse duration, and energy density (fluence).

4.5. Imprint Techniques

Imprint techniques, such as nano-imprinting, use a template to mechanically create an imprint pattern in a resist layer, after which resist cure and etch processing leaves features determined by the imprint mask. Initial efforts claimed a feature size of <100 nm [41]. The current feature size limitation is between 15 and 20 nm, but efforts are underway to accomplish feature size below 10 nm [42]. Imprint lithography sets itself apart from the previous techniques in that its spatial features are well controlled and can be designed randomly as well as in a well-organized manner. Careful control of the adhesion between the resist and the imprint template is required in order to achieve a defect-free nanopattern, although small defects in the nanopattern are not expected to affect adhesion force.

One of the disadvantages of all roughening processes in the frontend is that if no planarization is employed, all subsequent conformal films will have similar roughness, with matching contact surface features. Although the features will have broadened somewhat, the location of the asperities and matching dimples will have lesser impact in minimizing real contact area. An additional concern is that modifying surface roughness in the frontend could affect adhesion of subsequent thin film layers. Careful considerations and some experimental work is needed to determine the roughness effect on film adhesion.

Another consideration is the possibility that additional techniques could be employed to minimize adhesion forces. One of these techniques is the application of SAMs to increase the hydrophobic nature of the surfaces and, in general, to change the surface energy. Such SAMs could influence the contact mechanics substantially by absorbing impact energy and altering wear characteristics. Applications of these coatings fall outside the scope of this paper.

5. Conclusions

Pull-off force as an important indicator of device reliability is numerically studied in the current work based on a multi-asperity contact model. This model integrates a relatively accurate single-asperity contact model including adhesion and plasticity during both loading and unloading cycles. It also uses Pearson Type distribution to realize the representation of a rough surface with any combination of skewness and kurtosis. This model is expected to optimize the surface design of microdevices by lowering the pull-off force and contact area. The current status of surface texturing is also reviewed to evaluate the manufacturing possibility to realize the optimal surface texturing design.

Acknowledgement

The authors would like to thank Qualcomm MEMS Research and Innovation Center for rendering this work possible.

References

1. R. Maboudian and R. T. Howe, *J. Vac. Sci. Technol. B* **15**, 1 (1997).
2. W. A. de Groot, J. R. Webster, D. Felnhofer and E. P. Gusev, *IEEE Trans. Device Mater. Reliability* **9**, 190 (2009).
3. A. A. Polycarpou and I. Etsion, *Tribology Trans.* **41**, 531 (1998).
4. W. R. Chang, I. Etsion and D. B. Bogy, *ASME Trans. J. Tribology* **110**, 57 (1988).
5. R. S. Sayles, *Tribology Int.* **34**, 299 (2001).
6. L. Kogut and I. Etsion, *Tribology Trans.* **43**, 816 (2000).
7. B. Bhushan, *Tribology Lett.* **4**, 1 (1998).
8. G. Liu, Q. J. Wang and C. Lin, *Tribology Trans.* **42**, 581 (1999).
9. G. G. Adams and M. Nosonovsky, *Tribology Int.* **33**, 431 (2000).
10. N. Tas, T. Sonnenberg, H. Jansen, R. Legtenberg and M. Elwenspoek, *J. Micromech. Microeng.* **6**, 385 (1996).

11. X. Rottenberg, B. Nauwelaers, W. De Raedt and H. A. C. Tilmans, in: *Proceedings of 34th European Microwave Conference*, Amsterdam, pp. 77–80 (2004).

12. J. A. Greenwood and J. B. P. Williamson, *Proc. Roy. Soc. (London), Series A* **295**, 300 (1967).

13. P. Sahoo and A. Banerjee, *J. Phys. D: Appl. Phys.* **38**, 2841 (2005).

14. F. W. Delrio, M. P. de Boer, J. A. Knapp, E. D. Reedy Jr, P. J. Clews and M. L. Dunn, *Nature Materials* **4**, 629 (2005).

15. B. Lorenz and B. N. J. Persson, *J. Phys: Condens. Matter* **21**, 015003 (2009).

16. J. Song and D. J. Srolovitz, *Acta Materialia* **54**, 5305 (2006).

17. N. A. Fleck and J. W. Hutchinson, *J. Mech. Phys. Solids* **41**, 1825 (1993).

18. Y. Du, L. Chen, N. E. McGruer, G. G. Adams and I. Etsion, *J. Colloid Interface Sci.* **312**, 522 (2007).

19. L. Kogut and I. Etsion, *J. Colloid Interface Sci.* **261**, 372 (2003).

20. K. L. Johnson, K. Kendall and A. D. Roberts, *Proc. R. Soc. Lond. A* **324**, 301 (1971).

21. W. R. Chang, I. Etsion and D. B. Bogy, *ASME Trans. J. Tribology* **109**, 257 (1987).

22. Y. Du, G. G. Adams, N. E. McGruer and I. Etsion, *J. Appl. Phys.* **103**, 1 (2008).

23. Y. Kadin, Y. Kligerman and I. Etsion, *Int. J. Solids Struct.* **43**, 7119 (2006).

24. I. Etsion, Y. Kligerman and Y. Kadin, *Int. J. Solids Struct.* **42**, 3716 (2005).

25. K. Komvopoulos, *Wear* **200**, 305 (1996).

26. A. Majumdar and C. L. Tien, *Wear* **136**, 313 (1990).

27. K. Komvopoulos and N. Ye, *J. Tribology* **123**, 632 (2001).

28. J. I. McCool, *ASME Trans. J. Tribology* **108**, 380 (1986).

29. I. Etsion and M. Amit, *ASME Trans. J. Tribology* **115**, 406 (1993).

30. J. I. McCool, *Int. J. Machine Tools Manuf.* **32**, 115 (1992).

31. C. A. Kotwal and B. Bhushan, *Tribology Trans.* **39**, 890 (1996).

32. P. E. Elderton and L. J. Johnson, *System of Frequency Curves*. Cambridge University Press, London (1969).

33. N. Tayebi and A. Polycarpou, *Tribology Int.* **37**, 491 (2004).

34. N. Tayebi and A. Polycarpou, *J. Appl. Phys.* **98**, 1 (2005).

35. K. M. Hurst, W. R. Ashurst and C. B. Roberts, in: *Proceedings of the 9th International Symposium on Supercritical Fluids*, Arcachon, France, pp. 1–6 (2009).

36. K. M. Hurst, C. B. Roberts and W. R. Ashurst, in: *Proceedings of the ASME/STLE International Joint Tribology Conference*, San Antonio, TX, pp. 1–3 (2008).

37. S. T. Patton, J. M. Slocik, A. Campbell, J. Hu, R. R. Naik and A. A. Voevedin, *Nanotechnology* **19**, 405705 (2008).

38. K. S. Chen, A. A. Ayon, X. Zhang and S. M. Spearing, *J. Microelectromechanical Systems* **11**, 264 (2002).

39. W. Wei, M. Bachman and G.-P. Li, *Mater. Res. Soc. Symp. Proc.* **849**, KK8.6 (2005).

40. V. A. Dao, K. Hah, J. Heo, D. Kyeong, J. Kim, Y. Lee, Y. Kim, S. Jung, K. Kim and J. Yi, *Thin Solid Films* **517**, 3971 (2009).

41. S. Y. Chou, P. R. Krauss and P. J. Renstrom, *J. Vac. Sci. Technol. B* **14**, 4129 (1996).

42. M. D. Austin, H. Ge, W. Wu, M. Li, M. Yu, D. Wasserman, S. A. Lyon and S. Y. Chou, *Appl. Phys. Lett.* **84**, 5299 (2004).

A van der Waals Force-Based Adhesion Model for Micromanipulation

S. Alvo [a,b,*], **P. Lambert** [b,c], **M. Gauthier** [b] **and S. Régnier** [a]

[a] Institut des Systèmes Intelligents et Robotique (ISIR), Université Pierre et Marie Curie — Paris 6, UMR CNRS 7222, 4 place Jussieu, Boite Courrier 173, 75252 Paris Cedex 05, France
[b] FEMTO-ST Institute, Department AS2M, UMR CNRS 6174 — UFC/ENSMM/UTBM, 24 rue Alain Savary, 25000 Besançon, France
[c] Service des Systèmes Bio-, électromécaniques (BEAMS) CP 165/56, Université libre de Bruxelles, 50 Avenue Roosevelt, B-1050 Bruxelles, Belgium

Abstract

The robotic manipulation of microscopic objects is disturbed directly by the adhesion between the end-effector and the objects. In the microscale, no reliable model of adhesion is available and currently the behaviour of the micro-objects cannot be predicted before experiments. This paper proposes a new model of adhesion based on the analytical resolution of the coupling between the mechanical deformation of the micro-objects and van der Waals forces. In the nanoscale, the impact of the deformation can be neglected and the proposed model is thus similar to the classical expression for van der Waals forces. In the microscale, the deformation induces van der Waals forces to increase significantly and a new analytical expression is proposed. The limit of validity of this 'deformable van der Waals forces' is also discussed. This result can be used as an alternative to classical adhesion–deformation models in literature (Johnson–Kendall–Roberts (JKR) or Derjaguin–Muller–Toporov (DMT)), which have been validated at the macroscale but are not sufficient to describ the interaction forces in the microscale (typically from 100 nm to 500 μm).

Keywords

Pull-off force, deformation, adhesion, van der Waals forces, micromanipulation

1. Introduction

The study of micromanipulation consists of developing models and fabricating experimental tools for the individual manipulation and characterization of micro-components. The manipulator covers a large variety of microgrippers (mechanical and optical tweezers, capillary grippers...) actuated using numerous physical effects (thermal expansion, piezoelectricity, smart memory alloy...). Characterization mainly implies mechanical characterization of stiffness performed, for example,

---

[*] To whom correspondence should be addressed. Tel.: + 33(0)1 44 27 63 79; e-mail: sebastien.alvo@isir.upmc.fr

Adhesion Aspects in MEMS/NEMS

with an atomic force microscope. The major industrial perspective of micromanipulation is to develop reliable micro-assembly techniques, based on robotic assembly
or self-assembly. Both require adequate models to estimate the surface forces dis-
turbing the micromanipulation.

The goal of developing models at the microscale may be questioned for many
reasons:

1. The task is huge and the forces dominating at the micro- and nanoscales can
   be modeled only partially: for example, some of them cannot be modeled in a
   quantitative way (e.g., hydrogen bonds) suitable for robotics purpose, and most
   of proposed models are valid only at equilibrium (at least all the models based
   on the derivation of surface or potential energies);

2. It is impossible sometimes to know the parameters involved in the existing models, for example, the distribution of electrical charges on a dielectric oxide layer;

3. In the micro- and nanoscales, the physical measurements suffer from a very
   large experimental scatter, which makes the models' refinements questionable.
   According to individual's experience, experimental measurements are typically
   obtained with an error greater than some tens of percents.

Nevertheless the use of — even basic — models helps the microrobotician to
roughly describe the micro-objects behavior and design best grippers and tools.
Classical adhesion models like JKR [1], DMT [2] and Maugis–Dugdale [3] are usually proposed to study adhesion in micromanipulation [4]. These models are based
on the elastic deformation of two solids in contact (e.g., microcomponent/gripper
in micromanipulation) and cannot match the experimental data very well at the microscale.

We propose, therefore, a new analytical model taking into account the elastic
deformation and the van der Waals forces as the main source of adhesion [5, 6].
Capillary condensation is supposed to play a major role in humid environments [7],
but as a preliminary step in this paper, we will only consider dry environments in
which capillary condensation can be neglected. Moreover, materials used are supposed to be non-conductive and neutral so electrostatic forces are not considered.

This paper is organized in the following sections. Section 2 presents the coupled
problem of adhesion and deformation and the innovative iterative scheme used to
solve it. Section 3 recalls the basics of van der Waals (vdW) forces and presents the
vdW interaction between a truncated sphere and a plane, which is used in the iterative resolution. Section 4 describes the iterative scheme which provides an implicit
expression for the force. Two analytical approximations are also provided which
are valid in nano- and microscales, respectively. Finally, Section 5 suggests further
work and conclusions are drawn in Section 6.

2. Problem Description

2.1. The Lack of Adhesion Model for Micromanipulation

The main idea of this paper is to extend the existing model to the microscale. On the one hand, JKR, DMT or Maugis models are usually used to compute pull-off forces (force needed to split two objects in contact). They are computed with global energy calculation and their efficiency is well established at macroscale [8] but they are not correlated with experimental microscale measurements. These models take into account the impact of the deformation on the pull-off force but they are restricted to sphere/plane contact and cannot be applied to more complex geometries (e.g., parallelepipedic objects) usually used in microhandling [9].

On the other hand, atomic models based on the Hamaker computation could be extended from nanoscale to microscale. So van der Waals theory could be used to predict adhesion phenomena [6] for any kind of geometry (see Section 3.1). However, in the microscale, this theory underestimates pull-off forces. Moreover, it does not take into account the deformation which has a significant impact in microscale.

An Atomic Force Microscope (AFM) with a cantilever whose stiffness is 0.3 N/m has been used to measure pull-off forces. Measurements have been carried out in a temperature and humidity controlled environment to minimize the capillary condensation. For example, the measured pull-off force is 1.6 ± 0.6 µN when a glass sphere with 10 µm diameter contacts a glass substrate. JKR and DMT theories predict a larger force of 8.0 µN and 10.7 µN and the van der Waals theory predicts a smaller value of 0.6 µN. New investigation should be performed in order to predict pull-off forces better in the microscale.

So new models are necessary to understand the behavior of objects whose size is between 100 nm and 100 µm. This paper proposes a way to take into account the deformation of the object in the calculation of van der Waals forces.

2.2. A New Method for Adhesion Modeling

The sum of van der Waals forces applied by an object to another one depends on their contact surface and can also be considered as a global force which induces a deformation. This deformation increases the contact surface and the global van der Waals forces too. This coupled problem can be seen as an algorithm that uses sequentially two models (Fig. 1). The first one computes van der Waals forces according to the object shape. The other one computes deformation shape according to an external load. An iterative calculation is able to converge to the physical equilibrium.

This generic principle can be applied to calculate or simulate adhesion forces for a large variety of geometries. However, this paper focuses on the analytical expression applied to sphere/plane contact based on an analytical expression for the van der Waals forces and a deformation model *via* the Hertz theory. For more complex geometries including roughness [10], the calculation of the van der Waals forces

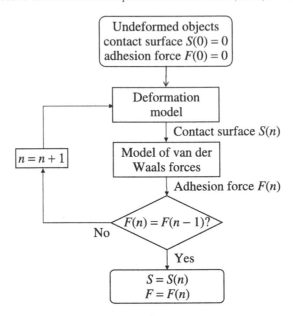

Figure 1. Algorithm proposed for calculating the adhesion force between two objects using the coupling between deformation and van der Waals forces.

should be replaced by a numerical calculation, and another deformation model should be considered (e.g., [11] for axisymmetric shapes) or modeled *via* finite element method.

3. Modeling Adhesion with van der Waals Forces for Simple Rigid Geometries

The first step in the calculation is to calculate the van der Waals forces on a deformed object. A pairwise summation of the energy given by the Lennard–Jones potential leads to classical expression for the van der Waals forces. Computation for simple shape objects attracted by a plane is carried out here as an example which can be applied to any kind of geometry.

3.1. The Lennard–Jones Potential and the Hamaker Approximation

The Lifshitz theory considers 'dance of charges' coupled with relativistic effects [12] to calculate electrodynamic free energy that is at the origin of van der Waals forces. It is commonly approximated using the Lennard–Jones potential:

$$\phi(r) = \varepsilon_0 \left(\frac{r_0}{r}\right)^{12} - 2\varepsilon_0 \left(\frac{r_0}{r}\right)^{6}. \tag{1}$$

This potential represents the interaction between two neutral atoms (or molecules) separated by a distance r (Fig. 2). r_0 is the equilibrium distance when the force $F = -\partial\phi/\partial r$ (Fig. 3) is null and ε_0 is the corresponding energy. The $1/r^{12}$ term

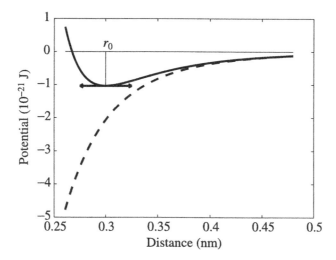

Figure 2. Comparison of Lennard–Jones potential (solid line) and $1/r^6$ potential (dashed line).

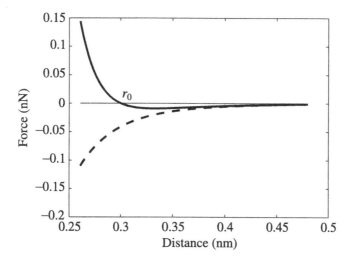

Figure 3. Comparison of forces derived from the Lennard–Jones potential (solid line) and from the $1/r^6$ potential (dashed line).

describes the strong repulsion that appears when the two atoms are closer than r_0. For $r > r_0$, the Lennard–Jones potential is usually approximated by the $1/r^6$ term (Fig. 2):

$$\phi(r) = -\frac{C}{r^6}, \quad \text{where } C = 2\varepsilon_0 r_0^6. \tag{2}$$

Adhesion forces are computed by a pairwise summation of the Lennard–Jones energies. Considering molecular densities of materials, the discrete summation can be replaced with an integral. This method is commonly called the Hamaker approximation [13] which is summarized very well in [14].

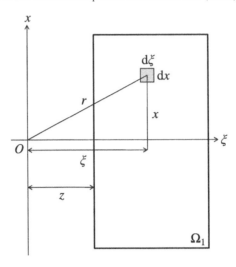

Figure 4. Notations used for the first step of van der Waals forces calculation: attraction force between a molecule and a rigid, infinite plane.

3.2. Interaction between a Molecule and an Infinite Plane

The first step in computation of the van der Waals forces is to sum every pairwise energy between the plane and a molecule of the second object. The van der Waals forces have a short cut-off radius so the plane can be considered as infinite without any influence on the result. Then, using notations of Fig. 4 and considering a molecular density β_1 for the plane Ω_1, the adhesion energy π_p between a molecule and a plane can be formulated as below:

$$\pi_p(z) = \beta_1 \int_{\Omega_1} \frac{-C}{r^6}\, d\Omega_1 = -\frac{\pi C \beta_1}{6z^3}. \tag{3}$$

3.3. Interaction between a Sphere and an Infinite Plane

Now the second part of the calculation can be performed by considering a molecular density β_2 for the second object (a sphere of radius R). For sphere and plane, the energy of interaction is:

$$\Pi_{sp} = \beta_1 \beta_2 \int_{\Omega_2} -\frac{\pi C}{6z^3}\, d\Omega_2. \tag{4}$$

The Hamaker constant $A = \pi^2 C \beta_1 \beta_2$ appears in the formulation.

The sphere studied is supposed to be larger than its distance from the plane (assumed to be $r_0 = 0.3$ nm). So for $R \gg r_0$ the force that is derived from the potential of equation (4) is:

$$F_{sp} = -\frac{AR}{6r_0^2}. \tag{5}$$

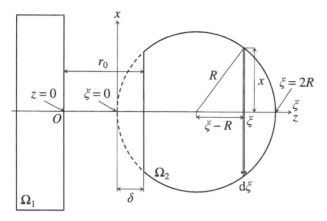

Figure 5. Integration of van der Waals forces between a truncated sphere and an infinite plane.

3.4. Interaction between a Truncated Sphere and an Infinite Plane

We consider a truncated sphere which represents the volume of a sphere of radius R where the cap with a δ height has been removed (see Fig. 5). The planar surface of the truncated sphere is at a distance r_0 from the plane. Integral of equation (4) has now to be computed with a modified Ω_2 described in Fig. 5. The volume Ω_2 extends from $\xi = \delta$ to $2R$ for the truncated sphere rather than from 0 for the full sphere. Thus, still considering non-atomic sized spheres ($R \gg r_0$) and little deformations ($R \gg \delta$), a new expression for the van der Waals forces can be formulated as:

$$F_{s'p} = -\frac{AR(r_0 + 2\delta)}{6r_0^3}. \tag{6}$$

The first term of this force is equal to the force between a full R-radius sphere and a plane seen in equation (5). Considering that $\delta \ll R$, the term $(AR\delta)/(3r_0^3)$ represents the force due to a cylinder whose radius corresponds to the truncated part of the sphere. The increase of the contact surface between a sphere and a truncated sphere induces an increase of the van der Waals forces despite the reduction of the volume.

The force computed in equation (6) is used in our coupled model to represent the van der Waals forces applied to a deformed object. The van der Waals forces computation consequently considers that the plane is rigid.

4. Sphere Deformation under van der Waals Forces

A deformation model is needed for the second phase of the algorithm presented in Fig. 1. In the case of a contact between a sphere and a plane, deformation according to load can be computed *via* the Hertz theory.

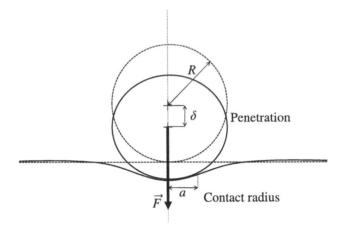

Figure 6. Sketch of a sphere/plane contact.

4.1. Hertz Contact Modeling

This model which defines the contact surface is the basis for the other models well known in adhesion modeling (JKR and DMT). It computes the radius of contact a and the penetration δ for two spheres pressed together with a force F (Fig. 6). Hertz model does not consider adhesion directly. This model is based on a geometrical and mechanical analysis under three assumptions:

- Contact radius a is small compared to the radii of spheres;

- There is no friction at the interface;

- There is no tensile stress in the contact area.

Elastic properties of the ith material are E_i (Young's modulus) and v_i (Poisson's coefficient). The modified Young's modulus E^* is defined as:

$$\frac{1}{E^*} = \frac{1 - v_1^2}{E_1} + \frac{1 - v_2^2}{E_2}. \tag{7}$$

So in case of a sphere/plane contact, the contact radius a and penetration depth δ can be calculated *via* Hertz theory:

$$a^3 = -\frac{3}{4}\frac{RF}{E^*}, \tag{8}$$

$$\delta = \frac{a^2}{R}. \tag{9}$$

4.2. Computation of a Model Including Deformation for a Sphere/Plane Contact

We assume that the deformed sphere can be seen as a truncated sphere which is described in Section 3.4. At every step of the algorithm, van der Waals forces F_n are calculated by equation (6) and the contact radius a_n is given by the Hertz theory (equation (8)). The penetration into the plane δ_n is linked to the contact radius a_n by equation (9). So equation (6) can be rewritten as:

$$F_n = -\frac{A(r_0 R + 2a_n^2)}{6r_0^3}. \tag{10}$$

At the initial step of the algorithm, the contact radius is considered as null, i.e., $a_0 = 0$. For all steps n, equation (8) implies:

$$a_{n+1} = -\left(\frac{3R}{4E^*}\right)^{1/3} F_n^{1/3}. \tag{11}$$

Considering equation (10), a sequence (a_n) can be defined:

$$a_{n+1} = \lambda^{1/3}(r_0 R + 2a_n^2)^{1/3}, \tag{12}$$

where

$$\lambda = AR/8r_0^3 E^*. \tag{13}$$

The sequence (12) converges to its unique positive fixed point a_∞. Fixed point is classically defined by $a_n = a_{n+1} = a_\infty$, so a_∞ is the unique real solution of equation (14):

$$a_\infty^3 - 2\lambda a_\infty^2 - \lambda r_0 R = 0. \tag{14}$$

This third-order equation can be solved analytically with the Cardan formula but the result is too complex to be exploited easily. The surface radius a_{dvdW} of the deformable van der Waals (dvdW) model can be obtained by numerical computation of equation (14) (see Fig. 7). Moreover, Sections 4.3 and 4.4 will show that an analytical solution exists for nano and microspheres.

4.3. Approximation of Micro-sized Spheres

This subsection presents an expression for the van der Waals forces in the case of micro-sized spheres. Equation (14) could be normalized by the $(2\lambda)^3$ term:

$$\left(\frac{a_\infty}{2\lambda}\right)^3 - \left(\frac{a_\infty}{2\lambda}\right)^2 - \frac{R_c}{R} = 0, \tag{15}$$

where

$$R_c = \frac{r_0 R^2}{8\lambda^2} = \frac{8r_0^7 E^{*2}}{A^2} \tag{16}$$

is a constant which depends only on properties of the sphere material.

If $R_c/R \ll (a_\infty/2\lambda)^3$, equation (15) can be rewritten as:

$$\frac{a_\infty}{2\lambda} \approx 1.$$

So for $R \gg R_c$, the algorithm for the deformation caused by van der Waals forces converges to a contact radius:

$$\tilde{a}_{dvdW} = 2\lambda = \frac{AR}{4r_0^3 E^*}. \tag{17}$$

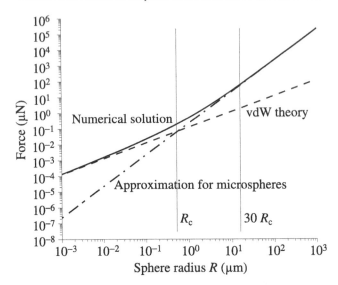

Figure 7. Comparison of forces computed with classical and deformable van der Waals theories at nano- and microscales (log scale). Dashed line: classical van der Waals theory. Dash-dotted line: approximation of microspheres (equation (18)). Solid line: numerical solution of the deformable van der Waals (dvdW) model (equation (14)) matches classical van der Waals (equation (5)) at nanoscale and analytical formula (18) at microscale.

Table 1.
Mechanical properties of glass

Young's modulus[a]	$E = 68$ GPa
Poisson's coefficient[a]	$\nu = 0.19$
Modified elastic modulus (see equation (7))	$E^* = 35.3$ GPa
Hamaker coefficient	$A = 6.5 \times 10^{-20}$ J
Minimum distance between atoms	$r_0 = 0.3$ nm

[a] Source: www.matweb.com for SiO_2 96%.

Applying this approximation to equation (10), a simplified force can be solved as below:

$$\tilde{F}_{\text{dvdW}} = -\frac{A^3 R^2}{48 r_0^9 E^{*2}}. \tag{18}$$

For example, in case of glass objects (see Table 1), R_c is 0.5 μm and this approximation ($R \gg R_c$) can be used for glass micro-sized spheres commonly used in micromanipulation.

4.4. Approximation of Nano-sized Spheres

This section is focused on the second asymptotic solution of equation (14) which represents the case of nano-sized spheres.

Another way to normalize equation (14) is to divide it by a $\lambda r_0 R$ term:

$$\left(\frac{a_\infty}{\sqrt[3]{\lambda r_0 R}}\right)^3 - \left(\frac{R}{R_c}\right)^{1/3}\left(\frac{a_\infty}{\sqrt[3]{\lambda r_0 R}}\right)^2 - 1 = 0. \tag{19}$$

So for $R/R_c \ll a_\infty/\sqrt[3]{\lambda r_0 R}$, equation (19) becomes:

$$\left(\frac{a_\infty}{\sqrt[3]{\lambda r_0 R}}\right)^3 \approx 1.$$

And for $R \ll R_c$, the contact radius and adhesion force become:

$$\tilde{a}_{\text{dvdW}} = \sqrt[3]{\lambda r_0 R} = \sqrt[3]{\frac{AR^2}{8r_0^2 E^*}}, \tag{20}$$

$$\tilde{F}_{\text{dvdW}} = -\frac{AR}{6r_0^2} = F_{\text{sp}}. \tag{21}$$

In the case of glass ($R_c = 0.5$ μm), this model matches the classical van der Waals theory at nanoscale ($R \ll R_c$). Moreover, the classical theory cannot predict the contact radius so the original expression (20) completes the current theory.

4.5. Discussion

The aim of this discussion is to estimate the error in the analytical expressions $\tilde{F}_{\text{dvdW}}$ proposed in (18) and (21) compared to an exact solution F_{dvdW} of equation (14). A coefficient Q_F is defined as a normalized force error quotient:

$$Q_F = \left|\frac{F_{\text{dvdW}} - \tilde{F}_{\text{dvdW}}}{F_{\text{dvdW}}}\right| = \left|\frac{\Delta F}{F}\right|. \tag{22}$$

Considering equation (10), F_{dvdW} can be written as:

$$F_{\text{dvdW}} = \lim_{n\to\infty} F_n = -\frac{A(r_0 R + 2a_{\text{dvdW}}^2)}{6r_0^3}. \tag{23}$$

The estimation of the error in the analytical expressions consists in estimating the error in the force Q_F as a function of the radius R of the sphere. The limit of validity of nano- and micro-sized spheres approximations depends on the normalized error ε between the analytical expression $\tilde{a}_{\text{dvdW}}$ for the contact radius and the exact contact radius a_{dvdW}:

$$\frac{a_{\text{dvdW}}}{\tilde{a}_{\text{dvdW}}} = 1 + \varepsilon, \tag{24}$$

where ε is supposed to be negligible compared to 1.

4.5.1. Estimation of Error in the Case of Micro-sized Spheres
For $R \gg R_c$, considering the definition (24) of ε and the value (17) of $\tilde{a}_{\text{dvdW}}$, equation (15) can be written as:

$$(1+\varepsilon)^3 - (1+\varepsilon)^2 - \frac{R_c}{R} = 0. \tag{25}$$

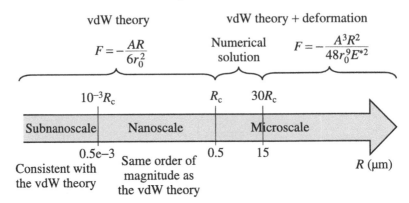

Figure 8. Adhesion models depending on sphere radius (critical radius R_c calculated for a glass sphere).

Considering that the normalized error ε is negligible compared to 1, equation (25) becomes:

$$\varepsilon \approx \frac{R_c}{R} \ll 1. \tag{26}$$

This result can be used to estimate Q_F:

$$\text{For } R \gg R_c, \quad Q_F \approx 3\frac{R_c}{R}. \tag{27}$$

Equation (27) gives an estimation of the error if the analytical formula (18) is used rather than a numerical solution of equation (14). This relation can be used to determine the validity domain of equation (18). We consider that the model (18) is valid only if the relative error Q_F is smaller than a criterion $1/k$. The parameter $1/k$ represents the maximum relative error of the model (typically $1/k = 10\%$) within its validity domain. The validity domain is thus defined by:

$$Q_F \leqslant 1/k. \tag{28}$$

According to equation (27) this inequality can be expressed as a function of the sphere radius R rather than the relative error Q_F:

$$Q_F \leqslant 1/k \quad \Longleftrightarrow \quad R \geqslant 3k R_c. \tag{29}$$

For glass, an error in force less than 10% can be observed for $R \geqslant 3 \times 10 R_c = 15$ µm (Fig. 8). So this approximation is clearly validated for micro-sized spheres.

4.5.2. Estimation of Error in the Case of Nano-sized Spheres

In the same way, using the nanosize approximation $R \ll R_c$, we can show that:

$$\varepsilon \approx \frac{1}{3\sqrt[3]{R_c/R} - 2} \ll 1, \tag{30}$$

and thus:

$$Q_F \leqslant 1/k \quad \Longleftrightarrow \quad R \leqslant 1/k^3 R_c. \tag{31}$$

For glass, an error in force less than 10% can be observed for $R \leqslant 1/10^3 R_c = 0.5$ μm. For such small objects, the hypothesis $R \gg r_0$ used for the Hamaker summation in Section 3.4 is no longer valid. This fact shows that the deformation cannot be neglected in the nanoscale either. However, the classical van der Waals expression found at nanoscale (21) gives an order of magnitude of the force. Indeed, the exact solution of equation (14) and the classical van der Waals expression (20) give the same order of magnitude for $k = 1$ ($|F_{dvdW}| \leqslant 2|F_{sp}|$), i.e., for $R \leqslant R_c = 0.5$ μm.

5. Future Work

In order to validate the proposed models, comparisons with experimental data and finite element simulation will be performed in the near future. In this way, the influence of the assumptions on deformation shape and on the rigid plane (for the integration of van der Waals forces) can be analyzed.

In a more general way, two similar studies that consider capillary condensation and electrostatic forces as the main causes of adhesion are also planned. These will help us to know which phenomenon is preponderant according to the environment characteristics and the object's material. These studies should be the way to build a simulator based on numerical integration applicable to more complex geometries. This numerical tool will be able to predict micro-objects interaction, which should be able to provide design rules in micromanipulation.

6. Conclusion

A new principle for adhesion forces computation has been proposed in this paper. The principle is based on the calculation of the coupling between van der Waals forces and the deformation of the object. Indeed, van der Waals forces induce a deformation that increases the contact surface, and at the same time the increased contact surface also increases the van der Waals forces. In order to solve this coupling problem, the proposed algorithm uses two independent models: Hertz model for the deformation and analytical expression for the van der Waals forces. In case of sphere/plane contact, analytical expressions for the deformable van der Waals forces have been proposed and their validity domains have been determined. The proposed algorithm can be extended to other more complex geometries using numerical computations.

This work has shown that the impact of the deformation of the object cannot be neglected especially in the microscale. In the nanoscale, the impact of the deformation is small and an order of magnitude of van der Waals forces can be found by neglecting the deformation.

Acknowledgements

This work is supported by the French National Project NANOROL ANR-07-ROBO-0003. Thanks to Dr. Wei Dong for reviewing the manuscript.

References

1. K. L. Johnson, K. Kendall and A. D. Roberts, *Proc. R. Soc. Lond. A* **324**, 301–313 (1971).
2. B. V. Derjaguin, V. M. Muller and Yu. P. Toporov, *J. Colloid Interface Sci.* **53**, 314–326 (1975).
3. D. Maugis, *J. Colloid Interface Sci.* **150**, 243–269 (1992).
4. D. Maugis, *Contact, Adhesion and Rupture of Elastic Solids*. Springer (2000).
5. M. Savia, Q. Zhou and H. N. Koivo, in: *Proceedings of 2004 IEEE/RSJ International Conference on Intelligent Robots and Systems*, Sendai, Japan, pp. 1722–1727 (2004).
6. F. W. DelRio, M. P. de Boer, J. A. Knapp, E. D. Reedy, P. J. Clews and M. L. Dunn, *Nature Materials* **4**, 629–634 (2005).
7. P. Lambert, *Capillary Forces in Microassembly, Modeling, Simulation, Experiments, and Case Study*. Springer-Verlag GmbH (2007).
8. E. Charrault, C. Gauthier, P. Marie and R. Schirrer, *Langmuir* **25**, 5847–5854 (2009).
9. B. Lopez-Walle, M. Gauthier and N. Chaillet, *IEEE Trans. Robotics* **24**, 897–902 (2008).
10. M. Sausse Lhernould, A. Delchambre, S. Régnier and P. Lambert, *Appl. Surface Sci.* **253**, 6203–6210 (2007).
11. I. N. Sneddon, *Inter. J. Eng. Sci.* **3**, 47–57 (1965).
12. V. A. Parsegian, *Van der Waals Forces. A Handbook for Biologists, Chemists, Engineers, and Physicists*. Cambridge University Press (2006).
13. H. C. Hamaker, *Physica* **4**, 1058–1072 (1937).
14. R. A. Sauer and S. Li, *J. Nanosci. Nanotechnol.* **8**, 1–17 (2007).

Part 2
Computer Simulation of Interfaces

Lattice Gas Monte Carlo Simulation of Capillary Forces in Atomic Force Microscopy

Joonkyung Jang [a,*] and George C. Schatz [b]

[a] Department of Nanomaterials Engineering, Pusan National University, Miryang 627-706, Republic of Korea
[b] Department of Chemistry, Northwestern University, 2145 Sheridan Rd, Evanston, IL 60208-3113, USA

Abstract

We review recent work concerned with lattice gas (LG) Monte Carlo (MC) simulations of the water meniscus formed between an atomic force microscope (AFM) tip and the surface in contact with the tip. Grand canonical MC simulations were performed to study the meniscus structure and capillary force, and this work allowed us to examine the mechanism of meniscus formation as a function of the tip–surface distance and humidity. It is found that the meniscus becomes unstable when it is narrower than the diameter of the tip–surface contact area. The calculations suggest that the ultimate size limit for a stable meniscus is five molecular diameters. We developed thermodynamic integration and perturbation methods to calculate the capillary force. The magnitude and humidity dependence of capillary force are significantly affected by the hydrophilicity of both the tip and surface. A mean field density functional theory (DFT) closely approximates the capillary forces calculated from the MC simulation. Changing the atomic scale roughness of the tip and surface drastically changes the capillary force. In particular, at low humidity, a slight roughening of the tip or surface leads to a drastic change in the force. The roughness effect persists even at 80% relative humidity. The capillary force is governed by the degree of confinement of the water and, therefore, increases as the free volume between the tip and surface decreases. The humidity dependence of the capillary force depends on the susceptibility of the meniscus width to tip retraction. For strongly hydrophilic tips at high humidity, the susceptibility is small so that the capillary force decreases as the humidity rises.

Keywords

Lattice gas, Monte Carlo simulation, water meniscus, atomic force microscopy, dip pen nanolithography, capillary force, capillary condensation, pull-off force

1. Introduction

When confined between two solid surfaces or in pores, gases naturally condense into a liquid at a pressure lower than the bulk saturation value. This so-called capillary condensation [1] plays a crucial role in imaging and patterning utilizing an

---

* To whom correspondence should be addressed. Tel.: +82-55-350-5277; Fax: +82-55-350-5653; e-mail: jkjang@pusan.ac.kr

Adhesion Aspects in MEMS/NEMS
© Koninklijke Brill NV, Leiden, 2010

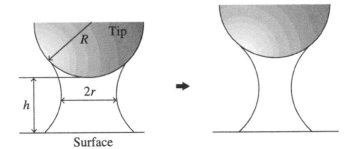

Figure 1. Diagram of the water meniscus formed in an AFM experiment. Shown on the left is a concave meniscus formed between a spherical tip and a flat surface. R is the tip radius, h the tip–surface distance and r the half width of the meniscus at the waist. Shown on the right is a new meniscus with an increased h and a decreased r resulting from the retraction of the tip. These figures are adapted from Ref. [13].

atomic force microscope (AFM) tip. Due to spatial confinement provided by the tip in close proximity to the surface, water molecules in the air condense to form a meniscus connecting the surface and tip (as illustrated in Fig. 1). This meniscus gives rise to a large capillary force on the AFM tip which governs the force (image) measured by the tip [2, 3]. The meniscus also serves as a transport channel for molecules that are deposited on the surface in dip-pen nanolithography (DPN) [4]. It has been speculated that the resolution of DPN is dependent on the width of the meniscus. Despite their importance in imaging and lithography using an AFM tip, the molecular properties of the meniscus and the capillary force are poorly understood. It would be interesting to know how the meniscus structure and the capillary force are controlled by the change in the tip–surface distance (mechanical control) or by change in humidity (thermodynamic control). It would be equally interesting to know how the meniscus and the force depend on the physicochemical properties (hydrophilicity) or geometrical shapes (roughness and curvature) of the tip and surface. Numerous experiments have been conducted which have reported that the capillary force sensitively depends on the humidity and the hydrophilicity of the tip [5–9]. A transparent interpretation of these experiments, however, is hampered by various unknown factors, such as tip geometry, surface corrugation, and contamination. For example, there has been controversy as to whether the capillary force for a mica surface should be a monotonically increasing function of humidity or not [6].

   Theoretical investigations, avoiding experimental uncertainties and complications, can provide clear insights into the properties of water meniscus and capillary force. In contrast to experimental observations, the conventional Laplace–Kelvin theory [10] predicts that dependence of the capillary force on humidity is very minor. By considering the effects of tip shape [8, 9] and liquid evaporation [11], this continuum theory can be significantly improved so as to reproduce the experimental results. However, at the nanometer scale, as in AFM experiments, a molecular description of the meniscus becomes more useful. Thus as an alternative to the continuum theory, we have been studying the use of lattice gas (LG) models and grand canonical Monte Carlo (GCMC) simulations to describe the meniscus and capillary force [12–20]. This approach provides molecular-level insight into the meniscus that is missing in the continuum description. The LG simulation has some nice features: many properties of the LG are known exactly, including the chemical po-tential at the bulk gas–liquid transition and the wetting transition temperature. Due to its computational simplicity, excellent statistics and simulation of large systems are possible in the LG model. It has been reported that the LG model success-fully explains the phase behavior of nanoscale water confined in a carbon nanotube [21]. As we will show later, the LG calculation reproduces the typical magnitude of the experimental pull-off force and its dependence on humidity. Molecular density functional theory (DFT) [22, 23] provides an alternative molecular approach that has been widely used to study phase behavior in confined spaces. This mean field theory, however, neglects long-range density correlations and is limited in its ability to describe fluctuations of the water meniscus [17].

Here, our previous LG model and GCMC simulation results are surveyed. In Sec-tion 2.1, we describe details of the LG model and the GCMC simulation methods. In Section 2.2, we describe how the capillary force is calculated using thermodynamic integration methods. A thermodynamic perturbation method is also described as an alternative, and we show how to combine the GCMC simulation with the DFT to calculate the capillary force. Section 3 presents the principal results obtained from the previous studies. In Section 3.1, the structural properties of the water meniscus are discussed. Section 3.2 presents results for the capillary force and shows how it is affected by the humidity and the roughness of the tip and surface. The microscopic relation between the humidity dependence of the capillary force and the meniscus structure is also presented. We conclude in Section 4 by summarizing our results.

2. Theory and Simulation Methods

2.1. Lattice Gas Model for the Water Meniscus

GCMC simulations were performed using a cubic lattice confined vertically be-tween a lower surface and an upper tip, as shown in Fig. 2. Molecules are allowed to exist only at lattice points.

Each molecule interacts with its nearest neighbor (NN) molecules with an in-termolecular attraction energy ε and has its own chemical potential μ. When a molecule is located at one of the NN sites of the tip and surface, it is subjected to binding energies, b_T and b_S, respectively. The effective Hamiltonian for the GCMC (the exponent of the Boltzmann-type statistical distribution) simulation is given by

$$H = -\varepsilon \sum_{i,j=\text{NN}} c_i c_j - b_T \sum_{\substack{i=\text{tip} \\ \text{bound.}}} c_i - b_S \sum_{\substack{i=\text{surf.} \\ \text{bound.}}} c_i - \mu N, \tag{1}$$

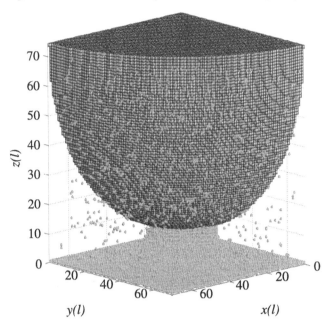

Figure 2. Tip and surface geometries and a snapshot of the water meniscus simulated by the GCMC method. The molecules are drawn as spheres and the tip boundary sites as cubes. Only the first quadrant of the system $(x, y \geqslant 0)$ is shown. The length and force, if not specified, are in units of l and ε/l, respectively (l = lattice spacing and ε = intermolecular attraction energy). Figure reprinted with permission from J. Jang, G. C. Schatz, and M. A. Ratner, *Phys. Rev. Lett.* **92**, 085504 (2004). Copyright (2009) by the American Physical Society. Readers may view, browse, and/or download material for temporary copying purposes only, provided these uses are for noncommercial personal purposes. Except as provided by law, this material may not be further reproduced, distributed, transmitted, modified, adapted, performed, displayed, published, or sold in whole or part, without prior written permission from the American Physical Society. http://link.aps.org/abstract/PRL/v92/p085504

where c_i is the occupation number (1 or 0) of the ith lattice site, and the first summation runs over NN pairs, the second is for the sites next to the tip, and the third for the sites next to the surface. N is the number of occupied sites ($N = \sum_i c_i$). We ran simulations by using the isomorphism of the LG model to an Ising model [24, 25]. Only the first quadrant of the system $(x, y > 0)$ is simulated, and the rest of the system is constructed by taking mirror images of the first quadrant. The simulation with this system leads to identical results to those obtained without invoking reflection symmetry. The relative humidity (RH) is defined as $\exp[(\mu - \mu_c)/k_B T]$, where μ_c ($= -3\varepsilon$) is the chemical potential at the bulk gas–liquid transition [26]. This definition of RH is the ideal gas limit expression for the system pressure relative to the bulk saturation pressure. The bulk critical temperature for the LG is given by $k_B T_C/\varepsilon = 1.128$ [26]. The identification of $T_C = 647.3$ K for water gives $\varepsilon = 4.8$ kJ/mol. The temperature is fixed at $T/T_C = 0.46$, corresponding to room temperature. If we use the typical lattice spacing $l = 0.37$ nm for the LG model of water [21], the force unit is $\varepsilon/l = 0.021$

nN. We systematically vary b_T and b_S to investigate the effects of the hydrophilic-ity of the tip and surface. An AFM tip in DPN is often coated with hydrophobic alkane thiols. However, in some experiments, these alkanethiols are functionalized with amine or carboxylic acid groups, which would tend to make the tips hy-drophilic. In view of this, we consider a range of tip properties, including strongly and weakly hydrophilic and hydrophobic tips. In the case where we compare the simulation with a specific experiment, the values of b_T and b_S must be speci-fied. For example, for a silicon nitride tip, b_T can be taken to be the heat of adsorption of a water molecule on the silicon nitride tip (50 kJ/mol) [27]. In the case of a mica surface, b_S can be taken from ab initio calculations (46 kJ/mol) [28].

We also performed simulations on a two-dimensional (2D) square lattice [19, 20]. The thermodynamic parameters for the 2D LG are exactly known as $\mu_c = -2\varepsilon$ and $k_B T_C/\varepsilon = (1/2)/\ln(1 + \sqrt{2})$ [24]. Therefore, the energy and force units in the simulation are given by $\varepsilon = 2.0$ kJ/mol and $\varepsilon/l = 0.0088$ nN, respectively.

2.2. Calculation of Capillary Force

The capillary force F for a given tip–surface distance h is given by [19]:

$$F(h) = -\left(\frac{\partial \Omega}{\partial h}\right)_{\mu,T} - p\left(\frac{\partial V}{\partial h}\right)_{\mu,T}, \tag{2}$$

where Ω and V are the grand potential and volume of the system, respectively, and p is the pressure of the bulk system with a chemical potential μ. Since Ω is not directly observable in the simulation, we adopt an indirect method called thermodynamic integration to determine either it or the force. In the T integration method [19], we utilize the following relation,

$$\left.\frac{\partial(\beta\Omega)}{\partial\beta}\right|_{\mu,h} = H, \quad \beta^{-1} = k_B T, \tag{3}$$

where H is the effective Hamiltonian, equation (1). In practice, starting from $\beta = 0$ (where Ω is known exactly), we run simulations for intermediate β values. At each β value, we evaluate equation (3) by using a standard GCMC simulation. By numerically integrating equation (3), we obtain the desired Ω. In the μ integration method [19], the capillary force is calculated by using the following relation,

$$\left(\frac{\partial F}{\partial \mu}\right)_{h,T} = \left(\frac{\partial N_{ex}}{\partial h}\right)_{\mu,T}, \tag{4}$$

where N_{ex} is the excess number of molecules with respect to the bulk number ($N_{ex} = N - N_{bulk}$). The capillary force $F(h)$ is calculated by integrating equa-tion (4) with respect to μ. Starting from a sufficiently low μ (which gives zero force), we numerically integrate equation (4) by evaluating $(\partial N_{ex}/\partial h)$ at interme-diate μ values. The T integration method has the advantage that it naturally divides the capillary force into its energetic and entropic contributions [19]. The μ integration method, on the other hand, is numerically more stable.

In combination with the GCMC simulation, we also used DFT for the calculation of the capillary force. The grand potential in the mean-field DFT Ω_{DFT} is given by [22]:

$$\Omega_{DFT}/V = k_B T \sum_i [\rho_i \log \rho_i + (1 - \rho_i) \log(1 - \rho_i)]$$

$$- (\varepsilon/2) \sum_{i,j=NN} \rho_i \rho_j + \sum_i (V_i - \mu) \rho_i, \qquad (5)$$

where ρ_i is the average occupancy of the ith site, and V_i is the potential field emanating from the surface and the tip. Using ρ_i obtained from the GCMC simulation, we calculated Ω_{DFT} using equation (5) and then the capillary force using equation (2). For the bulk system, we calculate Ω_{DFT} and the density ($\rho = N_{bulk}/V$) using a self-consistent numerical solution to equation (5) without the V_i term. In the calculation of these bulk properties, DFT gives virtually identical results to the GCMC simulation.

Thermodynamic integration methods are numerically robust, but require many extra simulations for the intermediate states. We previously proposed an alternative way to calculate the capillary force called the thermodynamic perturbation method [15]. The key quantity in the calculation of the capillary force is the derivative of Ω with respect to h. The derivative in the LG model is given by the difference in the grand potential, $\Omega(h + l) - \Omega(h)$, where $\Omega(h + l)$ and $\Omega(h)$ are the Ωs at the tip–surface distances of $h + l$ and h, respectively. The difference in Ω can be written in terms of partition functions as:

$$\Omega(h + l) - \Omega(h) = -k_B T \ln[Z(h)/Z(h + l)], \qquad (6)$$

where $Z(h)$ and $Z(h + l)$ are the partition functions with tip–surface distances of h and $h + l$, respectively. The formal expression for $Z(h)$ is:

$$Z(h) = \sum_{C_1=0}^{1} \sum_{C_2=0}^{1} \cdots \sum_{C_B=0}^{1} \exp[-\beta H(c_1, c_2, \ldots, c_B)], \qquad (7)$$

where B is the total number of lattice sites, and c_i ($i = 1, 2, \ldots, B$) is the occupation number. With some mathematical manipulation, we can derive an important relation,

$$\Omega(h + l) - \Omega(h) = k_B T[(B - B') \ln 2 + \ln(\langle \exp[\Delta H/k_B T]\rangle_{\mu,T,h+l})], \qquad (8)$$

where $\Delta H = H(c_1, c_2, \ldots, c_B, \ldots, c_{B'}) - H(c_1, c_2, \ldots, c_B)$, and $\langle \cdots \rangle_{\mu,h+l,T}$ means an ensemble average for a fixed μ, tip–surface distance $h + l$ and T. Equation (8) expresses the difference in Ω as an ensemble average of exponential of ΔH which, in principle, can be evaluated by running only a single simulation.

The accurate evaluation of equation (8), however, will require sampling a range of ΔH much broader than that obtained in the standard GCMC method. There are advanced MC techniques designed to meet such a requirement. The overlapping distribution method and umbrella sampling [29] are examples. This thermodynamic perturbation method, combined with the advanced MC techniques, is expected to be a computationally efficient approach to calculate the capillary force. The implementation of this perturbation method is left as future work.

3. Simulation Results

3.1. Structure and Stability of the Water Meniscus

In Fig. 2, we show a representative snapshot of the meniscus generated in a GCMC simulation. The tip is weakly hydrophilic and the surface is strongly hydrophilic (gold-like). The molecules form a meniscus (with a radius of about 19 molecular diameters) between the spherical tip and flat surface. In the following discussion, the length unit is the lattice spacing, l, which can be identified as the molecular diameter of water (roughly 0.37 nm [21]), and the forces are in units of ε/l by default, but are sometimes reported in physical units, such as nN. Note that the surface is completely covered (wet) by molecules (contact angle zero), while the tip is only partially covered and the meniscus forms a nearly 90° contact angle with the tip. The shape of the meniscus fluctuates significantly from snapshot to snapshot. By averaging over snapshots, as shown in Fig. 2, we can obtain the density profile of the meniscus, ρ_i. Typically, the average occupancy of each site is calculated from 10 000 snapshots and those sites with an average occupancy of more than a half are defined as liquid sites. Figure 3 illustrates the density profile, ρ_i, for a hydrophilic tip on a hydrophilic surface (located at $Y = 0$). The simulation is done using a 2D square lattice for $h = 10l$ and 70% RH. Those sites with average occupancy $\rho_i = 0.25$–0.5 are defined as a dense gas and are drawn as open circles. The gray and black filled circles represent the liquid densities with $\rho_i = 0.5$–0.75 (dilute liquid) and $\rho_i = 0.75$–1 (dense liquid), respectively. Instead of the sharp boundary expected for a macroscopic meniscus, the boundary of the nanoscale meniscus is rather fuzzy, indicating its intrinsic instability (see below).

As the tip withdraws from the surface, the meniscus, such as that shown in Fig. 2, shrinks in width and eventually evaporates. Figure 4 shows how this occurs for a hydrophilic tip on a hydrophilic surface at RH 30%. Shown in Fig. 4(a–d) are snapshots of the meniscus for various tip–surface distances. As the tip retracts from the closest distance (Fig. 4(a)), the meniscus narrows in width (Fig. 4(b) and 4(c)) and snaps off (Fig. 4(d)). For each tip–surface distance in Fig. 4(a–d), we inspected how the meniscus width varied from snapshot to snapshot. The probability distribution for the occurrence of each width is plotted in Fig. 4(e) for the distances shown in Fig. 4(a–c). The average widths in these cases are approximately $36l$ (Fig. 4(a)), $27l$ (Fig. 4(b)), and $11l$ (Fig. 4(c)). As the tip recedes from the surface, the dis-

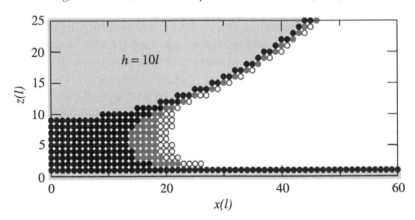

Figure 3. Density profile of the water meniscus condensed between a hydrophilic tip and a hydrophilic surface (located at $Z = 0$). The profile is obtained from a 2D GCMC simulation on the XZ plane. Only the right half of the system is shown. Lengths are in units of the lattice spacing l. The tip–surface distance h is $10l$ and the RH is 70%. The black, gray, and open circles represent the dense and dilute liquid sites and dense gas sites, respectively. The figure is adapted from Ref. [20].

tribution moves toward zero. At the distance just before snap-off, the distribution broadens notably and shows a bimodal distribution which peaks at around 15 and 0. The distribution located at zero width indicates that the meniscus is unstable. As a quantitative measure of stability, we evaluated the fluctuation, which is defined as the standard deviation of the meniscus width relative to its average [17]. After examining a range of RHs (0–80%) and tip–surface distances, we found that even the most stable meniscus had a fluctuation of more than 1%, in contrast to the virtually zero fluctuation in the thermodynamic limit.

We defined those menisci with fluctuations of less than 10% as stable ones. Among the stable menisci, we searched for the smallest meniscus possible by adjusting the humidity, the tip–surface distance, and the hydrophilicity of the tip. The smallest meniscus width is governed by the tip contact area (defined as the part of tip which contacts the surface at the shortest tip–surface distance), but is insensitive to the hydrophilicity of the tip. We investigated the correspondence between the smallest width and the tip contact diameter by varying the tip geometry. Specifically, we considered ellipsoidal tips with various aspect ratios, so that the tip contact diameter ranged from 1 to 16 molecular diameters. In Fig. 5, we plot the smallest meniscus width *vs* the tip contact diameter for a weakly and a strongly hydrophilic tip. The equivalence of the width and diameter is excellent for tips with contact diameters as small as 8 molecular diameters (note also that the smallest width is independent of whether the tip is weakly or strongly hydrophilic). Upon further decreasing the contact diameter down to just one molecular width, the smallest width deviates from the corresponding contact diameter and converges to 5 molecular diameters. This suggests that 5 molecular diameters is the ultimate size limit for a stable meniscus.

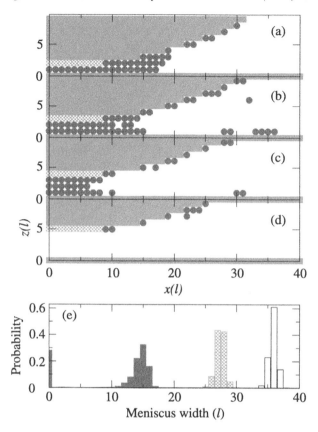

Figure 4. Shrinkage and breakage of the meniscus and the distribution of its width. (a)–(d) Meniscus snapshots at different tip–surface distances for a hydrophilic tip on a hydrophilic surface at 30% RH. Only the right half of the *XZ* cross-section of the system is shown in (a)–(d). Lengths are in units of the lattice spacing *l*. The molecules are drawn as circles and the tip and surface as gray areas. The tip contact area is defined as the part of tip closest to the surface (thus contacting the surface at the shortest tip–surface distance) and is drawn hatched in the figures. (e) Probability distribution of the meniscus width for (a) open bars, (b) hatched bars and (c) solid bars. The figures are adapted from Ref. [17].

We explored the transition from vapor to a meniscus as the humidity increased at a fixed tip–surface distance [20]. Before the meniscus formation, a hydrophilic surface or tip becomes covered with a layer of water. As the humidity reaches a certain saturation value, dense gas forms around the tip and surface, and a water meniscus results if the humidity is increased further. An accumulation of dense vapor precedes the formation of the meniscus, in accordance with the observation on nanoslits [30] in MD simulations. The periphery of the initial meniscus is diffuse due to its thermal instability. Once initially formed, the meniscus broadens continuously until it fills the entire system as humidity further increases. In Fig. 6, we show isotherms of the meniscus width for a hydrophilic tip separated from a hydrophilic

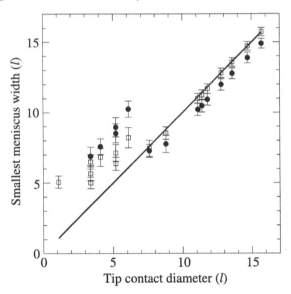

Figure 5. Smallest meniscus widths for various tip contact diameters. Lengths are in units of the lattice spacing l. The smallest possible width for a given tip geometry is determined by examining the stable menisci that form at RHs ranging from 10 to 80%. The smallest width is plotted as a function of the tip contact diameter for a weakly (circles) and a strongly (squares) hydrophilic tip (the error bars represent the standard deviation of the width). The surface is strongly hydrophilic. The line refers to the case where the smallest width is given by the tip contact diameter. The figure is adapted from Ref. [17].

surface by 4 (top), 10 (middle), and 16 (bottom) molecular diameters. For short tip–surface distances (top and middle), two consecutive transitions occur. The first jump in the width corresponds to the onset of meniscus formation and the second to the filling of the entire system with liquid. In between, the meniscus continuously broadens. As the tip–surface distance increases from $4l$ to $10l$, the initial meniscus width increases and the range of humidity in which continuous filling occurs decreases (middle). At the largest distance of $h = 16l$ (bottom), the two transitions collapse into a single gas–liquid transition characteristic of the conventional capillary condensation mechanism. The broad range of continuous filling for the shortest tip–surface distance has an interesting implication for thermodynamic control of the meniscus. The result shows that at a tip–surface distance of 1.3 nm, the width of the meniscus can be continuously tuned from 4.5 nm to 32 nm by controlling the humidity.

Wei and Zhao [31] reported that the formation of a meniscus involves water vapor condensation followed by the motion of the water film adsorbed on the surface. Unfortunately, the dynamics of meniscus growth is beyond reach of the present equilibrium MC simulation. As mentioned above, however, we did observe an accumulation of water vapor before the formation of a meniscus.

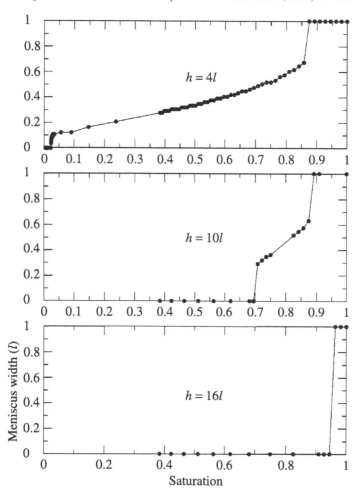

Figure 6. Isotherms of meniscus formation and broadening at several tip–surface distances (h). The meniscus width is plotted *vs* saturation (relative humidity) for a hydrophilic tip on a hydrophilic surface. The y axis refers to the meniscus width as a fraction of the tip diameter. The results are obtained from a 2D LG simulation. The figures are adapted from Ref. [20].

3.2. Capillary Force Arising from the Water Meniscus

In Fig. 7, the capillary force is plotted against the tip–surface distance for a hydrophilic (top) and hydrophobic (bottom) tip. The surface is hydrophilic in both cases. In each panel, the force–distance curves are drawn for three different RHs of 30% (filled circles), 50% (open squares), and 70% (filled triangles). With increasing humidity, the range of the force becomes larger in the case of the hydrophilic tip. The force for the hydrophobic tip vanishes at tip–surface distances above $2l$, regardless of the RH. The vanishing of the force is due to the evaporation of the meniscus as the tip recedes from the surface. For the hydrophilic tip, the menisci

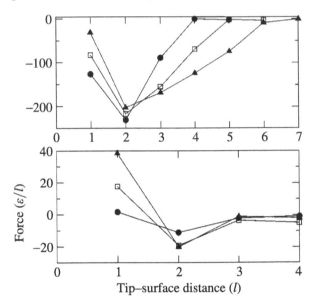

Figure 7. Capillary force *vs* tip–surface distance. For a hydrophilic (top) and a hydrophobic (bottom) tip, the force–distance curve is evaluated at RHs of 30% (filled circles), 50% (open squares), and 70% (filled triangles). The surface is hydrophilic for both tips. The force is attractive (repulsive) if it is negative (positive). Forces and distance are in units of nN and nm, respectively. Lines are drawn as a visual guide. The figures are adapted from Ref. [18].

have concave shapes and disappear at a longer tip–surface distance as the humidity rises. For the hydrophobic tip, however, the only meniscus that forms (regardless of the RH) is a monolayer of molecules sandwiched between the tip and surface. Thus, the force becomes zero when the tip–surface distance is longer than $2l$. When the tip touches the surface (a distance of $1l$), the force is attractive for the hydrophilic tip, but is repulsive for the hydrophobic tip. At this distance, the tip squeezes molecules out of the area where the tip comes into contact with the surface. For the hydrophilic tip, this squeezing out of molecules is compensated for by forming a broader meniscus: as the molecules become more confined, they experience a stronger interaction with the tip and surface and thus form a bigger (in volume) meniscus. As a result, the meniscus is more stable when the tip is in contact with the surface, giving an attractive force. For the hydrophobic tip, however, contacting the tip with the surface removes a portion of the monolayer from the surface. Because of the weak molecule-tip attraction in this case, the geometric confinement at the contact does not yield a wider meniscus as in the case of the hydrophilic tip. Therefore, this change is energetically unfavorable and thus gives rise to a repulsive force. Regardless of the tip hydrophilicity and humidity, the force is most attractive at the closest non-contact distance ($h = 2l$).

The T integration method (Section 3.2) enables the decomposition of the capillary force into its energetic and entropic components. Such a decomposition is

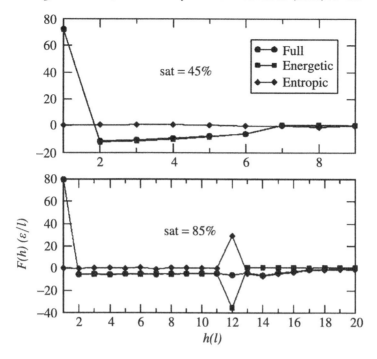

Figure 8. The capillary force and its decomposition into its energetic and entropic contributions. The capillary force *F* (drawn as circles) is plotted *vs* the tip–surface distance *h* at low (45%) and high (85%) RHs. Forces and lengths are in units of ε/l and *l*, respectively. Both the tip and surface are hydrophilic. The energetic and entropic contributions to the force are drawn as squares and diamonds, respectively. The lines serve as a visual guide. The figures are adapted from Ref. [19]. Figure reprinted with permission from J. Jang, G. C. Schatz, and M. A. Ranter, *Phys. Rev. Lett.* **90**, 156104 (2003). Copyright (2009) by the American Physical Society. http://link.aps.org/abstract/PRL/v90/p156104

done for a hydrophilic tip on a hydrophilic surface, as shown in Fig. 8 (2D simulation). At most tip–surface distances and humidities, the entropic force is negligible and the energetic contribution dominates the capillary force. Only in the case of a strongly hydrophilic tip at the snap-off distance for high humidity ($h = 12l$ for 85% RH, bottom of Fig. 8) is the entropic force comparable to the energetic force. As the meniscus snaps off, the tip experiences a repulsive entropic force, which drives the transition from liquid meniscus (ordered) to a gas state (disordered). This repulsive entropic force is almost perfectly cancelled by the attractive force arising from the loss of adhesion energy, which resists energetic destabilization due to the snap-off, such that the net force varies smoothly with separation distance.

The force necessary to detach an AFM tip in contact with a surface is called the pull-off force, and is often measured experimentally as a function of relative humidity. To calculate the pull-off force in the simulation, we first calculated the distance dependence of the force $F(h)$, as shown in Figs 7 and 8. We then identified the magnitude of the most attractive force in the $F(h)$ curve as

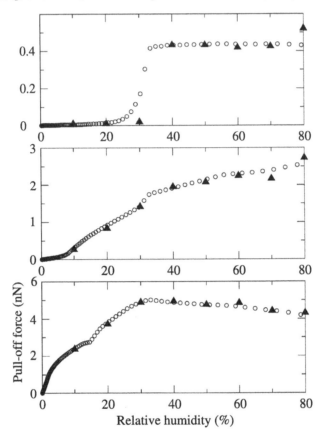

Figure 9. Humidity dependence of the pull-off force for different tip hydrophilicities. The forces calculated by the μ integration method (open circles) and DFT (filled triangles) are plotted for the hydrophobic (top), weakly hydrophilic (middle) and strongly hydrophilic (bottom) tips on a hydrophilic surface. The figures are adapted from Ref. [18].

the pull-off force. We plot in Fig. 9 the pull-off force as a function of the humidity for tips with different hydrophilicities. The humidity dependence of the pull-off force varies drastically as the tip hydrophilicity changes. The average pull-off forces are roughly 0.3, 1.7, and 3.9 nN for the hydrophobic (top), weakly hydrophilic (middle), and strongly hydrophilic (bottom) tips, respectively. Also plotted is the pull-off force calculated using the mean-field DFT (drawn as filled triangles). The DFT results closely approximate the thermodynamic integration results.

AFM tips and surfaces would be expected to have some degree of roughness, especially on the atomic scale. Here we describe our investigations of the effect of tip roughness on the capillary force. Figure 10 illustrates the four different tips that were studied. The tip labeled A is a smooth spherical tip, while Fig. 10(B–D) correspond to rough tips. The root-mean-square (RMS) roughnesses for tips B, C,

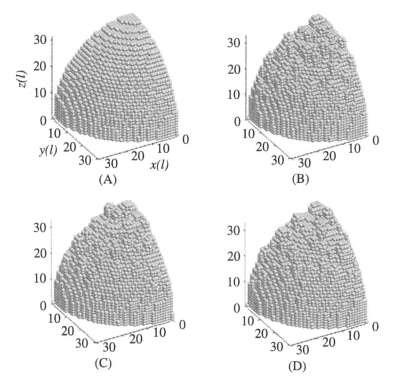

Figure 10. AFM tips with atomic-scale roughness. Lengths are in units of l. The X, Y and Z labels of (A) apply to (B—D). The figure shows a smooth spherical tip A and 3 rough tips B, C and D. The tip sites are represented as cubes. The tip radius is 11 nm. Only the 1st quadrant ($x \geqslant 0$, $y \geqslant 0$) of the tip is shown. The figures are adapted from Ref. [16].

and D are 0.22 nm, 0.21 nm and 0.19 nm, respectively. Due to the small scale of the roughness, the tips are not very different from each other.

In Fig. 11, the pull-off force of each tip in Fig. 10 is plotted as a function of humidity. The energetic parameters, $b_T/\varepsilon = 2.68$ and $b_S/\varepsilon = 2.47$, are chosen to mimic a silicon nitride tip interacting with a mica surface. For the various tip shapes, the dependence of the pull-off force in Fig. 11 on the humidity is quite different. This is remarkable, because the shapes of the tips are not very different from each other. Interestingly, the MD simulation of Luan and Robbins [32] showed that atomic scale roughness greatly influences the contact mechanics of a cylindrical tip interacting with a flat surface in vacuum (without a meniscus). The contact area and friction of tip are drastically changed by the small roughness of the tip. As the RH rises from zero, the pull-off force of the smooth tip (Fig. 10(A)) quickly increases to a maximum at RH 6.5% and then gradually decreases. For the rough tips (Fig. 10(B–D)), we also observe a rapid increase of the force as the RH increases from 0 to 6.5%. As the RH further increases from 6.5%, the pull-off force for the rough tips slowly increases, instead of decreasing as in the case of the smooth tip. For tips B and C, the pull-off force shows alternating maxima and

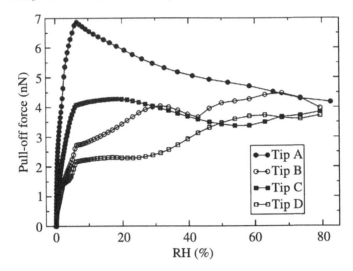

Figure 11. Pull-off force *vs* RH for the smooth tip (tip A) and three rough tips (tips B, C and D) in Fig. 10. The lines serve as a visual guide. The figure is adapted from Ref. [16].

minima with rising humidity. The pull-off force of tip D is mostly a monotonically increasing function of humidity. The pull-off force shown in Fig. 11 is of the same order of magnitude as the experimental force ranging from 2 to 12 nN. Due to the small size of tip, the pull-off force in the simulation is, on average, smaller than the experimental value [6, 33]. Considering our coarse-grained lattice model, the overall agreement is impressive. Intriguingly, with slight changes in the tip shape, our simulations qualitatively reproduce both the monotonic [6] and non-monotonic [33] dependences of the pull-off force on the humidity observed experimentally for a silicon nitride tip on a mica surface. A quantitative comparison with experiment, however, seems difficult considering the coarse-grained nature of our lattice gas model and the unknown experimental details (such as tip shape and surface roughness). Overall, the effect of the roughness is greater at low humidity and decreases as the humidity increases. For example, at RH 6.5%, the pull-off force of the smooth tip (6.9 nN) is 3 times larger than that of tip D (2.3 nN). This is due to the fact that the size and shape of the small-scale meniscus at low humidity are sensitive to the atomic-scale roughness of the tip. By and large, the pull-off forces of the rough tips are smaller than that of the smooth tip. As the RH approaches 80%, the pull-off force of each tip becomes similar, but not exactly the same. This means that even at this high humidity, the water meniscus can sense the roughness of the AFM tip.

What would be the effects of roughness of the surface on pull-off force? To evaluate this, we examined six rough surfaces and one flat surface. The six rough surfaces drawn in Fig. 12 are generated by introducing random bumps onto the flat surface. Surfaces (a), (b) and (c) have RMS roughnesses of 0.24, 0.22, and 0.22 nm, respectively. Surfaces (d), (e) and (f) have RMS roughnesses of 0.59, 0.54, and 0.60 nm, respectively, which are about three times rougher than surfaces (a)–(c).

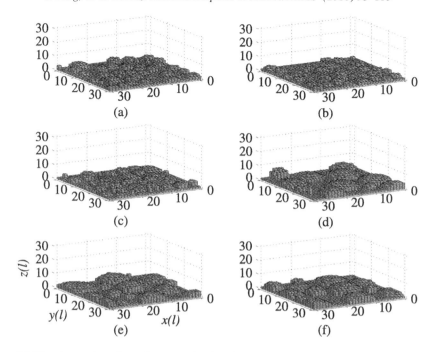

Figure 12. The six rough surfaces that were simulated. Lengths are in units of *l*. The axis labels of (e) apply to the rest of figures. The surface (hydrophilic) sites are represented as cubes. The figures are adapted from Ref. [14].

In Fig. 13, we plot the humidity dependent pull-off force for a smooth tip on the various surfaces shown in Fig. 12. As with tip roughness, we see that surface roughness profoundly influences the pull-off force. The roughness effect is especially large at low RHs (<20%). The pull-off force of the rough surface can be nearly 5 nN smaller than that of the smooth surface. The difference in the pull-off forces between the rough and smooth surfaces becomes small as the RH approaches 80%. At low humidity, the meniscus is narrow and is sensitive to little bumps on the surface. At high humidity, the meniscus becomes broad and covers up these small bumps, so that the atomic details of the surface become less important. Even at the highest RH, however, the pull-off force differs for the different surfaces. The size of the meniscus does not reach the macroscopic limit and, therefore, the pull-off force reflects the atomic-scale morphology of the surface to some extent. Notice that the pull-off force of the rough surface can be greater than that of the smooth surface, especially at high RHs. This can be explained as follows. A rough surface has gorges between the bumps on the surface. When the water meniscus fills in these gorges, the molecules experience a stronger binding force from the surface than from a smooth surface (because the molecules in these gorges are more confined by the surface walls), and the stronger molecular binding to the surface gives rise to a stronger pull-off force.

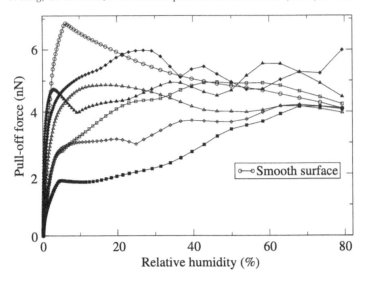

Figure 13. Effect of surface roughness on the pull-off force. We plot the pull-off force *vs* RH for a smooth surface (drawn as open circles) and for the six rough surfaces shown in Fig. 12. The open symbols represent surfaces (a)–(c) of Fig. 12. The filled symbols correspond to surfaces (d)–(f) in Fig. 12. The figure is adapted from Ref. [14].

The capillary force originates from water condensation which is, in turn, due to the molecular confinement between the tip and surface. It is thus reasonable to expect that increased confinement would lead to a bigger meniscus and a stronger pull-off force. We quantified the confinement by the average vertical distance between the tip and surface: the larger the average distance, the less confined is the system. In Fig. 14, the overall pull-off force is plotted *vs* the average tip–surface distance. The overall pull-off force is the pull-off force averaged over the different RHs. The error bars represent the RMS deviation of the overall pull-off force with respect to the RH. All in all, the overall pull-off force decreases as the average tip–surface distance increases. There is a small exception to this trend at an average distance of 4.4 nm, where the overall pull-off force is higher than that at 4 nm. This deviation is ascribed to the presence of gorges in the rough surface. As described above, the molecules confined in these gorges are more strongly attracted to the surface than those on a flat surface. These molecules can give extra strength to the pull-off force, thereby overriding the decrease in vertical confinement. In other words, the average vertical distance alone cannot account for the presence of these gorges and the enhanced confinement due to them, but this effect is rather small.

As the humidity rises, the meniscus becomes wider. It seems plausible that a wider meniscus would give rise to a larger capillary force. However, the experiments and simulations show that the capillary force can become smaller as the humidity rises. The quantity of interest is the derivative of the pull-off force with respect to RH, $s = \exp[(\mu - \mu_c)/k_B T]$. Defining the pull-off distance $h_{\min}$ as the tip–surface distance at which $F(h)$ is minimal (most attractive), from equa-

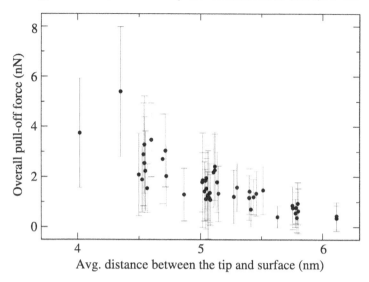

Figure 14. Correlation between the pull-off force and the degree of confinement. The overall pull-off force is plotted *vs* the average vertical distance between the tip and surface. The error bars represent the RMS deviation (with respect to the RHs ranging from 0 to 80%) of the overall pull-off force. The figure is adapted from Ref. [14].

tion (4) we obtain the expression: $-\partial F(h_{min})/\partial s = -(k_B T/s)(\partial N_{ex}/\partial h)$ where $-\partial N_{ex}/\partial h$ is evaluated at $h = h_{min}$. A positive (negative) value of $\partial N_{ex}/\partial h$ at $h = h_{min}$ represents a decreasing (increasing) pull-off force with respect to increasing humidity. Since the bulk contribution to N_{ex} is negligible ($N \gg N_{bulk}$), N_{ex} can be identified as N. Then, $\partial(\text{Pull-off force})/\partial s \approx -(k_B T/s)(\partial N/\partial h)$. If the meniscus density is constant (which is true for the LG model), N is proportional to the meniscus volume, V_b. Therefore, the dependence of the pull-off force on the humidity is determined by $-\partial V_b/\partial h$, i.e., the susceptibility of the meniscus volume to an increase in the tip–surface distance,

$$\left(\frac{\partial(\text{Pull-off force})}{\partial s}\right)_T \approx -(\rho_b k_B T/s)\left(\frac{\partial V_b}{\partial h}\right)_T, \qquad (9)$$

where ρ_b is the number density of the meniscus and the derivative with respect to h is evaluated at $h = h_{min}$. According to equation (9), the pull-off force is an increasing (decreasing) function of humidity if V_b decreases (increases) with increasing h. There are two competing contributions to the change in V_b as the tip retracts. As h increases, the height of the meniscus increases (see Fig. 1) and, therefore, its volume expands. At the same time, an increase in h narrows the meniscus width $2r$ and, thus, reduces its volume. This competition can be clearly seen for a cylindrical meniscus with a radius r and a tip–surface distance h. dV_b/dh for the cylindrical meniscus is given by [13]:

$$(1/\pi Rr)(dV_b/dh) = r/R - (-2dr/dh)(h/R + 1 - \sqrt{1 - r^2/R^2}). \qquad (10)$$

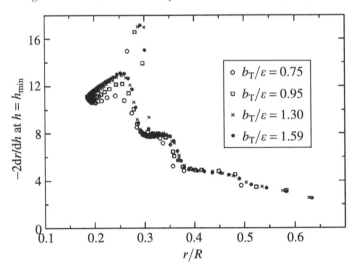

Figure 15. Susceptibility of the meniscus width to an increase in the tip–surface distance. For four energetically different tips $b_T/\varepsilon = 0.75, 0.95, 1.30$ and 1.59 on a hydrophilic surface, we calculated $-2dr/dh$ at the pull-off distance h_{min} by varying the RH from 0 to 80%. The calculated $-2dr/dh$ is plotted as a function of the half-width of the meniscus relative to the tip radius, r/R. The figure is adapted from Ref. [13].

In equation (10), the first term on the right-hand side, r/R, represents the increase in V_b as h increases. On the other hand, the second term including $-2dr/dh$ is negative and corresponds to the decrease in V_b during the retraction of the tip from the surface. Therefore, dV_b/dh is determined by the sum of these two competing terms. As the humidity increases, both the first and second terms on the right-hand side in equation (10) increase in magnitude. It is thus the value of $-2dr/dh$ that determines the sign of dV_b/dh. $-2dr/dh$ is the susceptibility of the meniscus width to an increase in the tip–surface distance. A large $-2dr/dh$ means that the corresponding meniscus shrinks rapidly as the tip is retracted. A meniscus with a small $-2dr/dh$, on the other hand, is stable with respect to the withdrawal of the tip.

In Fig. 15, we plot $-2dr/dh$ as a function of r/R for various tip hydrophilicities quantified by b_T/ε. Overall, the susceptibility, $-2dr/dh$, varies from 2 to 18 and becomes smaller as the meniscus becomes wider (r/R increases). Note that a narrow meniscus ($r/R < 0.3$) is sensitive to the withdrawal of the tip and, thus, its susceptibility significantly depends on the tip hydrophilicity b_T/ε. In contrast, a wide meniscus is relatively stable with respect to the retraction of the tip and its susceptibility does not depend much on the tip hydrophilicity (one can see the converging behavior of the susceptibility for $r/R > 0.3$). Near $r/R = 0.3$, the susceptibility is especially sensitive to changes in b_T ($r/R = 0.3$ corresponds to an RH of 29% for $b_T/\varepsilon = 1.59$ in our simulation).

In most cases, the susceptibility, $-2dr/dh$, is large, so that the second term in equation (10) dominates the first term to give a negative value for dV_b/dh. As a result, the pull-off force is an increasing function of humidity. For a wide meniscus with a small $-2dr/dh$ value, however, the second term can be smaller in magnitude than the first term. The pull-off force in that case decreases with increasing humidity. For a weakly hydrophilic tip, as shown in the middle of Fig. 9, the pull-off force always increases with increasing humidity. This is due to the fact that the water meniscus rapidly shrinks in width if the tip is retracted (a large $-2dr/dh$). By contrast, the pull-off force for a strongly hydrophilic tip (bottom of Fig. 9) increases up to a certain RH (~30%), but decreases thereafter. In this case, the meniscus at high humidity is wide (large r/R in equation (10)) and is not strongly dependent on the tip–surface distance (small $-2dr/dh$ in equation (10)). This leads to a decrease in the pull-off force with increasing humidity. Note also that in the bottom of Fig. 9 the slope of the pull-off force with respect to humidity becomes smaller as the RH increases. This is ascribed to the inverse humidity term $(1/s)$ in equation (9).

4. Conclusion

The LG model is a simple molecular description of the water meniscus and the capillary force in AFM experiments. Being a molecular model, the LG model lends itself to modern statistical mechanical methods such as GCMC simulation, the density functional theory, and the thermodynamic integration and perturbation methods, to name a few. Compared to other molecular models based on continuous intermolecular potentials, the LG model provides great efficiency and excellent statistics in MC simulation and various calculations. It is, therefore, possible to study a much wider range of experimental parameters (e.g., humidity, tip geometry and temperature) than is possible for other molecular models. It is also encouraging that the LG model reproduces the essential features of capillary force found in AFM experiments. Given all these benefits, the LG model seems an attractive alternative to a fully molecular model of the capillary force in AFM.

The capillary force in AFM experiment sensitively depends on humidity while the Kelvin equation predicts a little humidity dependence. This is not surprising because the continuum theory incorrectly assumes that the water meniscus shape can be described by two principal radii, and its volume remains unchanged as the tip is retracted. It is possible to elaborate on the continuum theory by considering the existence of a surface water film and the non-spherical shape of the tip. Nevertheless, Xiao and Qian [9] concluded that the continuum theory is not able to reproduce the experimental behavior of the capillary force with respect to humidity. The fact that a meniscus is not a smooth continuum but a collection of discrete molecules is manifested by the significant fluctuation in structure for a small meniscus (few molecules wide). A molecular theory is definitely needed for such a case. With the LG model, one can investigate the fluctuation of meniscus over a wide range of humidities and tip geometries. The LG model can be used to study other issues beyond the scope of continuum theory, such as the effects of the atomic scale roughness on the capillary force. The molecular insights gained from the LG simulation will be invaluable in fundamental research on AFM as well as in tip-based nanolithographies such as DPN.

Acknowledgements

JJ was supported by a Korea Research Foundation Grant funded by the Korean Government (MEST) (No. 2009-0071412). GCS was supported by NSF Grant CHE-0843832.

References

1. L. Gelb, K. E. Gubbins, R. Radhakrishnan and M. Sliwinska-Bartkowiak, *Rep. Prog. Phys.* **62**, 1573 (1999).
2. J. N. Israelachvili, *Intermolecular and Surface Forces*. Academic Press, London (1992).
3. H. Choe, M.-H. Hong, Y. Seo, K. Lee, G. Kim, Y. Cho, J. Ihm and W. Jhe, *Phys. Rev. Lett.* **95**, 187801 (2005).
4. K. Salaita, Y. Wang and C. A. Mirkin, *Nature Nanotechnol.* **2**, 145 (2007).
5. M. He, A. S. Blum, D. E. Aston, C. Buenviaje, R. M. Overney and R. Luginbuhl, *J. Chem. Phys.* **114**, 1355 (2001).
6. D. L. Sedin and K. L. Rowlen, *Anal. Chem.* **72**, 2183 (2000).
7. D. B. Asay and S. H. Kim, *J. Chem. Phys.* **124**, 174712 (2006).
8. H.-J. Butt, M. Farshchi-Tabrizi and M. Kappl, *J. Appl. Phys.* **100**, 024312 (2006).
9. X. Xiao and L. Qian, *Langmuir* **16**, 8153 (2000).
10. L. R. Fisher and J. N. Israelachvili, *J. Colloid Interface Sci.* **80**, 528 (1981).
11. C. Gao, *Appl. Phys. Lett.* **71**, 1801 (1997).
12. J. Jang, J. Jeon and S. Hwang, *Colloids Surfaces A* **300**, 60 (2007).
13. J. Jang, M. Yang and G. Schatz, *J. Chem. Phys.* **126**, 174705 (2007).
14. J. Jang, J. Sung and G. C. Schatz, *J. Phys. Chem. C* **111**, 4648 (2007).
15. S. W. Han and J. Jang, *Bull. Korean Chem. Soc.* **27**, 31 (2006).
16. J. Jang, M. A. Ratner and G. C. Schatz, *J. Phys. Chem. B* **110**, 659 (2006).
17. J. Jang, G. C. Schatz and M. A. Ratner, *Phys. Rev. Lett.* **92**, 85504 (2004).
18. J. Jang, G. C. Schatz and M. A. Ratner, *J. Chem. Phys.* **120**, 1157 (2004).
19. J. Jang, G. C. Schatz and M. A. Ratner, *Phys. Rev. Lett.* **90**, 156104 (2003).
20. J. Jang, G. C. Schatz and M. A. Ratner, *J. Chem. Phys.* **116**, 3875 (2002).
21. L. Maibaum and D. A. Chandler, *J. Phys. Chem. B* **107**, 1189 (2003).
22. P. B. Paramonov and S. F. Lyuksyutov, *J. Chem. Phys.* **123**, 084705 (2005).
23. M. J. De Oliveira and R. B. Griffiths, *Surface Sci.* **71**, 687 (1978).
24. T. L. Hill, *Statistical Mechanics*. McGraw-Hill, New York, NY (1956).
25. K. Binder (Ed.), *Application of the Monte Carlo Method in Statistical Physics*, 2nd ed. Springer, Berlin (1987).
26. R. Pandit, M. Schick and M. Wortis, *Phys. Rev. B* **26**, 5112 (1982).
27. B. Fubini, M. Volante, V. Bolis and E. Giamello, *J. Mater. Sci.* **24**, 549 (1989).
28. M. Odelius, M. Bernasconi and M. Parrinello, *Phys. Rev. Lett.* **78**, 2855 (1997).
29. D. Frenkel and B. Smit, *Understanding Molecular Simulation*. Academic Press, San Diego, CA (2002).
30. W. J. Stroud, J. E. Curry and J. H. Cushman, *Langmuir* **17**, 688 (2001).

31. Z. Wei and Y.-P. Zhao, *J. Phys. D: Appl. Phys.* **40**, 4368 (2007).
32. B. Luan and M. O. Robbins, *Nature* **435**, 929 (2005).
33. J. Hu, X. D. Xiao, D. F. Ogletree and M. Salmeron, *Science* **268**, 267 (1995).

Large Scale Molecular Dynamics Simulations of Vapor Phase Lubrication for MEMS

Christian D. Lorenz [a,*], **Michael Chandross** [b] **and Gary S. Grest** [b]

[a] Department of Physics, King's College London, London WC2R 2LS, UK
[b] Sandia National Laboratories, Albuquerque, NM 87185, USA

Abstract

While alkylsilane monolayers reduce both adhesion and friction in MEMS, experiments and simulations have shown that they are easily damaged by momentary contact even at low loads. Vapor phase alcohols appear to provide a potential solution to this problem, reducing friction in MEMS with no noticeable wear, and allowing devices to run for billions of cycles without failure. The underlying mechanisms behind both the reduction in friction as well as the healing of damage are however unclear. We report on the results of large scale molecular dynamics simulations aimed at understanding the tribology of vapor phase alcohols in contact with amorphous silica substrates. The healing mechanism is investigated by simulating asperity contact with a model AFM tip in contact with a monolayer of propanol on an amorphous silica substrate. We find that because of the low vapor pressure, alcohol molecules removed by shear contact remain close to the substrate, moving around the contact region to replenish molecules removed from the damage site. For comparison, the tribology of propanol and water confined between two opposing flat silica surfaces is also studied.

Keywords

Molecular dynamics simulations, vapor phase lubrication, MEMS, nanotribology, AFM

1. Introduction

The impact of microelectromechanical systems (MEMS) on all sectors of manufacturing has been enormous. MEMS are currently in use in car airbags, digital light projectors, ink jet printers, accelerometers in video games, cell phones and digital cameras, and many other consumer and industrial products. There are still, however, no commercial products with MEMS devices that contain contacting, moving parts. The major, unresolved issue is the frictional contact between the silica substrates that are generally used to fabricate MEMS.

---

* To whom correspondence should be addressed. Tel.: +44 0207 848 2639; Fax: +44 0207 848 2932; e-mail: chris.lorenz@kcl.ac.uk

Adhesion Aspects in MEMS/NEMS
© Koninklijke Brill NV, Leiden, 2010

MEMS can be (and indeed have been) made out of many different materials, but the most popular by far is silicon, both because of the far-reaching infrastructure already in place for manufacturing and processing silicon, but also because of the attractive potential of including the actuation from a MEMS device on the same chip as the logic circuit that controls it. Silicon, however, has many undesirable material characteristics. Silicon oxidizes upon exposure to air, generating a highly reactive, hydrophilic oxide layer (hereafter referred to as silica), which is primarily responsible for the adhesion and friction issues that are so problematic in MEMS devices with contacting, moving surfaces. Because the surface-to-volume ratio is small in MEMS devices, surface forces dominate. At a separation of approximately 10 nm, van der Waals interactions between two smooth μm^2 pieces of silicon are approximately two orders of magnitude greater than the typical restoring forces that MEMS actuators can provide [1].

Silicon MEMS initially had failure issues due to parts sticking upon release (stiction) because of the presence of residual water in the system. This adhesion problem in MEMS was essentially solved with the use of alkylsilane self-assembled monolayers (SAMs), which passivate the silica surface and provide hydrophobic coatings. This allowed parts to be freely separated, and even MEMS with intermittent contact could operate effectively. While alkylsilane SAMs are also reasonably effective boundary lubricants [2], providing a far better frictional response than bare silica, the unfortunate reality is that they are an exceptionally fragile lubricant layer, and are easily damaged with both normal and shear contacts of even nominal loads.

Recent experimental work with a MEMS tribometer has shown that changes to the physical and chemical nature of surfaces occur almost immediately after both normal and sliding contacts [3]. With repeated impact loading at a contact velocity of 2 mm/s at 750 nN applied load, the coating was severely damaged within 50 000–100 000 cycles after an initial run-in period that lasted from 0 to 100 000 cycles. For sliding contacts degradation was seen to occur extremely quickly, after just 800 cycles with the same applied load.

Our previous simulation work has demonstrated that normal loads on the order of 50 nN are sufficient to remove alkylsilane chains from the monolayer with a single insertion of model atomic force microscope (AFM) tip with radius of curvature of 10 nm. Under shear, the performance is worse with less than 10 nN applied normal load necessary to cause severe damage to the monolayer [4]. This damage has been shown both by these simulations and by experiments [5] to be irreversible. Damage provides preferential sites for water adsorption and penetration, which quickly leads to further debonding of monolayer chains [6], indicating that SAMs are not an adequate long-term solution for tribological coatings in MEMS devices.

Recent work, however, has demonstrated that gas-phase alcohols may provide a solution to the friction problem, in that when MEMS devices are exposed to low vapor pressures of short chain alcohols, devices with contacting, moving parts can run for billions of cycles without failure and with no visible wear [7]. While this appears to be a successful lubrication strategy, in order to move towards applications, a number of questions still need to be addressed. Perhaps, most importantly, the fact that there appears to be some mechanism of damage/wear recovery at work in these alcohol systems is still an open issue. Additionally, the mechanism behind the frictional properties of the alcohols is still unknown, and in particular the fact that short chain systems are expected to lead to disordered monolayers with poor coverage does not explain the good tribological response. Experimental studies have also found evidence of long-chain hydrocarbon species forming on the surface in the contact region [7]. It is as yet unknown whether these reactant products are beneficial or even important to the frictional response.

The shear response of confined fluids has been the focus of several simulations studies over the course of the past 20 years. Thompson and Robbins [8, 9] first observed stick–slip motion in atomically thin, fluid films confined between two solid plates. Thompson *et al.* [10] were the first to observe stick–slip behavior for short chain polymers. Since these pioneering studies, many other studies have been conducted. Bitsanis *et al.* [11] found that fluid inhomogeneity in a Lennard–Jones liquid causes the shear stress and effective viscosity to be smaller than for the homogeneous fluid. Khare *et al.* [12] found that confined bead-spring polymer fluids exhibit a much stronger tendency for slip at the wall/fluid interface than simple fluids and the amount of slip increased with shear rate. They also found that the amount of slip increases with chain length until it reaches an asymptotic value at approximately the entanglement length of the polymer. Stevens *et al.* [13] and Jabbarzadeh *et al.* [14] both studied the shear behavior of a united-atom model of hexadecane and found that as the strength of the wall/fluid interaction increases the amount of slip decreases but the slip always occurs at the wall/fluid interface. Koike and Yoneya [15] studied the effect of molecular chain length on frictional behavior of confined liquid films of bead-spring polymers under shear. They found that at high pressure an interfacial slip appears between the wall and the film, and the shear stress does not depend on chain length, while at low pressure they observed interlayer slip within the polymer film for long chain lengths and that shear stress does depend on chain length. Gao *et al.* [16] have found that surface roughness reduces the ordering found in thin hexadecane confined films, which suppresses the solvation forces and results in liquid-like dynamic and response characteristics. They also found that at high shear rates the interfacial layers of the hexadecane films stick to the rough surface causing partial slip within the film, while with flat surfaces the slip occurs at the film boundary with the surface, and there are small shear stresses in the film. Cui *et al.* [17] found that dodecane films confined between mica surfaces display shear-thinning starting at shear rates that are orders of magnitude lower than the transition point for bulk fluids. In addition, they found that the relaxation time of the confined liquid is seven orders of magnitude larger than that of the bulk fluid. Priezjev and Troian [18] found that at low shear rates, the slip length calculated from velocity profiles of confined bead-spring polymers is strongly correlated to the slip length calculated from Green–Kubo analysis. Also, they found that molecular weight dependence of the slip length of chains of length greater than 10 is dominated strongly by the bulk viscosity. Priezjev [19] studied the shear rate dependence of the slip length in thin bead-spring polymer films confined between atomically flat surfaces and found that there is a gradual transition from no-slip to steady-state slip flow that is associated with the faster relaxation of the polymer chains near the wall.

While all of the previously mentioned studies have focused on the shear response of confined polymers, there have also been several recent simulation studies of the behavior of nanoconfined water under shear. Leng and Cummings [20] have found that hydration layers on mica surfaces with thickness ranging from 0.92 nm to 2.44 nm demonstrate fluidic shear responses, but for what they call 'bilayer ice' which has a thickness of 0.61 nm they report significant shear enhancement and shear-thinning over a wide range of shear rates. They report that the water molecules in the 'bilayer ice' have a rotational relaxation time of 0.017 ms and an effective shear viscosity at least 10^6 times larger than the bulk value. Sendner *et al.* [21] have studied the dynamics and structure of water under shear near hydrophobic and hydrophilic diamond surfaces. They found that near the hydrophobic surface there is a finite surface slip and near the hydrophilic surface there is no slip but the water viscosity is found to be increased within a thin surface layer. There have been other studies of the shear response of ultrathin films of confined water between crystalline substrates [22–25]. While the main focus of these studies was the structure of the water film under shear, they have also found extremely low friction coefficients for the water, and they have observed stick–slip motion within their systems.

Here we have performed molecular dynamics (MD) simulations of a model AFM tip in contact with an amorphous silica substrate initially coated with a propanol monolayer. Our results show that even for low applied loads, the alcohol molecules are squeezed out of the contact region. This is partially a result of the lack of chemical bonds between the alcohol molecules and the substrate. The low vapor pressure of the alcohol results in the alcohol molecules strongly favoring adsorption on the substrate. Therefore, molecules that are squeezed out of the contact region by the shearing of the tip move around the tip itself and re-form the monolayer to 'heal' the damage where molecules had been removed. Because our model tips are small and have perfect curvature, it is essentially impossible to trap any small lubricant molecules in the contact region, even for extremely low loads. It is, therefore, difficult to measure the frictional properties of the propanol molecules using these tip-based simulations as even nominal contact pressures lead to the removal of all lubricant as described above. For larger tips it is possible to trap the alcohol molecules between the tip and the substrate, which could lead to different results. To study this case we also performed MD simulations of propanol confined between two flat, amorphous silica substrates and compared the results to water, a prototypical short chain, non-hydrocarbon molecule that strongly adsorbs on the silica substrate. We find that the propanol molecules adsorb on the silica in much the same manner as alkylsilane or alkoxyl molecules, albeit without a chemical bond, and provide a reduction in friction with a similar friction coefficient. The calculated friction coefficient of the water system is generally lower than that of the alcohol system, possibly due to the orientation of the water molecules upon shear. We have also recently performed shear simulations of water confined between hydrophilic self-assembled monolayers and found that, because of the low viscosity of the water even under confinement, it is necessary to perform shear simulations at high velocities (in the range of 100 m/s) in order to measure any appreciable shear stresses and friction forces [40]. A similar mechanism is relevant for the low friction forces measured in the simulations here.

In Section 2, we briefly review the model and simulation methodology. In Section 3 we present our results for a model AFM tip in contact with an amorphous silica substrate coated with a monolayer of propanol. In Section 4 we present our results for propanol and water confined between two flat substrates. In Section 5 we briefly summarize our results and discuss future work.

2. Model and Methodology

For the simulations of a model AFM tip, a propanol monolayer coated amorphous silica substrate was contacted with a 10 nm radius of curvature tip, as described in our previous work [26]. The amorphous silica substrate was produced using a previously reported procedure [27–29]. We began with bulk α-quartz crystalline silica, which was then melted and quenched before two surfaces were created by cleaving the system in the z-dimension. The resulting substrate had dimensions 250 Å × 250 Å × 15 Å. The substrate was annealed until we achieved a coverage of silicon and oxygen defects at the interfaces of 4.6 OH/nm^2, which is in agreement with the coverage observed experimentally [30]. These defects were terminated with –OH and –H groups, respectively. For the tip, we started with the same bulk α-quartz crystalline silica that had been melted, quenched and cleaved, and then carved out a tip with a radius of curvature of 100 Å. The tip was cut such that it had a thickness of 15 Å and hollowed out in order to reduce the computational burden. Finally, the tip was annealed in order to obtain the desired coverage of silanol groups, which in this case rendered the tip hydrophobic. We hold OH groups on the outer edge of the substrate and the inside of the tip fixed in order to allow for the application of normal loads and tangential shear velocities. The propanol monolayer was prepared by placing propanol molecules at a surface density of 4 chains/nm^2 above the silica substrate such that the alcohol group was approximately 3 Å from the silica substrate. The monolayer and substrate were allowed to thermalize for 100 ps before being placed under the AFM tip. The combined system consists of the substrate, propanol molecules and AFM tip consisted of 197 000 atoms. Using this system, we have conducted both compression and shear simulations.

For comparison, we also studied the tribological properties of two monolayers of propanol and water molecules between amorphous silica substrates. The amorphous silica substrate has dimensions 55.2 Å × 55.8 Å × 27.3 Å and is made as described above. A layer of OH groups that are attached to the outer edges of each substrate are held rigid for the application of normal pressures and tangential shear velocities. The substrate was placed in contact with either a propanol monolayer containing 155 molecules, such that there were 5.03 propanol molecules/nm^2, or a water monolayer containing 375 water molecules (12.2 water molecules/nm^2). The monolayer systems (i.e., without the tip) were mirrored in the z-dimension, and were placed into contact with one another such that the two monolayers of liquid (310 propanol molecules or 750 water molecules in the system total) were positioned between the two silica substrates. These systems were compressed at a velocity of 2 m/s. Although these compression rates are high compared to experiment, our previous work demonstrated that for similar systems the velocity does not have a strong effect on the contact force [28]. Then starting from configurations at four different separations with average loads of 2.0 nN, 7.1 nN, 18.7 nN and 29.8 nN for water and 2.6 nN, 6.5 nN, 16.7 nN and 33.5 nN (Note: For our system with an area of 55.2 Å × 55.8 Å, a force of 1 nN is equivalent to a pressure of 32.5 MPa) for propanol we carried out shear simulations in which we displaced the fixed OH groups on the two silica substrates in opposite directions with a shear velocity v_x of 1 m/s, resulting in a relative shear velocity of 2 m/s. Due to the low viscosity of water, we have also conducted shear simulations at relative shear velocities of 20 m/s and 200 m/s in order to resolve shear forces larger than the statistical errors of our measurements.

Figure 1 shows the propanol and water systems at average loads of 2.6 nN and 2.0 nN, respectively, and at zero shear. Each system was equilibrated at constant volume for 1 ns before we began shearing the systems. Shear velocities in molecular dynamics simulations are high because of the necessity of accurately capturing bond vibrations in the numerical integration. This leads to time steps on the order of 1 fs, which directly affects the accessible shear velocities. When comparing to experimental friction studies performed with the Atomic Force Microscope (AFM), as is often the case, the difference in velocities is approximately six orders of magnitude. MEMS devices, on the other hand, tend to operate at velocities in the cm/s regime, making the comparison more favorable. The systems were sheared for 5–20 ns, during which we collected the data used in the analysis, which is summarized in Section 3.

All of the simulations were carried out using the LAMMPS molecular dynamics code [31]. The propanol molecules and silica substrate are modeled using the OPLS force field [32, 33]. The inter- and intra-molecular interactions of the water are described using the TIP3P model [34]. The temperature was maintained at 298 K by a Langevin thermostat with a damping coefficient of 100 fs. The SHAKE algorithm was used to constrain all hydrogen bonds in the system and the angle of the water molecules [35]. A 1 fs time step was used with the velocity Verlet integrator. The van der Waals interactions and short-range Coulomb interactions were cut off at 1.0 nm and a slab version of the PPPM algorithm was used to compute the long-range Coulomb interactions [36].

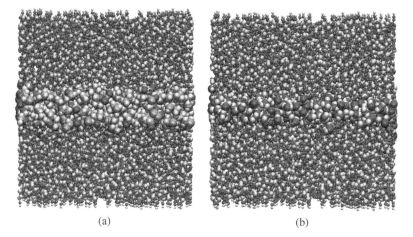

(a)   (b)

Figure 1. Snapshots of (a) propanol and (b) water between two amorphous silica substrates at loads of 2.6 nN and 2.0 nN, respectively. Colors: silicon — yellow, oxygen — red, hydrogen — white, carbon — cyan.

3. AFM Tip

In order to simulate damage due to asperity contact, we subject our adsorbed propanol monolayer to normal and shear contacts from a model AFM tip. For all applied normal loads, we find that as the tip approaches the substrate, all the propanol molecules between the tip and substrate are squeezed out. There are no alcohol molecules remaining in the contact region. This is evident in Fig. 2(a) where we show a top-down view of the propanol molecules on the substrate, with all substrate and tip atoms removed for clarity. There is a hole in the monolayer evident in the center of the figure where the tip contacts the substrate, and no propanol molecules are present.

Most of the propanol molecules that were squeezed out of the contact region do not enter the gas phase but remain on the substrate. Figure 2(b) shows the same point in the simulation as Fig. 2(a), but with all atoms present, including tip and substrate atoms. The molecules visible in the vapor phase of Fig. 2(b) are an indication that while a low pressure vapor is supported by this system, it is energetically favorable for most of the propanol molecules to remain physisorbed on the silica substrate, even when being forcibly squeezed out of the contact region. The fact that we find very few propanol molecules in the vapor phase is consistent with both measured isotherms [41] and friction studies which found that at low vapor pressure a monolayer remains on the substrate [39]. The partial pressure necessary to maintain a monolayer of propanol on amorphous silica is approximately 2.12 Torr [39]. For size of the simulation cell studied here, an ideal gas at this pressure would have, on average, less than one molecule in the vapor phase.

In Fig. 2(c), we show the same simulation as in Fig. 2(a) and 2(b), but after shearing for 2 ns at a relative velocity of 2 m/s. In Fig. 2(a) and 2(c), molecules that are initially in front of the tip (in the direction of shear) are colored orange,

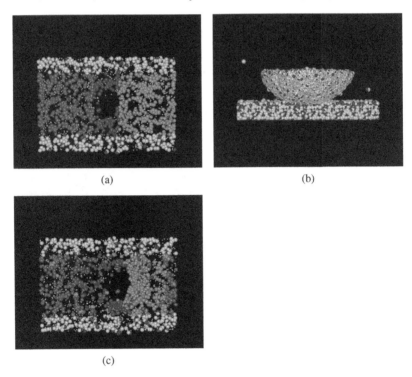

(a)   (b)

(c)

Figure 2. Snapshots of shear simulations of propanol monolayer with AFM tip. The snapshots show the damage done to the propanol monolayer before shear (only the monolayer is shown in (a), while the entire system is shown in (b)) and the healing mechanism of the propanol molecules, which are displayed after shearing for 2 ns in (c). The atoms in the substrate and tip are colored such that silicon atoms are yellow, oxygen atoms are red and hydrogen atoms are white. The propanol chains are colored based on their position relative to the tip, such that the chains initially in front of the tip are orange, those initially on the side of the tip are purple, those initially behind the tip are green and those outside of the zone of influence of the tip are yellow.

those initially to the side of the tip (perpendicular to the shear direction) are colored purple, and those initially behind the tip (in the shear direction) are colored green. Molecules that are out of the zone of influence of the tip are colored yellow. After shearing it is clear that a number of molecules initially in front of the tip have moved to the side of the tip, and those molecules initially to the side of the tip have moved to fill the gap behind the tip left by molecules initially behind the tip that have been plowed out of the way. Note that none of these specially colored molecules have mixed with those molecules that are further from the contact region.

We thus speculate that this is the mechanism behind the damage healing. Molecules are plowed from the contact region by asperity contacts. Because of the low vapor pressure, it is more energetically favorable for these molecules to remain on the substrate or at least relatively close to other alcohol molecules in the monolayer. These molecules therefore remain in close contact with the tip and other alcohol molecules in a region of high density at the leading edge of the tip. The sides of the tip have areas of lower density, and the density in the area behind the tip, where very few molecules remain after plowing, is even lower still. These density gradients lead to diffusive motion of the alcohol molecules to the damage sites, effectively replenishing the lubricating monolayer. Thus as long as the tip (or asperity) does not come in direct contact with the substrate the propanol molecules are able to move away from the contact resulting in low friction. Direct contact of the tip with the substrate would, however, be expected to give rise to high friction. The intermediate case in which some propanol molecules are trapped in the contact region is examined in the next section.

4. Confined Fluids

Figure 3 shows plots of the normal force $F_\perp$ as a function of the separation between two silica substrates for the confined water and propanol systems. In the initial configurations, the silica surfaces are separated by approximately 25 Å, and each substrate has an equilibrated monolayer adsorbed on it. The terminal groups of the two monolayers are separated by approximately 10–15 Å. The distance between the silica surfaces is decreased by moving the two substrates toward one another, each at a velocity of 1 m/s. The separation is calculated by finding the minimum distance between any two atoms on opposing silica surfaces.

As the separation between the two silica substrates is decreased, $F_\perp$ remains negligible until it becomes slightly negative due to the attractive van der Waals interactions in both systems. Following the attractive region, $F_\perp$ increases at small separations where repulsive forces dominate. $F_\perp$ begins to increase at a separation of ~5 Å in the confined water system. In the propanol system, $F_\perp$ begins to increase at a larger separation of ~8 Å which is consistent with the overlapping of the two monolayers which have thickness of ~4.5 Å. The increase in $F_\perp$ for the propanol

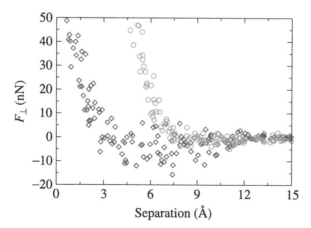

Figure 3. Normal force $F_\perp$ as a function of the separation between the silica substrates for confined water (blue diamonds) and propanol (red circles) systems. Both systems were compressed at a velocity of 2 m/s.

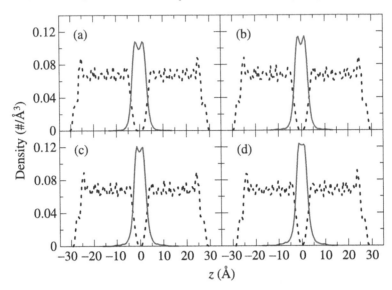

Figure 4. Number density of the water molecules (blue solid line) and silica substrate (black dashed line) as a function of the z-dimension for loads of (a) 2.0 nN, (b) 7.1 nN, (c) 18.7 nN and (d) 29.8 nN.

system is steeper than that observed in the repulsive region of the water system. The water molecules are more mobile due to their relative small size compared to the propanol molecules and, therefore, are more able to move and relieve stress that builds up within the film. Also, the water molecules can be forced into small voids that exist on the amorphous silica surface, while the larger propanol molecules remain near the silanol groups at the substrate surface. This can be seen in Figs 4 and 5, which show the number densities of the water and propanol systems at four different loads ranging from ∼2.0 nN to ∼30 nN.

In contrast to the previous shear simulations, larger tips, or more flat, multi-asperity contacts are able to trap molecules, and can be effectively lubricated in the contact region. We, therefore, performed simulations in which pairs of propanol monolayers were confined between infinitely flat silica substrates. In this geometry, it is impossible for molecules to be removed from the contact region, regardless of applied load, and we can therefore effectively measure adhesion and friction. In addition to the alcohol molecules, we have measured the adhesion and friction of two layers of water between the same amorphous silica substrates. While the calculated friction coefficients from previous shear simulation studies of ultrathin films of confined water [22–25] have, under certain circumstances, been extremely low, the models used have been simplified (with either atomically flat, crystalline, or rigid substrates), making direct comparisons difficult.

Figure 6 shows the measured friction force as a function of the applied load $F_\perp$ for the water and propanol systems sheared at a relative velocity of 2.0 m/s. The friction force and applied load values are determined by averaging each force for 5 ns of steady state behavior. From these data, we determine a microscopic coeffi-

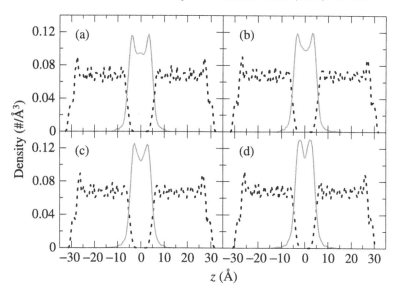

Figure 5. Number density of the propanol molecules (red solid line) and silica substrate (black dashed line) as a function of z-dimension for loads of (a) 2.6 nN, (b) 6.5 nN, (c) 16.7 nN and (d) 33.5 nN.

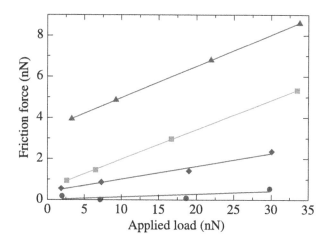

Figure 6. Friction force as a function of applied load for the water system at a relative shear velocity of 2.0 m/s (blue circles), 20.0 m/s (blue diamonds) and 200.0 m/s (blue triangles) and propanol system a relative shear velocity of 2.0 m/s (red squares).

cient of friction μ by considering that the friction depends upon both the load and the area, i.e., $F = \alpha A + \mu F_{\perp}$, where F is the friction force, α is a constant, A is the contact area and $F_{\perp}$ is the applied load. The slope of the plot in Fig. 6 gives μ. The typically measured or 'macroscopic' coefficient of friction encompasses a number of issues (i.e., microscopic roughness and asperity contact) that we do not treat in these parallel slab simulations.

The coefficient of friction measured from the data for the propanol systems is 0.14. This value is slightly smaller than the friction coefficients we have previously measured for alkylsilane $Si(OH)_3(CH_2)_{10}CH_3$ SAMs ($\mu = 0.19$) [27] that are also physisorbed on the surface. The measured friction forces in the propanol system are also essentially identical to the values that were previously measured at the same loads and velocities for the hydrocarbon alkylsilane SAMs [27]. This agreement is in one sense remarkable, given the entirely different nature of the head group of alcohol molecule when compared to an alkylsilane, but also unsurprising given our previous findings that it is the disorder at the sliding surface (i.e., at the interface between the propanol monolayers themselves) that dominates the frictional behavior [28]. It is likely that the difference in head groups is responsible for the minor differences in both the friction force and coefficient. With only one hydroxyl group (as opposed to three in the alkylsilanes) the alcohol has fewer available binding sites and is more mobile on the substrate and more able to accommodate applied loads. We have measured the orientation of the propanol molecules with respect to the substrate, and found that the majority of molecules are aligned with their head groups near the silica surface while the tails have approximately a 30° cant away from the normal to the surface. This is similar to what was seen in our previous studies of alkylsilane monolayers [28], and helps to explain why the calculated friction coefficient is similar, especially when compared to the alkoxyl monolayers.

As a comparison, we have measured the friction of confined water between the same silica substrates, as water is a small, non-hydrocarbon molecule that is expected to adsorb strongly and competitively with the propanol, yet is generally not considered to be a good lubricant for silica. For propanol, as for our previous work on SAMs, the microscopic friction is high enough that it is possible to measure significant shear stresses at the slowest sliding velocities normally accessible to MD simulations, normally about 1 m/s. However, as we will show below and as we have found in other recent work on the viscosity and tribological response of water confined between hydrophilic SAMs [40], even small amounts of water reduce the friction enough that shear stresses are too low to measure reliably, and simulation velocities must be increased by an order of magnitude or more in order to extract friction forces.

In Fig. 6 we show the calculated friction force for the two monolayers of confined water at relative velocities of 2 m/s, 20 m/s and 200 m/s. At the lowest velocity of 2 m/s the friction force is essentially too low to measure and although we can extract a friction coefficient from this graph, when statistical errors are taken into account these numbers cannot be considered reliable. As the shear velocity is increased to 20 m/s, both the friction force and coefficient (0.06) increase as the water shows some effects of the confinement. This effect can also be seen in the increased viscosity of the water as will be discussed below. As the velocity is further increased to 200 m/s, the friction forces and coefficient (0.15) increase further.

Placing this work in the context of the available experiments is difficult. Qian *et al.* measured friction coefficient between silica at various levels of humidity and an

AFM tip made from Si_3N_4 [38] and found that at extremely low levels of humidity μ was high, but decreased as humidity increased. Previously, Bingelli and Mate [43] had measured the friction coefficient of silica as a function of humidity with a much larger tungsten tip and found that the friction coefficient starts slightly above $\mu = 0.5$ and decreases with increasing humidity. The comparison to these results is difficult because while we only have two monolayers of water, our surfaces are also fully covered. The experiments, on the other hand, consist of a single silica layer exposed to a humid environment probed by an AFM tip. The amount of water on the substrate can be gleaned from an experimental isotherm [42], but when these layers interact with a probe tip the formation of a meniscus can cause differences in adhesion forces, which may strongly influence the measured friction. Additionally, we find that both our calculated friction forces and coefficients are strongly velocity dependent. This coupled with the fact that our velocities are orders of magnitude higher than typical AFM velocities makes any direct comparison less than straightforward. We are, however, able to perform detailed analyses that are out of the reach of experiments. Examination of the conformation of the water shows that the majority of molecules tend to lie parallel to the substrate to maximize bonding with the silica. This may provide a smoother sliding surface for water molecules farther away from the surface.

Since both films are fluid, we have also investigated the velocity of the molecules in the shear direction as a function of location in the film. Figures 7 and 8 show the velocity profiles of the water and propanol systems respectively at the various loads studied. The velocities were averaged over approximately 15 ns for the water sys-

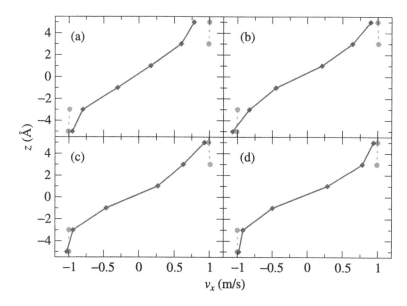

Figure 7. Velocity profile of the water molecules (blue solid line) and the interfacial silica atoms (red dashed lines) as a function of separation for loads of (a) 2.0 nN, (b) 7.1 nN, (c) 18.7 nN and (d) 29.8 nN.

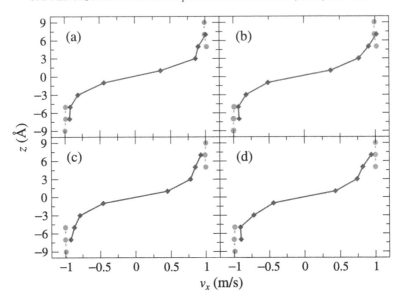

Figure 8. Velocity profile of the propanol molecules (blue solid line) and the interfacial silica atoms (red dashed lines) as a function of separation for loads of (a) 2.6 nN, (b) 6.5 nN, (c) 16.7 nN and (d) 33.5 nN.

tems and 8 ns for the propanol systems and over all fluid atoms within 2 Å thick slabs across the z-dimension. In doing so, we observe that the fluid molecules that are within the 2 Å of the silica surface move at nearly the same velocity as the silica substrate. However, at a distance of 2–3 Å from the silica surface, we observe that the average velocity of the water/propanol molecules is significantly less than the relative velocity of the silica substrate. This is approximately the same distance from the surface where the peaks in the density occur in each of the films. The magnitude of the average velocity of the fluid molecules continues to decrease further from the silica substrate, eventually reaching a value of zero half way between the two silica surfaces, which is in good agreement with a Couette-like velocity profile.

A major benefit of our simulation technique is that it allows us to probe multiple aspects of the dynamics of the confined water simultaneously. As discussed in the Introduction, the shear response and dynamics of confined water has been a topic of debate for many years. In addition to the friction force shown in Fig. 6 we have also measured the viscosity of the confined water systems to investigate the shear response. In Fig. 9 we show the calculated viscosity of the same water systems. The method used has been described in our previous work [40], but briefly the viscosity is proportional to the calculated shear stress divided by the shear rate, which can be calculated from the velocity profiles, discussed above. We are unable to extract a reliable value for the viscosity at a velocity of 2 m/s because the calculated shear stresses are comparable to the statistical uncertainty at this velocity. At 20 m/s the water does show an increase in viscosity due to the confinement, with an increase by as much as a factor of 6 at high pressures. Note that the zero shear rate viscosity for

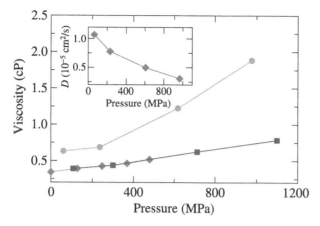

Figure 9. Viscosity of the confined water as a function of applied pressure. Data are shown for relative velocities of 20 m/s (red circles) and 200 m/s (blue squares). Also shown are data for the zero shear rate viscosity of bulk TIP3P water from Ref. [40] (green diamonds). Lines are shown as guides to the eye. Calculated two-dimensional mean-squared diffusion coefficients for the confined water systems are shown in the inset as a function of applied pressure. For comparison, bulk TIP3P water at 20 MPa has a diffusion coefficient of 6.0×10^{-5} cm^2/s [44].

the TIP3P model (also shown in Fig. 9) is only 0.35 cP at zero pressure, compared to the experimental value of 1.0 cP. At 200 m/s, the water shows significant shear-thinning behavior. In all cases, the calculated viscosity shows that the water remains liquid-like. We have also measured the two-dimensional diffusion coefficient of the water molecules (shown in the inset of Fig. 9), and verified that while the diffusion is slightly suppressed due to the confinement, there is still no indication of the formation of ice-like layers.

5. Summary and Conclusions

Our simulations indicate that vapor phase alcohols can indeed provide a robust lubrication solution for MEMS devices. While the introduction of alkylsilanes, in general, had previously solved the adhesion problem in MEMS, they have been shown to be too fragile and easily damaged by asperity contact [26]. Such contact removes chains from the monolayer and these chains are not redeposited. The vapor phase alcohols, on the other hand, provide a similar reduction in adhesion and friction but have the added benefit of redeposition after any damage, likely because of the low vapor pressure of alcohol molecules, as indicated by our simulations.

Our work is currently unable to encompass all these effects in a single simulation demonstrating both the tribological properties as well as the response to damage, so instead we focused on a two-prong approach in order to study the practicality of vapor phase alcohols as lubricants for MEMS devices. We studied monolayer damage from asperity contacts through the use of an ideal AFM tip contacting the surface. Rather than a liability, the fact that no lubricant molecules remain trapped in the contact region can, in some ways, be considered an advantage of this aspect of the simulation as it allows us to model the most severe scratch tests possible with total removal of all lubricant molecules. Even in this case we find that because of the extremely low saturation pressure of the propanol, molecules do not enter the vapor but instead accumulate in regions of increased density due to the plowing of molecules from the shearing contact. Density gradients lead to diffusive motion of molecules around the contact region to areas of lower density, which replenishes molecules at the damage sites, allowing for lubrication of future contacts.

The second aspect of our simulation method involves infinite flat plate simulations in which the lubricant molecules are unable to escape the contact region. These simulations allow us to measure the tribological properties of an ideal contact without the added complications of damage or wear. In this case we find that, much like the longer alkylsilane SAMs [28], short chain alcohols provide quality lubrication with a microscopic friction coefficient of 0.14. Even though these short chains are disordered, our previous findings on the irrelevance of chain length (and the stronger dependence on monolayer order) hold [28], likely because the physical makeup of the two systems is similar in that the head group of the alcohol molecule is located close to the silica substrate while the hydrophobic tails point away, with disorder at the sliding surface.

Experiments have found indication of the formation of long-chain hydrocarbon species upon repeated shearing of silica substrates treated with alcohol vapor [7]. While our simulations are unable to investigate these processes, our initial results tend to indicate that such a process is unnecessary for either the friction or healing mechanisms. Further simulations with reactive potentials would be necessary to study this process and elucidate both the formation and effects of these long-chain hydrocarbon molecules. In addition to the inclusion of chemistry, future directions for this work should include increasing the length of the alcohol molecules, varying the thickness of the confined fluid layers, and combining water and alcohol in the same system to study the effects of competitive adsorption on the tribological response.

Acknowledgements

CL acknowledges the KCL Division of Engineering Start-Up funds for supporting this project. This work is supported by the Laboratory Directed Research and Development Program at Sandia National Laboratories. Sandia is a multiprogram laboratory operated by Sandia Corporation, a Lockheed Martin Company, for the United States Department of Energy under contract DE-AC04-94AL85000.

References

1. R. Maboudian, *Surf. Sci. Reports* **30**, 207 (1998).
2. I. Szulfarska, M. Chandross and R. W. Carpick, *J. Phys. D: Appl. Phys.* **41**, 123001 (2004).
3. D. A. Hook, S. J. Timpe, M. T. Dugger and J. Krim, *J. Appl. Phys.* **104**, 034303 (2008).
4. M. Chandross, C. D. Lorenz, M. J. Stevens and G. S. Grest, *J. Manuf. Sci. Eng.* **132**, 030916 (2010).

5. X. D. Xiao, J. Hu, D. H. Charych and M. Salmeron, *Langmuir* **12**, 235 (1996).
6. J. M. D. Lane, M. Chandross, C. D. Lorenz, M. J. Stevens and G. S. Grest, *Langmuir* **24**, 5734 (2008).
7. D. B. Asay, M. T. Dugger, J. A. Ohlhausen and S. H. Kim, *Langmuir* **24**, 155 (2008).
8. P. A. Thomspon and M. O. Robbins, *Science* **250**, 792 (1990).
9. M. O. Robbins and P. A. Thomspon, *Science* **253**, 916 (1991).
10. P. A. Thomspon, G. S. Grest and M. O. Robbins, *Phys. Rev. Lett.* **68**, 3448 (1992).
11. I. Bitsanis, J. J. Magda, M. Tirrell and H. T. Davis, *J. Chem. Phys.* **87**, 1733 (1987).
12. R. Khare, J. J. de Pablo and A. Yethiraj, *Macromolecules* **29**, 7910 (1996).
13. M. J. Stevens, M. Mondello, G. S. Grest, S. T. Cui, H. D. Cochran and P. T. Cummings, *J. Chem. Phys.* **106**, 7303 (1997).
14. A. Jabbarzadeh, J. D. Atkinson and R. I. Tanner, *J. Chem. Phys.* **110**, 2612 (1999).
15. A. Koike and M. Yoneya, *J. Phys. Chem. B* **102**, 3669 (1998).
16. J. Gao, W. D. Luedtke and U. Landman, *Tribology Lett.* **9**, 3 (2000).
17. S. T. Cui, C. McCabe, P. T. Cummings and H. D. Cochran, *J. Chem. Phys.* **118**, 8941 (2003).
18. N. V. Priezjev and S. M. Troian, *Phys. Rev. Lett.* **92**, 018302 (2004).
19. N. V. Priezjev, *Phys. Rev. E* **80**, 031608 (2009).
20. Y. Leng and P. T. Cummings, *J. Chem. Phys.* **125**, 104701 (2006).
21. C. Sendner, D. Horinek, L. Bocquet and R. R. Netz, *Langmuir* **25**, 10768 (2009).
22. M. Paliy and O. M. Braun, *Tribology Lett.* **23**, 7 (2006).
23. A. Pertsin and M. Grunze, *Langmuir* **24**, 135 (2008).
24. A. Pertsin and M. Grunze, *Langmuir* **24**, 4750 (2008).
25. A. V. Khomenko and N. V. Prodanov, *Condensed Matter Physics* **11**, 615 (2008).
26. M. Chandross, C. D. Lorenz, M. J. Stevens and G. S. Grest, *Langmuir* **24**, 1240 (2008).
27. C. D. Lorenz, E. B. Webb III, M. J. Stevens, M. Chandross and G. S. Grest, *Langmuir* **21**, 11744 (2005).
28. M. Chandross, E. B. Webb III, M. J. Stevens, G. S. Grest and S. H. Garofalini, *Phys. Rev. Lett.* **93**, 166103 (2004).
29. C. D. Lorenz and A. Travesset, *Phys. Rev. E* **75**, 061202 (2007).
30. R. K. Iler (Ed.), *The Chemistry of Silica*. Wiley, New York, NY (1979).
31. S. J. Plimpton, *J. Computational Phys.* **117**, 1 (1995).
32. W. L. Jorgensen, D. S. Maxwell and J. Tirado-Rives, *J. Am. Chem. Soc.* **118**, 11225 (1996).
33. W. L. Jorgensen, private communication (2003).
34. W. L. Jorgensen, J. Chandrasekhar, J. D. Madura, R. W. Impey and M. L. Klein, *J. Chem. Phys.* **79**, 926 (1983).
35. J. P. Ryckaert, G. Ciccotti and H. J. C. Berendsen, *J. Computational Phys.* **23**, 327 (1977).
36. P. S. Crozier, R. L. Rowley and D. Henderson, *J. Chem. Phys.* **114**, 7513 (2001).
37. A. Luzar and D. Chandler, *Phys. Rev. Lett.* **76**, 928 (1996).
38. L. Qian, F. Tian and X. Xiao, *Tribology Lett.* **15**, 169 (2003).
39. K. Strawhecker, D. B. Asay, J. McKinney and S. H. Kim, *Tribology Lett.* **19**, 17 (2005).
40. C. D. Lorenz, J. M. D. Lane, M. Chandross and G. S. Grest, *Modelling Simulat. Mater. Sci. Eng.* **18**, 034005 (2010).
41. A. L. Barnette, D. B. Asay, M. J. Janik and S. H. Kim, *J. Phys. Chem. C* **113**, 10632 (2009).
42. D. B. Asay and S. H. Kim, *J. Phys. Chem. B* **109**, 16760 (2005).
43. M. Bingelli and C. M. Mate, *Appl. Phys. Lett.* **65**, 415 (1994).
44. C. D. Lorenz, J. M. D. Lane, M. J. Stevens, M. Chandross and G. S. Grest, *Langmuir* **25**, 4535 (2009).

Atomistic Factors Governing Adhesion between Diamond, Amorphous Carbon and Model Diamond Nanocomposite Surfaces

Pamela L. Piotrowski [a,*], **Rachel J. Cannara** [b,*], **Guangtu Gao** [a], **Joseph J. Urban** [a], **Robert W. Carpick** [c] **and Judith A. Harrison** [a,**]

[a] United States Naval Academy, Chemistry Department, 572 Holloway Road, Annapolis, MD 21402, USA

[b] National Institute of Standards and Technology, Center for Nanoscale Science and Technology, 100 Bureau Dr., Gaithersburg, MD 20899, USA

[c] Dept. of Mechanical Engineering & Applied Mechanics, University of Pennsylvania, 220 S. 33rd St., Philadelphia, PA 19104-6315, USA

Abstract

Complementary atomic force microscopy (AFM) measurements and molecular dynamics (MD) simulations were conducted to determine the work of adhesion for diamond (C)(111)(1 × 1) and C(001)(2 × 1) surfaces paired with carbon-based materials. While the works of adhesion from experiments and simulations are in reasonable agreement, some differences were identified. Experimentally, the work of adhesion between an amorphous carbon tip and individual C(001)(2 × 1)–H and C(111)(1 × 1)–H surfaces yielded adhesion values that were larger on the C(001)(2 × 1)–H surface. The simulations revealed that the average adhesion between self-mated C(001)(2 × 1) surfaces was smaller than for self-mated C(111)(1 × 1) contacts. Adhesion was reduced when amorphous carbon counterfaces were paired with both types of diamond surfaces. Pairing model diamond nanocomposite surfaces with the C(111)(1 × 1)–H sample resulted in even larger reductions in adhesion. These results point to the importance of atomic-scale roughness for adhesion. The simulated adhesion also shows a modest dependence on hydrogen coverage. Density functional theory calculations revealed small, C–H bond dipoles on both diamond samples, with the C(001)(2 × 1)–H surface having the larger dipole, but having a smaller dipole moment per unit area. Thus, charge separation at the surface is another possible source of the difference between the measured and calculated works of adhesion.

Keywords

Adhesion, diamond, nanocomposite diamond, amorphous carbon, diamond-like carbon, nanotribology, work of adhesion, pull-off force, atomic force microscopy, friction force microscopy, molecular dynamics, *ab initio* studies, density functional theory

---

[*] These authors contributed equally to this work.

[**] To whom correspondence should be addressed. E-mail: jah@usna.edu

Adhesion Aspects in MEMS/NEMS

1. Introduction

As a result of its unique electrical, mechanical and tribological properties, diamond has been the focus of great interest both as an object of scientific study and as an ideal material for applications, ranging from cutting tool coatings and waste water purifiers to chemical sensors, electronic devices and micro- and nanoelectromechanical systems (M/NEMS). Due to its high fracture strength and chemical robustness, it can withstand exposure to harsh environments and resist mechanical wear long after most other materials fail. Thus, diamond is a potential candidate for replacing silicon for M/NEMS applications [1–3]. Silicon has a relatively high surface energy, is very brittle, and tends to fracture easily. The native oxide on silicon produces a hydrophilic surface that encourages water to form bridges between micro- and nanoscale objects in close proximity, often causing device failure. Hence, silicon M/NEMS surfaces typically require special coatings (e.g., self-assembled monolayers) to protect these devices from the devastating effects of adhesion [4, 5]. With these issues in mind, the adhesion properties of diamond at the micro- and nanoscale must be investigated to determine whether diamond is superior to silicon.

The adhesion properties of any pair of surface materials can depend strongly on the atomic termination of the surfaces. *Ab initio* studies have shown that the presence of hydrogen on diamond surfaces forming an interface with a metal can reduce the work of adhesion by one order of magnitude, depending on the metal [6, 7]. Density functional theory (DFT) calculations revealed that terminating the diamond or silicon surface with hydrogen produces lower surface energies than termination by oxygen or hydroxyl groups [6, 8, 9]. Moreover, in the presence of water or a mixture of water and hydrogen gas, adhesion can be reduced by causing the surfaces to be terminated with $-H$ as opposed to terminating with $-OH$ or $=O$ [10]. In an atomic force microscopy (AFM) experiment, Kaibara *et al.* compared pull-off forces on H- and O-terminated patterned regions of a diamond (001) film and measured higher adhesion forces for the oxygenated region [11]. The difference was more than two-fold in air due to the higher hydrophilicity of the O-terminated regions. Indeed, the contrast decreased substantially when the sample was vacuum-annealed and then measured in an inert argon atmosphere, due to the desorption and subsequent absence of water.

Hydrogen surface termination and bulk content are also important for adhesion and tribological properties of diamond-like carbon (DLC) [12] and nanocrystalline diamond films [13, 14]. Molecular dynamics (MD) simulations have shown that the removal of hydrogen atoms from self-mated diamond contacts leads to increased chemical bond formation, or adhesion [15, 16]. In that case, covalent bonds were formed across the interface, and separation of the surfaces led to fracture. Similarly, removing the hydrogen termination from a single-crystal diamond counterface in sliding contact with DLC increased friction by increasing the formation of chemical bonds between the counterface and the DLC [17]. More recently, the incorporation of hydrogen into DLC films was shown to reduce friction in self-mated DLC-DLC contacts [18].

By definition, the work of adhesion (W) is the energy per unit area required to separate two semi-infinite surfaces from their equilibrium separation to infinity. For two dissimilar materials, this quantity is equivalent to a sum of the surface energies, γ_1 and γ_2, of the two surfaces (because, effectively, the separation process creates the two surfaces), minus the interfacial energy, γ_{12}. Hence, $W = \gamma_1 + \gamma_2 - \gamma_{12}$, where W is also known as the Dupré energy of adhesion. To control adhesion forces between two materials, the work of adhesion can be varied by changing the surface termination. However, this assumes that the two surfaces are perfectly flat. If one or both of the surfaces is rough, then the actual area of contact will be less than the apparent contact area [19], even at the atomic scale [20–22]. Taller asperities (or protruding atoms) at the interface will prevent some of the smaller features (or other atoms) from closely approaching each other, increasing the separation between different regions of the surface and, therefore, reducing the energy of interaction. While the extent to which nano- or even atomic-scale roughness affects adhesion measurements is not fully understood [23], recent MD simulations indicate that the effects could be dramatic [20, 21].

At present, both modeling and experiment have been used to investigate adhesion at the nanometer scale. In an AFM experiment, adhesion is measured by recording pull-off forces (L_C) between an AFM tip and a sample surface in a controlled or ambient environment. The work of adhesion is calculated from the measured pull-off force using a continuum model of contact mechanics [24]. The calculation requires assumptions or measurements made about the nature of the contact, including the range of interaction and the tip shape and size [25–28]. For linear elastic materials where the tip shape is well-described by a paraboloid, the pull-off force is directly proportional to the product of the work of adhesion and the radius of curvature of the tip. The proportionality constant depends on the range of adhesion interactions compared with the elastic deformations caused by adhesion [26]. This, in turn, depends on the elastic properties of both materials. These properties may be different at a surface or for small tip shapes compared with bulk values. Moreover, small-scale roughness may go unobserved and yet completely alter the nature of the measurement. Delrio *et al.* used both experiment and numerical analysis to show that an increase in microscale surface roughness (from 2 nm to 10 nm) can decrease adhesion forces for nominally identical surface chemistries [29]. Luan and Robbins used MD simulations to examine the effects of tip shape and atomic-scale roughness on the work of adhesion between a tip and an atomically flat surface [21]. In that work, W decreased by a factor of two when the tip and surface were incommensurate and by another factor of two when atomic-scale roughness was introduced. Hence, surface roughness is an important parameter over a wide range of length scales, so it is important to understand its effects down to the atomic scale.

In this work, we present experimental and theoretical studies of the work of adhesion between H-terminated diamond samples of a given orientation and diamond samples paired with other carbon-based materials. For these investigations, we combined MD simulations, *ab initio* calculations, and AFM experiments. Unlike the previous *ab initio* studies that compared surface energies for H-terminated C(001)(1×1) (unreconstructed) and C(111)(1×1) [30] (in addition to those mentioned in the above discussion of surface termination), the MD simulations used here simulate the (2 × 1)-reconstructed C(001)–H surface. Both density functional theory [9] and the potential energy function used here [31] predict that the reconstructed C(001)(2 × 1)–H surface is a more energetically favorable structure than the C(001)(1 × 1)–H surface [32]. Experimental results are consistent with this as well. In addition, the MD simulations performed here account for long-range van der Waals interactions and consider the work of adhesion, as opposed to only considering surface energy. We also present *ab initio* calculations to determine the presence of bond dipoles of single-crystal diamond surfaces. The work of adhesion values obtained by MD were compared with AFM data recorded from force *vs* displacement (L *vs* Z) and friction measurements. The friction measurements provided information about the contact mechanics of the interface, namely, the load-dependence of the contact area, and works of adhesion could be derived from the pull-off forces measured from both the sliding experiments and the quasistatic L *vs* Z measurements. The AFM friction measurements used to extract the adhesion and pull-off data were reported previously in a separate paper [33] describing similar MD and AFM comparisons of orientation-dependent friction in dry nitrogen between microcrystalline diamond (MCD) surfaces and tips made of diamond and amorphous carbon. The latter tips were formed by electron beam-induced deposition (EBID) in a transmission electron microscope. Here, we glean further information from those AFM experiments for comparison with new MD simulations of adhesion and DFT calculations of surface dipoles. One result reported in the work by Gao *et al.* [33] showed that both the pull-off force L_C and the sliding work of adhesion (W_{FL}) were always higher on the (001) surface than the (111) surface for both of the tips used in the AFM experiments. Because L_C and W_{FL} were both larger for the smaller tip (of radius 45 nm), with the smaller tip exhibiting roughly twice the adhesion of the larger (150 nm) tip, it was speculated that the tips had chemical differences, and any roughness differences were not considered.

Since then, Brukman *et al.* performed temperature-dependent studies in ultrahigh vacuum of the friction on diamond (001) and (111) single-crystal surfaces between 25 K and 225 K using two different polycrystalline diamond tips of radii 30 nm (Tip 1) and 60 nm (Tip 2) [34]. Sliding was along the [11$\bar{2}$] direction on the (111) and the [010] on the (001) crystal, and sample surfaces were not deliberately H-terminated. In those studies, the measured W_{FL} was lower on the (001) surface both for Tip 2 at all temperatures and for Tip 1 between 90 K and 180 K. Adhesion values were very close for Tip 1 at other temperatures, and the difference in adhesion between (111) and (001) increased for Tip 2 at temperatures below 135 K. Similar to Gao *et al.* [33], substantially different behavior was observed depending on which tip was used. However, for both tips, a similar trend and ratio of L_C

values for (001) and (111) samples was observed as compared with the behavior of the corresponding W_{FL}, indicating that the contact mechanics behavior (which relates W to L_C) was similar for the two surfaces and independent of tip shape and chemistry. Auger electron spectroscopy showed that the (001) sample surface had slightly greater oxygen coverage than (111) (0.21 monolayers *vs* 0.16 monolayers, respectively). It is possible that the additional O- (in place of H-)termination gives rise to the difference in W_{FL} for (001) *vs* (111), but that interpretation opposes the result of Kaibara *et al.* [11].

These previous measurements and those of Ref. [25] confirm that the nature of the tip has a strong effect not only on the magnitude of adhesion but also on the manner in which adhesion depends on the surface orientation and temperature. In addition, there is some indication that the work of adhesion (W_{LZ}) obtained from L *vs* Z pull-off measurements differs significantly from the dynamic sliding case. In this work, previous results will be reviewed and compared with additional W_{LZ} data recorded during the experiments reported in Ref. [25]. MD simulations will also be employed to isolate the possible mechanisms responsible for these experimental observations. MD was used in Refs [25] and [26] to study temperature-dependent friction on (001) and (111) surfaces, but adhesion forces were neglected in those simulations based on the assumption that they would only add an offset to the friction *vs* load data. Moreover, the goal was to deduce whether there were intrinsic differences in the shear properties of the two surfaces in the absence of chemical differences. In the study presented here, the investigation of diamond samples is broadened by focusing on adhesion. By combining MD simulations and *ab initio* DFT calculations with AFM experiments, we aim to build a more complete picture of the physics and mechanics at interfaces with diamond samples.

2. MD Simulation Methods

In the MD simulations, W is calculated by bringing two surfaces into perpendicular contact and generating an L *vs* Z curve. The energy of this interaction is obtained by using the trapezoidal rule to numerically integrate the area enclosed by the attractive portion of the force curve. The work of adhesion is obtained by dividing the energy of interaction by the area of the computational cell (Table 1). Self-mated C(111)(1 × 1) and C(001)(2 × 1) contacts were examined. The interactions of diamond crystals with H-termination, model diamond nanocomposite (MDN) surfaces, and several amorphous carbon surfaces were also examined. A complete set of the counterface–sample pairs examined with MD is given in Table 1. The nature of MD simulations also allowed for an examination of the influence of variables, which are not easily changed in an AFM experiment, including H-termination, temperature, and counterface–surface alignment on W.

A sample MD simulation system composed of two infinitely flat C(111)(1 × 1)–H surfaces is shown in Fig. 1. In this system, the counterface (upper surface) was placed approximately 10 Å away from the lower surface. Both diamond surfaces

Table 1.
MD system details

Counterface + sample pair	Counterface atoms C:H	Surface atoms C:H	$sp^3 : sp^2 : sp$ (%) (amorphous systems)	X (Å)	Y (Å)	Z (Å)	RMS roughness (Å)	Density (kg/m³)
C(111)(1 × 1)–H + C(111)(1 × 1)–H	1008:C 144:H	1008:C 144:H		30.256[a]	26.202[b]	14.75		
C(001)(2 × 1)–H + C(001)(2 × 1)–H	1296:C 144:H	1296:C 144:H		30.256[a]	30.256[c]	16.5		
sp^2-amorphous + C(111)(1 × 1)–H	4008:C 0:H	1008:C 144:H	12.5 : 85.3 : 2.2	30.177[a]	26.135[b]	41.5	0.57	2750
sp^3-amorphous + C(111)(1 × 1)–H	2123:C 118:H	1008:C 144:H	94.4 : 5.6 : 0	30.177[a]	26.135[b]	23.8	0.48	2860
sp^2-amorphous + C(001)(2 × 1)–H	4608:C 0:H	1152:C 144:H	13.3 : 80.6 : 6.0	30.256[a]	30.256[c]	45	0.68	2660
C(111)(1 × 1)–H + MDN1	5460:C 780:H	13179:C 461:H		65.555[d]	65.506[e]	27.85	1.18	
C(111)(1 × 1)–H + MDN2 on C(111)	5460:C 780:H	12887:C 461:H		65.555[d]	65.506[e]	29.4	1.47	

[a] This corresponds to 12 unit cells in the X direction.
[b] This corresponds to 6 unit cells in the Y direction.
[c] This corresponds to 12 unit cells in the Y direction.
[d] This corresponds to 26 unit cells in the X direction.
[e] This corresponds to 15 unit cells in the Y direction.

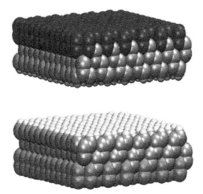

Figure 1. Two, hydrogen-terminated C(111)(1 × 1)–H surfaces. Large and small spheres represent carbon and hydrogen atoms, respectively. The bottom-most, middle, and top-most spheres atoms are fixed, thermostatted, and have a constant velocity applied to them, respectively.

were composed of seven layers of carbon atoms and one layer of hydrogen atoms, and each surface contained 144 atoms. The dimensions of the computational cell were 30.2 Å by 26.1 Å, which corresponds to twelve diamond unit cells repeated along the [$\bar{1}$10] lattice direction and six unit cells repeated along the [$\bar{1}\bar{1}$2] lattice direction, respectively. Each surface was divided into one rigid and one thermostatted region. The counterface was moved toward the surface by giving the rigid atoms a constant velocity of 100 m/s (1 Å/ps). The rigid atoms in the lower surface were held fixed. Unless otherwise indicated, the remaining atoms were maintained at a temperature of 300 K using the Langevin thermostat [35]. Newton's equations of motion were then integrated using a time step of 0.25 fs. Nearly identical results can be achieved when obtaining the *L vs Z* curve, either by applying the thermostat to the gray atoms in Fig. 1 or by equilibrating the system to 300 K and then integrating the gray atoms without constraints while obtaining the *L vs Z* curve.

As listed in Table 1, seven systems were studied by MD: (i) self-mated C(111)(1 × 1)–H; (ii) self-mated C(001)(2 × 1)–H; (iii)–(iv) C(111)(1 × 1)–H against two different amorphous carbon counterfaces; (v) C(001)(2 × 1)–H against an amorphous carbon counterface; and (vi)–(vii) C(111)(1 × 1)–H against two different model diamond nanocomposites (MDNs): MDN1 and MDN2. When paired with a C(111) counterface, the amorphous carbon samples were attached to C(111) substrates so that the periodic boundary conditions of the counterface–sample pair would be equal. One amorphous carbon counterface is composed of 94.4% sp^3-hybridized carbon, and the interface region is passivated with hydrogen [36]. The second surface is hydrogen-free and is mostly (85.3%) sp^2-hybridized carbon [18]. Both of these amorphous carbon samples have been used in our previous MD simulations of friction [18, 37, 38]. The amorphous carbon sample attached to C(001) is hydrogen-free and contains 80.6% sp^2-hybridized carbon (Table 1). The MDN surfaces were generated by inserting sixteen H-terminated diamond (111) and (110) grains into an amorphous carbon matrix [39]. Although the identity and number of

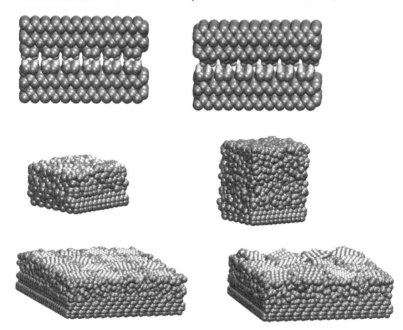

Figure 2. Sample configurations of the MD systems examined here. Large and small spheres represent carbon and hydrogen atoms, respectively. In the top left and right panels are two C(001)(2 × 1)–H surfaces in the aligned and shifted configurations, respectively. In the middle left and right panels are the sp^3-hybridized and sp^2-hybridized amorphous carbon surfaces, respectively. In the bottom left and right panels are the model diamond nanocomposite surfaces MDN1 and MDN2, respectively. The numbers of atoms and dimensions for each surface are given in Table 1.

each type of grain embedded in the matrix were the same, the location and tilt varied so that films of different roughnesses were created. The average root-mean-square (RMS) roughness on MDN2 and MDN1 are given in Table 1 [39]. Snapshots of the samples are shown in Fig. 2, and the total number of atoms and dimensions of each counterface–surface pair are summarized in Table 1.

The term counterface refers to the upper sample, which was always fully H-terminated, while varying degrees of hydrogen coverage on the lower samples were examined. Periodic boundary conditions were applied to the x- and y-directions (the contacting plane) for all simulations. The adaptive intermolecular reactive empirical bond-order (AIREBO) potential [31] was used to calculate the forces within the MD simulations. The AIREBO potential contains torsional terms, intermolecular terms (van der Waals interactions), and terms for modeling covalent bonding based on the second-generation REBO potential [40]. The terms in addition to the REBO potential increase the computational time significantly. With this in mind, the simulation systems are typically designed to minimize the number of atoms used, while not adversely influencing the simulation results. Because the AIREBO potential is based on the REBO potential, it is one of a few empirical potentials with long-range forces that can model chemical reactions. Nonetheless, it should be noted that because carbon and hydrogen have similar electronegativities, partial charges were not included in the REBO or the AIREBO potentials. Both the REBO and the AIREBO potentials have been used successfully to model mechanical properties of filled [41, 42] and unfilled [43] nanotubes, properties of clusters [44], indentation and contact [15, 16, 22, 39, 45], the tribology of diamond [22, 33, 38, 46–48] and DLC [17, 18, 37, 49], stress at grain boundaries [50], and friction and polymerization of model self-assembled monolayers [51–55].

Average W values for the self-mated diamond contacts were calculated by averaging data from contact simulations with independent starting configurations obtained by shifting the counterface in the contacting plane. Due to the periodicity of each system, the counterface was shifted incrementally a fraction of the lattice constant in the x and y directions. A total of five and seven independent starting configurations that span the surface unit cells were examined for the C(111)(1 × 1) and the C(001)(2 × 1) samples, respectively. Both the L vs Z data and W values were obtained from each of these independent starting configurations. The average W corresponds to the average of all of the starting configurations for a given counterface–sample pair. In an AFM friction vs load (F vs L) experiment, the tip is rastered over the sample (i.e., each scan line is obtained by incrementing the displacement perpendicular to the sliding direction). Thus, measurements are obtained over a two-dimensional range of relative positions between the tip and sample atoms. Averaging over different starting configurations approximates the line-averaged response from an AFM experiment in sliding contact.

3. MD Results

To calculate W, load vs separation (L vs Z) force curves were generated for each counterface–sample pair used in the MD simulation (Fig. 3). As the counterface moved toward the sample, the force on each atom within the lower sample was calculated and summed to give a total load at each timestep. The separation is the difference in the average z positions of atoms on the sample and those on the counterface closest to the interface. For the diamond samples, the average z position of the atoms closest to the interface is the average position of the terminal hydrogen atoms (or carbon atoms if the hydrogen is completely removed). The roughness of the amorphous carbon and MDN samples makes defining the surface separation more challenging. For the amorphous carbon and the MDN samples, the atom closest to the opposing sample is found. The vertical position of all atoms belonging to this sample that are within a vertical distance of this atom that is two times the RMS roughness (Table 1) are then averaged to find the average z position of the surface atoms.

Examination of the L vs Z curves in Fig. 3 reveals that when the counterface–sample pair is commensurate, the alignment (or relative position) of the two surfaces dramatically influences the force curve. For example, when the contacting pairs are both C(111)(1 × 1)–H, the samples are prevented from reaching small

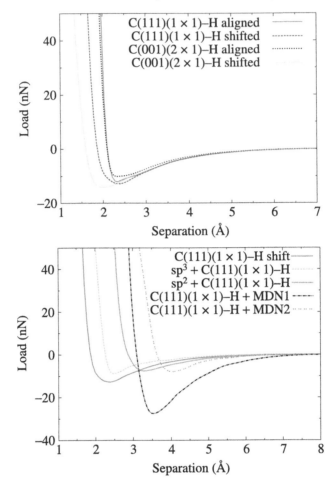

Figure 3. Load *vs* separation calculated from the MD simulations of the counterface–surface pair indicated in the legend. The separation is the distance between the average z positions of the lowest interface atoms on the counterface and the highest interface atoms on the surface. In the upper panel, two starting configurations of self-mated diamond surfaces are shown. Both the counterface and the surface are completely H-terminated. In the lower panel, the counterface–surface pairs are shown in the legend.

separations when the hydrogen atoms on opposing samples are directly above and below one another. This is referred to as the "aligned" configuration. Shifting the counterface slightly so that the hydrogen-atoms are not directly above and below one another allows the counterface and the sample to get closer together, shifting the repulsive portion of the force curve to smaller separations and enlarging the attractive region of the L *vs* Z curve, both of which ultimately impact W. The "shifted" configuration that ultimately leads to the highest W corresponds to the one where H atoms on opposing samples are equidistant from each other in the $x–y$ plane. These are the shifted L *vs* Z curves in Fig. 3. This phenomenon is also apparent in the force curves for two opposing $C(001)(2 \times 1)$–H surfaces (Fig. 3). Due to the larger lattice constants of the $C(001)(2 \times 1)$ surface, the repulsive portion of the force curve for the "shifted" configuration that leads to the highest W is at a smaller separation than it is for the self-mated $C(111)(1 \times 1)$–H contacts. Because W depends on surface alignment, force curves were generated for several different initial configurations for both the self-mated $C(111)(1 \times 1)$ and $C(001)(2 \times 1)$ contacts.

Substituting an amorphous carbon sample (the DLC) for one of the diamond samples makes the contacting pair incommensurate which is analogous to the incommensurate contact between the AFM tips and the diamond surface. However, the geometry of the tip-sample contact in the simulation and the experiment remains dissimilar. Using MDN surfaces also makes the interface incommensurate and adds an additional component of roughness. The force curves for the interaction of two amorphous carbon counterfaces and two MDNs with $C(111)(1 \times 1)$–H are also shown in Fig. 3. Comparison of data from the systems containing the two amorphous carbon samples reveals that the mostly sp^3-hybridized surface comes into closer proximity with the $C(111)$ sample surface than the mostly sp^2-hybridized surface. That is, the repulsive portion of the force curve is shifted to smaller separations. This is largely due to the fact that the mostly sp^3-hybridized film is hydrogen terminated, so hydrogen–hydrogen non-bonded interactions dominate, while the mostly sp^2-hybridized film is hydrogen-free so carbon–hydrogen non-bonded interactions dominate.

The two MDN samples were constructed to have different roughnesses, with MDN1 being smoother than MDN2 (Table 1). Because the MDN1 surface is smoother, the repulsive wall of the potential is shifted to smaller separations and a larger number of surface atoms interact with the diamond counterface at a given separation. Thus, the attractive portion of the force curve for the MDN1 surface is significantly deeper than it is for MDN2 surface. Care must be taken when attempting a direct comparison of the force curves of systems containing the MDN samples and those containing the amorphous samples, because the simulation size (contact area) differs in these systems.

The W is calculated by numerical integration of the area enclosed in the attractive portion of the force-separation curves (Fig. 3) and then dividing by the area of the computational cell in the contacting plane ($x–y$ plane, Table 1). The work of adhesion values for the "aligned" and one of the shifted configurations of both self-mated diamond systems as a function of H-coverage are shown in Fig. 4. For each of these samples, the differences in the force curves (Fig. 3) brought about by the relative positioning of the hydrogen atoms on the opposing diamond samples is reflected in the values of W shown in Fig. 4. Shifting the counterface so that its hydrogen atoms were laterally equidistant to the hydrogen atoms on the opposing surface produced the largest values of W. In this configuration, referred to as the "shifted" configuration (Fig. 4), the hydrogen atoms on opposing samples can draw nearer causing changes in the L vs Z curves (Fig. 3). The average W was calculated from the W obtained from each of the independent starting config-

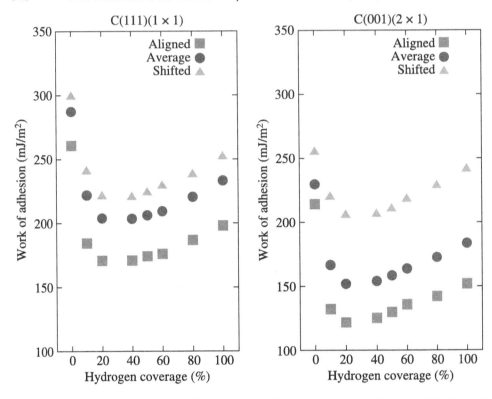

Figure 4. Work of adhesion W calculated from the MD simulations of self-mated $C(111)(1 \times 1)$ and $C(001)(2 \times 1) \times 1)$ contacts as a function of hydrogen coverage. The counterface is completely H-terminated, and the hydrogen termination is varied on the lower surface. The terms "aligned" and "shifted" are defined in the text. The shifted configuration shown yielded the largest work of adhesion for that counterface–surface pair.

urations. For all hydrogen coverages, the average W for $C(111)(1 \times 1)$ was larger than for $C(001)(2\times1)$. The fact that $C(111)(1 \times 1)$–H has a larger calculated average W than $C(001)(2 \times 1)$–H is because there is a higher density of atoms on the (111) surface (0.181 $\mathring{A}^{-2}$) than on the (001) surface (0.157 $\mathring{A}^{-2}$). Thus, there are more interactions contributing to the depth of the attractive portion of the force curve. Maximum and minimum values of W were obtained from independent starting configurations and used as uncertainty estimates from which the error bars in Fig. 5 were constructed. Based on these uncertainty estimates obtained from the different alignments, the MD simulations reveal no statistically significant difference in the adhesion of self-mated $C(111)(1 \times 1)$ and $C(001)(2 \times 1)$ contacts.

The values of W for all the counterface–sample pairs examined by MD are shown in Fig. 5 as a function of surface hydrogen coverage. For all hydrogen coverages, the MDN surfaces exhibit lower values of W than the average values of W and are outside the error bars calculated for both self-mated $C(001)(2 \times 1)$ and $C(111)(1 \times 1)$ contacts. In addition, MDN2 has a lower W than MDN1 for all hydrogen coverages. The roughness of MDN2 is larger than it is for MDN1 and prevents it from

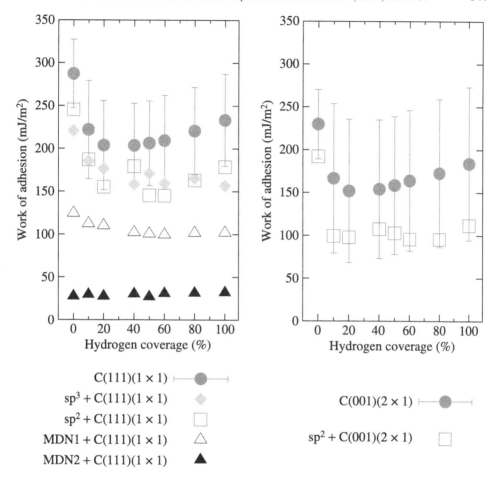

Figure 5. Work of adhesion calculated from the MD simulations for all the counterface–surface pairs examined in this work. The error bars represent an uncertainty in the average value, which is determine by shifting the lateral alignment of the counterface with respect to the surface.

coming into close proximity with the opposing diamond sample. As a result, the attractive portion of the force curve is very shallow (Fig. 3), which ultimately translates into small values of W compared to the MDN1 surface. In fact, roughness seems to be the dominant influence on W for both MDN samples and removing the H-termination from the MDN samples has little impact on the calculated values of W.

To examine adhesion between an incommensurate counterface–sample pair, amorphous carbon counterfaces were brought into contact with both C(111)(1 × 1) and C(001)(2 × 1) samples. Two different amorphous carbon (DLC) samples with different sp^2-to-sp^3 ratios were paired with C(111)(1 × 1) and one amorphous sample was paired with C(001)(2 × 1). The values of W as a function of hydrogen coverage of the diamond-sample surface are shown in Fig. 5. For all hydrogen cov-

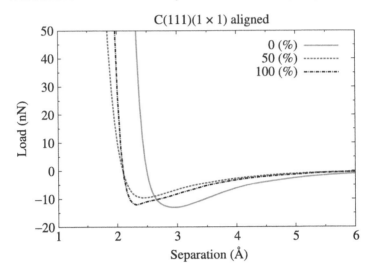

Figure 6. Load *vs* separation calculated from the MD simulations for the C(111)(1 × 1) for three different hydrogen coverages of the surface. The hydrogen atoms on opposing diamond surfaces are in the "aligned" configuration (directly above and below one another). The separation is the distance between the lowest interface atom on the counterface and the highest interface atom on the surface.

erages and all amorphous samples examined, the values of W were smaller than W for self-mated diamond contacts.

The values of W for the self-mated diamond contacts exhibit small minima at hydrogen coverages around 30–40%. The origin of this minimum can be understood by considering the interaction between two C(111)(1 × 1) samples. When the surfaces are both fully H-terminated, hydrogen–hydrogen non-bonded interactions dominate. As hydrogen atoms are removed from the lower surface, there are fewer hydrogen–hydrogen non-bonded interactions for a given separation. Carbon–hydrogen non-bonded interactions remain while hydrogen–hydrogen non-bonded interactions as hydrogen is removed; however, the carbon atoms are farther from the hydrogen atoms on the opposing sample surface. Thus, the attractive region of the force curve decreases slightly (Fig. 6). When most of the hydrogen atoms have been removed from the surface of one sample, the specimens can now further reduce their separation, and so carbon–hydrogen non-bonded interactions dominate. This causes the well depth of the force curve to increase and shift to larger separations (Fig. 6). Because the well depth has increased, the value of W increases.

While variable-temperature measurements are difficult to achieve experimentally, the temperature can be varied easily in MD simulations. The work of adhesion for the aligned configuration of fully-terminated, self-mated C(001)(2 × 1)–H and C(111)(1 × 1)–H contacts as a function of temperature reveals that increasing the temperature from 100 K to 700 K monotonically decreases the value of W by a small amount ($\sim$5 mJ/m^2 for the aligned configuration). The decrease is likely due to the effect of thermal expansion, leading to slightly larger average bond separations. Brukman *et al.* [34] observed experimentally in variable temperature ultra-high vacuum (UHV) AFM studies that on both (001) and (111) single crystals for both tips used, the work of adhesion decreased by 23–44% when increasing the temperature from 135 K to 220 K, whereas contrasting trends with temperature were seen in the experiment for different tips in the range of 50–135 K. In the MD simulations, the extrapolated change from 135 K to 220 K is 155.3 mJ/m^2 to 154.5 mJ/m^2 and 200.8 mJ/m^2 to 200.1 mJ/m^2 for the C(001)(2 × 1)–H and C(111)(1 × 1)–H self-mated contacts, respectively. This variation is far smaller than the AFM results.

4. DFT Methods and Results

Because electrostatics is not included in the MD simulations, DFT was used to examine the partial atomic charges at the C(111)(1 × 1)–H and C(001)(2 × 1)–H surfaces and determine whether differences arising from the electrostatics could be a contributing factor to the differences in adhesion that were observed in the AFM experiments. DFT calculations were carried out with the Gaussian03 suite of programs [56] using periodic boundary conditions. These calculations require only the coordinates for the contents of a single unit cell and the cell's dimensions. The unit cells were hydrogen-terminated on the upper and lower z-surfaces. The PBEPBE functional was employed with the 6-31G(d) basis set. Two charge derivation schemes were tested (as implemented in Gaussian03): (1) the Mulliken population analysis [57] and (2) fits to molecular electrostatic potential using the CHELPG method [58]. The Mulliken method was found to be more reliable; therefore, we report those results.

Consistency was checked with a sub-set of the simulation results by comparing with the B3LYP functional (for the 001 surface), and by using an explicit atom approach, where DFT calculations were performed on diamond segments (for the (111) surface). In the latter case, two system sizes, 102 atoms and 182 atoms initially, were considered. The explicit atom calculations were single point energy calculations at the B3LYP/6-31G(d) level of theory. Overall, the results were consistent between the two methods within 5–15%.

The results are summarized in Table 2. The hydrogen atom on the surface of the diamond sample is slightly positive for both samples, with the (001) having slightly more charge. The hydrogen-bearing carbon is slightly negative for both surfaces, again with the (001) surface having slightly more charge. The quaternary carbon is nearly neutral for both samples. The difference in charge across the bond is proportional to the bond moment. For the C–H bond, the difference in the number of electrons is 0.266 and 0.286 for the (111) and (001) surfaces, respectively. For the C–CH bond, the difference in number of electrons is −0.214 and −0.266 for the (111) and (001) surfaces, respectively. Thus, there are charge fluctuations in the z-direction for the (001) surface. However, the net charge difference per unit area depends on the density of the charge-bearing sites and corresponding C–H

Table 2.
Mulliken charges on diamond surface atoms (in electron units)

Atom[a]	Charge[b] (electrons)	Charge (electrons) per unit area for the topmost H atom (nm^{-2})	Charge energy difference per unit area for topmost layer (eV/nm^2)
Diamond (111)			
H	0.106	1.93	4.83
C (H)	−0.160		
C	0.055		
Diamond (100)			
H	0.113	1.78	4.49
C (H)	−0.173		
C	0.093		

[a] Atoms at the surface of diamond. C (H) refers to the hydrogen-bearing carbon. C refers to the quaternary carbon beneath the hydrogen-bearing carbon.
[b] Periodic boundary condition density functional theory (PBEPBE/6-31G(d)/auto) on unit cell.

bonds. The latter is 18.16 nm^{-2} and 15.73 nm^{-2} for the (111) and (001) surfaces, respectively. This produces a difference in electrons per unit area of 4.83 nm^{-2} and 4.49 nm^{-2} for the (111) and (001) surfaces, respectively, at the very surface (i.e., from the C–H bond polarization). The C–H bond length is nearly identical for both surfaces; therefore, the electrostatic dipole per unit area produced by the C^-–H^+ bonds is stronger for the (111) surface than the (001) surface. The C–C bonds in the layer below also produce a charge difference, but of opposite sign. Although this dipole reduces the effect of the C–H bond dipole, the C–C bonds are further from the interface. Therefore, the DFT calculations predict a slightly stronger electrostatic field at the (111) surface compared to (001), which would increase adhesion with neutral but polarizable groups on the counterface. It would also increase (decrease) adhesion with dipoles whose direction vector has a component parallel (anti-parallel) to the C–H bonds, essentially due to the local interaction between the charges of the outermost atoms. The AIREBO potential [31], since it has been fit to experimental data for hydrocarbon systems, captures the effect of the C–H dipole to some extent although without further investigation it is not clear how much discrepancy there is between the AIREBO potential and the DFT calculation for diamond surfaces. Furthermore, the number of electrons due solely to the outermost H atoms is 1.93 nm^{-2} and 1.78 nm^{-2} for the (111) and (001) surfaces, respectively. These charges will exert a local electrostatic repulsion on positively charged H atoms on the counterface. This effect is not captured by the AIREBO potential and would tend to make surfaces with protruding H atoms repel one another more strongly than MD simulations predict, thus reducing the work of adhesion. The DFT calculation predicts the effect is slightly larger for the (111) surface compared to the (001) surface.

5. AFM Methods and Results

Achieving precise lateral alignment of tip atoms with respect to the surface periodicity for adhesion or friction measurements is extremely challenging in an AFM. Doing so requires accurate control over the tip structure as well as knowledge of the local surface topology. This has been achieved only in exceptional cases, such as rotational alignment of a flake of graphite attached to an AFM tip, where the sample could then be rotated [59]. Lateral alignment is particularly challenging on diamond surfaces. From separate measurements on single-crystal diamond, we observed periodic stick–slip friction showing the surface lattice. Figure 7 shows lateral force images of C(001)(2 × 1)–H dimer domains, whose orientation can change on the scale of a few nanometers to tens of nanometers, consistent with previous scanning tunneling microscopy measurements [60, 61]. Aligning tip and surface atoms laterally within an individual domain requires higher precision and stability than is typically available in an AFM, as well as knowledge of the atomic structure of the tip itself, which is not possible in all but a few scanning probe systems. Thus, it is very difficult to determine and control local tip–sample lattice alignment in AFM L vs Z experiments, and we rely on the MD simulations for information about any dependence of adhesion on lateral tip–sample alignment.

As mentioned above, the AFM experiments included two separate measurements of adhesion (pull-off) forces, L_C. In the previous study [33], L_C values recorded during F vs L measurements on individual MCD (001) and (111) crystallites, which were grown by hot filament chemical vapor deposition, were reported. These crystallites were characterized by Raman spectroscopy, X-ray photoelectron spec-

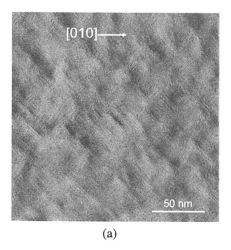
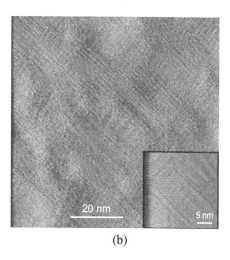

(a)   (b)

Figure 7. Lateral force images of a C(001)(2 × 1)–H single-crystal surface in dry nitrogen using a 45 nm EBID-coated amorphous tip. Here, the dimers are resolved, showing crisscross patterns at 45° to the [010] sliding direction. The stick–slip periodicity is approximately 0.5 nm (twice the diamond (001)(1 × 1) lattice constant). The image scan size in (a) is 200 nm by 200 nm, and the scan sizes in (b) are 80 nm by 80 nm and 20 nm by 20 nm (inset).

troscopy (XPS), elastic recoil detection analysis (ERDA), and X-ray absorption near-edge spectroscopy (XANES). The Raman spectrum showed a single sharp sp^3 peak at 1332 cm^{-1} and the absence of an sp^2 peak near 1580 cm^{-1} indicative of diamond character; the XPS exhibited a strong reduction in the oxygen signal after H-termination; ERD analysis showed the reduction of water adsorbed on the surface after H-termination (and air exposure); finally, the XANES spectra demonstrated diamond character and H-termination. In XANES, a sharp exciton feature and the second bandgap of diamond at 289.15 eV and ≈302 eV, respectively, demonstrate the high quality diamond character. The XANES spectra also demonstrated the presence of a small amount of π-bonded carbon due to a peak at 285 eV, likely due to sp^2 bonds at grain boundaries, and the presence of C–H bonds due to a shoulder near 287 eV. For all variable-load experiments, the lateral displacement of the tip along the surface that occurs in response to an applied load was compensated with piezo motion to keep the tip above the same location on the surface [62]. This tilt-compensation scheme was necessary because while the RMS roughesses of the surfaces was 1–3 nm over a 500 nm × 500 nm x–y range, the surfaces consisted of 10–50 nm diameter islands of ≈0.3 nm root-mean-square (RMS) roughness. Thus, it was important to confine the tip to individual islands during each measurement. Using the convenient Carpick–Ogletree–Salmeron (COS) approximation for fitting F vs L data with the Maugis–Dugdale transition model [25, 26], the sliding work of adhesion, referred to as W_{FL} here, was calculated. The fitting procedure for the transition model determines the behavior of a parabola-on-flat interface along a spectrum that spans the Derjaguin–Müller–Toporov [63] (DMT) and Johnson–Kendall–Roberts [64] (JKR) contact mechanical regimes described in Ref. [25]. By fitting the F vs L data using this procedure, the way in which the contact area varies with load was determined and, thus, a value for the contact area, A_C, at the pull-off point during the experiment was obtained. The work of adhesion is then proportional to the pull-off force L_C, assuming that sliding does not induce early pull-off.

In addition to the F vs L measurements that yielded W_{FL}, L vs Z measurements were performed both before and after individual sets of F vs L measurements on the same locations as the friction measurements. From the L vs Z curves, L_C values were obtained for comparison with those obtained from the F vs L experiments. Values for W_{LZ} were then calculated from these L_C values, using COS transition parameters to determine the contact area at pull-off. In the transition model, the work of adhesion is related to the pull-off force according to the equation,

$$W = \frac{-L_C}{\chi \pi R},\tag{1}$$

where R is the tip radius, and χ is a dimensionless parameter determined from the transition model fitting procedure. Note that L_C is generally a negative quantity, but here we will treat the L_C–W relationship and discussions about L_C in terms of its magnitude, and all values of L_C reported here are positive. The values of χ range from 1.5 to 2, corresponding to the JKR and DMT regimes, respectively.

Both the W_{FL} and the W_{LZ} obtained from the two types of AFM measurements (*F vs L* and *L vs Z*) are reported in Table 3. The experimental uncertainties in Table 3 correspond to the relative combined standard uncertainty of the reported value [65]. The use of continuum methods to model nanoscale contacts presents some uncertainty in the methodology [20], and quantitative values for adhesion are important for real applications. The range of adhesion values that represent possible results from our measurements is estimated by calculating W based on two different approaches for determining W from the measured L_C. The first approach, mentioned above, uses the COS transition parameter, χ, to identify the contact behavior with respect to the JKR–DMT spectrum. The second approach uses a physically reasonable estimate for the dimensionless Tabor parameter (μ_T), which is a measure of the ratio of the amount of deformation caused by adhesion forces relative to the counterface–surface equilibrium separation (z_0) [66]. Thus, μ_T is yet another indicator of the appropriate contact mechanics regime (assuming a continuum model applies). The exact equation for Tabor's parameter is:

$$\mu_T = \left(\frac{16RW^2}{9K^2z_0^3}\right)^{1/3},\tag{2}$$

where $K = \frac{4}{3}(\frac{1-\nu_{tip}^2}{E_{tip}} + \frac{1-\nu_{surface}^2}{E_{surface}})^{-1}$ is the combined elastic modulus, and ν and E are the Poisson's ratio and Young's modulus, respectively. Combining equations (1) and (2) yields

$$\mu_T = \left(\frac{16L_C^2}{9\pi^2RK^2\chi^2z_0^3}\right)^{1/3}.\tag{3}$$

Low (<0.1) and high (>5) values of Tabor's parameter correspond to DMT and JKR contacts, respectively. Low values indicative of the DMT regime, like those observed by Brukman *et al.* [34], would be expected for diamond surfaces.

While it is not clear what μ_T should be for an amorphous tip on diamond, it is expected to be more JKR-like than a diamond–diamond interface because of its lower modulus and possibly higher surface energy. With that in mind, an upper bound for μ_T for each AFM tip-sample pair was calculated by assuming JKR behavior in equation (1) (i.e., $\chi = 1.5$) and a relatively low value for E_{tip} of 100 GPa. Average bulk values for the elastic constants of diamond ($E_{(111)} = 1208$ GPa, $E_{(001)} = 1054$ GPa, $\nu_{(111)} = 0.047$ and $\nu_{(100)} = 0.105$) were used [67–69]. The z_0 values were taken from the MD simulations of corresponding, self-mated diamond pairs reported here. For C(001) and C(111), $z_0 = 0.19$ and 0.18 nm, respectively. In comparison with the amorphous counterfaces, the diamond–diamond interfaces represented the smallest possible values of z_0 and, therefore, gave an upper bound for μ_T, which was 0.015–0.023 for C(111)(1 × 1)–H and 0.015–0.035 for C(001)(2 × 1)–H, depending on the tip. A new χ parameter was then calculated [26]. Table 3 displays the W_{LZ} values that were calculated using the estimated μ_T (and new χ values). In general, the work of adhesion calculated in this way rep-

Table 3.
AFM data

Surface	Sliding direction	Tip radius (nm)	L_C from F vs L (nN)	W_{FL} from COS (J/m^2)	L_C from L vs Z (nN)	W_{LZ} from COS (J/m^2)	W_{LZ} from est. μ_T (J/m^2)
(111)(1 × 1)–H	[112]	45	29.7 ± 3.3	0.131 ± 0.018	23.9 ± 1.1	0.101 ± 0.008	0.093 ± 0.008
	[110]		25.3 ± 1.5	0.103 ± 0.009			
		150	27.6 ± 1.7	0.036 ± 0.003	23.1 ± 1.2	0.030 ± 0.003	0.026 ± 0.002
(001)(2 × 1)–H	[−110] or [110]	45	53.6 ± 2.4	0.258 ± 0.022	32.6 ± 1.3	0.157 ± 0.012	0.129 ± 0.010
		150	45.0 ± 1.7	0.064 ± 0.005	29.0 ± 1.0	0.041 ± 0.003	0.033 ± 0.002
	[010]		48.1 ± 1.6	0.066 ± 0.005			

resented an approximately 10–20% reduction in the work of adhesion compared with those calculated from the COS parameter, χ, and it moved the W_{LZ} values even further away from the corresponding W_{FL} values. The (very conservative) Tabor estimate produced more JKR-like contact behavior, but the difference was not large. Note that this result depends only slightly on E_{tip}, and inserting even the largest z_0 yields only minor changes.

6. Discussion

In the AFM experiments, two amorphous hydrocarbon-coated tips with different radii were used to examine the adhesion of individual diamond grains on an MCD surface in dry N_2 at room temperature. The work of adhesion was calculated from pull-off forces during force–displacement curves and while performing friction *vs.* load scans. While the magnitudes of W and the pull-off force depended on the measurement method, both methods yielded values of W and pull-off force that were larger for the C(001)(2 × 1)–H sample than for the C(111)(1×1)–H sample. For both diamond self-mated contacts, the MD simulations produced values of W that were within the range of values measured experimentally. However, the *average* values of W calculated from the MD simulations were larger for the C(111)(1 × 1)–H self-mated sample. In addition, due to the range of values associated with different starting configurations, the simulations produced no statistically relevant difference between the adhesion of diamond C(111)(1 × 1)–H compared to C(001)(2 × 1)–H. Despite the fact that simulation and experimental conditions were designed to correspond as closely as possible, the remaining differences must be responsible for the disagreement between AFM and MD. Possible explanations include the different contact geometries (parabola-on-flat in AFM *vs* two infinitely flat surfaces in MD), differences in atomic-scale roughness of the two surfaces in the experiment, the presence of partial charges, and the presence of defects or impurities.

Determining the contact area from AFM data remains one of the most challenging aspects of extracting materials properties from the data for which we rely on continuum mechanics models. Recent MD simulations of Luan and Robbins have shown that continuum contact models underestimate the contact area of nanoscale single-asperity contacts, calling into question the validity of continuum models at the nanoscale [20, 21]. Subsequent MD simulations revealed that this same trend exists when contact between a curved diamond tip and single-crystal diamond surfaces is simulated [33]. Although the results from continuum mechanics fits may be prone to quantitative error, a comparison of different surfaces using the same tip remains meaningful as long as roughness is constant. In addition, in simulations with finite-sized tips, it is possible to use various definitions of contact [20, 33, 52, 70]. Nonetheless, given the sensitivity of W to the contact area, infinitely flat surfaces were used in the MD simulations here.

The MD simulations showed that the relative position of the hydrogen atoms on opposing diamond samples results in a large variation in the calculated values of W. Substituting an amorphous carbon counterface for a diamond counterface removes the complication of averaging over counterface–sample alignments and has the benefit of making the counterface more similar to the AFM tip. It should be noted, however, that the sp^2:sp^3 ratio in the EBID-produced AFM tip was not measured [33]. With that in mind, counterfaces with different sp^2:sp^3 ratios were used in the simulations. All the amorphous, incommensurate counterfaces lowered the calculated values of W compared to self-mated diamond contacts. In addition to being incommensurate with the diamond samples, the amorphous carbon counterfaces were not atomically flat. Indeed, the simulations showed that progressive roughening of one of the contacting bodies, i.e., by progressing from C(111) to an amorphous sample and, ultimately, to an MDN sample, reduces the work of adhesion significantly.

The previous MD simulations of Luan and Robbins [21] and the simulations presented here demonstrate that roughness reduces the work of adhesion. It is possible that the larger AFM tip has a greater surface roughness than the smaller tip; in other words, the larger extent of the contact zone formed by the larger tip means that any rough features at the apex are more likely to prevent adjacent atoms from contacting the surface [71]. This effect may explain why the large AFM tip produced consistently lower values of W and pull-off force than the smaller AFM tip when paired with a given sample (Table 3). Likewise, although the apparent, nanoscale roughness was the same when measured by the two tips, any differences in atomic-scale roughness between the C(001) and C(111) samples could lead to the different W values reported here. As seen in the scanning electron microsopy (SEM) image in Fig. 8, the (001) surface is known to grow *via* layer-by-layer growth [72] whereas the (111) surface grows *via* multiple nucleation processes [73]. It is not well understood exactly how this affects atomic-scale roughness, but it is possible that growth dominated by individual nucleation sites produces an atomic-scale roughness that leads to lower adhesion for the (111) surface. On the other hand, the dimer rows on the (001) surface produce their own atomic-scale "roughness" of an ordered nature. Nonetheless, it is notable that the Tabor estimate produced a larger difference in comparison with the COS calculation for the (001) surface compared with the (111) surface. This discrepancy indicates that continuum mechanics models could be less valid for the (001) surface, even though from a growth standpoint the greater roughness of the (111) surface should lead to a greater deviation from the continuum model.

Care was taken to H-terminate the diamond samples used in the AFM experiments [33]. Thus, there should be no unsaturated carbon atoms on the sample. However, the concentration of dangling bonds on the tips is unknown. AFM measurements on ultrananocrystalline (UNCD) surfaces [14] and MD simulations of the contact of diamond and DLC surfaces have shown that the presence of unsaturated bonds increases adhesion [15–18]. In the simulations, this occurred *via* the forma-

Figure 8. SEM image of the microcrystalline diamond film. The terraced and triangular features are representative of the different growth mechanisms of the (001) (large central square) and (111) (triangular on right) crystallites, respectively.

tion of covalent bonds between the two contacting bodies. When both surfaces are fully H-terminated, the DFT calculations presented here show that positive charge exists at the surface due to the expected charge transfer from the H atom to the C atom. The calculated charge and bond moment for the C(001)(2×1)–H sample agrees with recently published DFT data [10]. As noted earlier, the AIREBO potential does not contain terms to model partial charges that arise from electronegativity differences. Thus, any changes in work of adhesion that arise from charges on both surfaces in proximity to one another is not captured in the MD simulations.

The addition of different atom types, such as oxygen, to the diamond surfaces could complicate the chemical reactivity and the charge landscape of the surfaces. Brukman *et al.* [34] used a variable-temperature, UHV AFM to measure the pull-off force and W between polycrystalline diamond tips and single-crystal C(111)(1×1) and C(001)(2 × 1). As mentioned earlier, Auger analysis revealed the presence of less than a monolayer of oxygen on the (111) and the (001) surfaces. While the way the oxygen is bound has not been resolved, recent DFT calculations reveal that added oxygen to the C(001)(2 × 1) sample increases the magnitude of the bond moments at the surface. When the diamond is –OH and =O terminated, the oxygen is negative and the bond moments are 0.99e and 0.5e, respectively [10]. The existence of these partial charges is likely to influence the work of adhesion measured by Brukman *et al.* In that case, the work of adhesion measured for both C(111)(1 × 1) and C(001)(2×1) surfaces with the smaller polycrystalline tip were indistinguishable near room temperature (225 K). In contrast to the trends reported here, when a larger polycrystalline tip was used, the C(111)(1 × 1) surface exhibited higher adhesion. In an effort to isolate the effect of partial charges on the adhesion of H-terminated diamond, MD simulations that utilize a potential energy function with Coulombic interactions are currently underway.

The MD simulations predict a slight temperature dependence of the work of adhesion such that it decreases as the temperature increases. Clear differences exist between previous variable-temperature AFM experiments [34] and the present simulations, and, therefore, it is not possible to conclude that the trends in the experimental and calculated work of adhesion with temperature are governed by the same phenomena.

7. Conclusion

The origins of adhesion forces and the work of adhesion were studied by complementary atomic force microscopy (AFM) and molecular dynamics (MD) simulations. AFM measurements, using amorphous hydrocarbon tips, found that the work of adhesion was larger on the C(001)(2 × 1)–H diamond surface by 27–55% compared to the C(111)(1 × 1)–H surface. Although the absolute values of the works of adhesion for experiment and simulation are in reasonable agreement, the simulations found that the average work of adhesion between two flat C(001)(2 × 1) surfaces was smaller than for self-mated C(111)(1 × 1) contacts for all hydrogen coverages examined, in contradiction to the AFM results. However, the relative alignment of the opposing surfaces was found to significantly affect the adhesion, such that incommensurate alignment of atoms strongly reduces adhesion. As this factor is not known or easily controlled in the AFM experiments, it could contribute to the different trends in adhesion for C(001)(2 × 1) *vs* C(111)(1 × 1) observed by the experiments and simulations. Indeed, the dependence of the work of adhesion on tip size is consistent with this idea: lower work of adhesion was observed with the larger tip, and could be due to the fact that the larger tip encounters more roughness, increasing the mean separation between the surfaces at equilibrium and correspondingly decreasing the work of adhesion.

The calculated W values show a modest dependence on hydrogen coverage, whereby an optimal coverage is found which is intermediate to fully terminated and fully exposed. Although fully H-terminated surfaces have a lower surface energy, removing an optimal number of H atoms reduces the work of adhesion by producing a larger mean separation between the counterface and the topmost atoms, which now include C atoms. The MD simulations predict a small dependence of work of adhesion on temperature. Previous AFM measurements see the same trend over a limited temperature range but the magnitude of the effect is larger in the experiments.

DFT calculations performed on hydrogen-terminated, single-crystal diamond quantified the small positive charges at the very surface due to H atoms, with the C(111)(1 × 1)–H surface having slightly more charge per unit area. This could act to repel a countersurface with protruding H atoms. This is another possible source of the difference between the experiment and MD simulations for the work of adhesion.

While further work is needed to resolve the sources of discrepancy between AFM and MD measurements, it is clear that atomic roughness and the corresponding mean separation of the surfaces at equilibrium is a critical factor influencing adhesion at the nanometer scale, which presents both a challenge and an opportunity for understanding and controlling adhesion at small scales.

Acknowledgements

P. L. P., G. G. and J. A. H. acknowledge support from AFOSR grant numbers F1ATA-09-086G002 and F1ATA-09-086G003 (and as part of the Extreme Friction MURI). J. A. H. also acknowledges support from the ONR grant # N00014-09-WR20155. P. L. P. and J. A. H. acknowledge support from NSF grant number CMMI-0825981. R. W. C. acknowledges support from AFOSR grant # FA9550-08-1-0024 and NSF grant number CMMI-0826076. J. A. H. and P. L. P. gratefully acknowledge J. D. Schall for helpful discussions. R. J. C. and R. W. C. gratefully acknowledge A. V. Sumant for helpful discussions and invaluable technical expertise. We also thank G. D. Wright and D. Whyte for the elastic recoil experiments, as well as D. S. Grierson and A. R. Konicek for the X-ray absorption spectroscopy measurements.

References

1. O. Auciello, S. Pacheco, A. V. Sumant, C. Gudeman, S. Sampath, A. Datta, R. W. Carpick, V. P. Adiga, P. Zurcher, Z. Ma, H. C. Yuan, J. A. Carlisle, B. Kabius, J. Hiller and S. Srinivasan, *IEEE Microwave Magazine* **8**, 61 (2007).
2. O. Auciello, J. Birrell, J. A. Carlisle, J. E. Gerbi, X. C. Xiao, B. Peng and H. D. Espinosa, *J. Phys.: Condens. Matter* **16**, R539 (2004).
3. I. S. Forbes and J. I. B. Wilson, *Thin Solid Films* **420**, 508 (2002).
4. M. P. de Boer and T. M. Mayer, *MRS Bull.* **26**, 302 (2001).
5. R. Maboudian, W. R. Ashurst and C. Carraro, *Sensors Actuators A* **82**, 219 (2000).
6. Y. Qi and L. G. Hector, *Phys. Rev. B* **69**, 235401 (2004).
7. X. G. Wang and J. R. Smith, *Phys. Rev. Lett.* **87**, 186103 (2001).
8. G. Kern, J. Hafner and G. Kresse, *Surf. Sci.* **366**, 445 (1996).
9. S. J. Sque, R. Jones and P. R. Briddon, *Phys. Rev. B* **73**, 085313 (2006).
10. G. Zilibotti, M. C. Righi and M. Ferrario, *Phys. Rev. B* **79**, 075420 (2009).
11. Y. Kaibara, K. Sugata, M. Tachiki, H. Umezawa and H. Kawarada, *Diam. & Rel. Mater.* **12**, 560 (2003).
12. A. Erdemir, *Surf. Coat. Technol.* **146**, 292 (2001).
13. A. V. Sumant, D. S. Grierson, J. E. Gerbi, J. Birrell, U. D. Lanke, O. Auciello, J. A. Carlisle and R. W. Carpick, *Adv. Mater.* **17**, 1039 (2005).
14. A. V. Sumant, D. S. Grierson, J. E. Gerbi, J. A. Carlisle, O. Auciello and R. W. Carpick, *Phys. Rev. B* **76**, 235429 (2007).
15. J. A. Harrison, R. J. Colton, C. T. White and D. W. Brenner, *Mater. Res. Soc. Symp. Proc.* **239**, 573 (1992).

16. J. A. Harrison, D. W. Brenner, C. T. White and R. J. Colton, *Thin Solid Films* **206**, 213 (1991).
17. G. T. Gao, P. T. Mikulski and J. A. Harrison, *J. Am. Chem. Soc.* **124**, 7202 (2002).
18. J. D. Schall, G. Gao and J. A. Harrison, *J. Phys. Chem. C* **V114**, 5321–5330 (2010).
19. D. L. Liu, J. Martin and N. A. Burnham, *Appl. Phys. Lett.* **91**, 043107 (2007).
20. B. Q. Luan and M. O. Robbins, *Nature* **435**, 929 (2005).
21. B. Q. Luan and M. O. Robbins, *Phys. Rev. E* **74**, 026111 (2006).
22. Y. F. Mo, K. T. Turner and I. Szlufarska, *Nature* **457**, 1116 (2009).
23. T. S. Chow, *Phys. Rev. Lett.* **86**, 4592 (2001).
24. B. Cappella and G. Dietler, *Surf. Sci. Rep.* **34**, 1 (1999).
25. R. W. Carpick, D. F. Ogletree and M. Salmeron, *J. Colloid Interface Sci.* **211**, 395 (1999).
26. D. S. Grierson, E. E. Flater and R. W. Carpick, *J. Adhesion Sci. Technol.* **19**, 291 (2005).
27. D. Maugis, *Langmuir* **11**, 679 (1995).
28. U. D. Schwarz, *J. Colloid Interface Sci.* **261**, 99 (2003).
29. F. W. Delrio, M. P. De Boer, J. A. Knapp, E. D. Reedy, P. J. Clews and M. L. Dunn, *Nature Materials* **4**, 629 (2005).
30. A. A. Stekolnikov, J. Furthmüller and F. Bechstedt, *Phys. Rev. B* **65**, 115318 (2002).
31. S. J. Stuart, A. B. Tutein and J. A. Harrison, *J. Chem. Phys.* **112**, 6472 (2000).
32. K. Bobrov, A. Mayne, G. Comtet, G. Dujardin, L. Hellner and A. Hoffman, *Phys. Rev. B* **68**, 195416 (2003).
33. G. T. Gao, R. J. Cannara, R. W. Carpick and J. A. Harrison, *Langmuir* **23**, 5394 (2007).
34. M. J. Brukman, G. T. Gao, R. J. Nemanich and J. A. Harrison, *J. Phys. Chem. C* **112**, 9358 (2008).
35. S. A. Adelman, *Adv. Chem. Phys.* **44**, 143 (1980).
36. G. T. Gao, P. T. Mikulski, G. M. Chateauneuf and J. A. Harrison, *J. Phys. Chem. B* **107**, 11082 (2003).
37. G.-T. Gao, P. T. Mikulski, G. M. Chateauneuf and J. A. Harrison, *J. Phys. Chem. B* **107**, 11082 (2003).
38. J. A. Harrison, G. Gao, J. D. Schall, M. T. Knippenberg and P. T. Mikulski, *Philos. Trans. Royal Soc. A* **366**, 1469 (2008).
39. J. D. Pearson, G. Gao, M. A. Zikry and J. A. Harrison, *Computational Mater. Sci.* **47**, 1 (2009).
40. D. W. Brenner, O. A. Shenderova, J. A. Harrison, S. J. Stuart, B. Ni and S. B. Sinnott, *J. Phys. C* **14**, 783 (2002).
41. B. Ni, S. B. Sinnott, P. T. Mikulski and J. A. Harrison, *Phys. Rev. Lett.* **88**, 205505 (2002).
42. H. Trotter, R. Phillips, B. Ni, Y. H. Hu, S. B. Sinnott, P. T. Mikulski and J. A. Harrison, *J. Nanosci. Nanotechnol.* **5**, 536 (2005).
43. B. I. Yakobson, C. J. Brabec and J. Bernholc, *Phys. Rev. Lett.* **76**, 2511 (1996).
44. D. W. Brenner, B. I. Dunlap, J. A. Harrison, J. W. Mintmire, R. C. Mowrey, D. H. Robertson and C. T. White, *Phys. Rev. B* **44**, 3479 (1991).
45. J. A. Harrison, S. J. Stuart, D. H. Robertson and C. T. White, *J. Phys. Chem. B* **101**, 9682 (1997).
46. J. A. Harrison and D. W. Brenner, *J. Am. Chem. Soc.* **116**, 10399 (1994).
47. J. A. Harrison, C. T. White, R. J. Colton and D. W. Brenner, *Phys. Rev. B* **46**, 9700 (1992).
48. M. D. Perry and J. A. Harrison, *J. Phys. Chem.* **99**, 9960 (1995).
49. S. L. Zhang, G. Wagner, S. N. Medyanik, W. K. Liu, Y. H. Yu and Y. W. Chung, *Surf. Coat. Technol.* **177–178**, 818 (2004).
50. O. A. Shenderova and D. W. Brenner, *Phys. Rev. B* **60**, 7053 (1999).
51. G. M. Chateauneuf, P. T. Mikulski, G. T. Gao and J. A. Harrison, *J. Phys. Chem. B* **108**, 16626 (2004).

52. M. T. Knippenberg, P. T. Mikulski, B. I. Dulnap and J. A. Harrison, *Phys. Rev. B* **78**, 235409 (2008).

53. P. T. Mikulski, G.-T. Gao, G. M. Chateauneuf and J. A. Harrison, *J. Chem. Phys.* **122**, 024701 (2005).

54. P. T. Mikulski and J. A. Harrison, *J. Am. Chem. Soc.* **123**, 6873 (2001).

55. A. B. Tutein, S. J. Stuart and J. A. Harrison, *Langmuir* **16**, 291 (2000).

56. M. J. Frisch, G. W. Trucks, H. B. Schlegel, G. E. Scuseria, M. A. Robb, J. R. Cheeseman, J. A. Montgomery Jr, T. Vreven, K. N. Kudin, J. C. Burant, J. M. Millam, S. S. Iyengar, J. Tomasi, V. Barone, B. Mennucci, M. Cossi, G. Scalmani, N. Rega, G. A. Petersson, H. Nakatsuji, M. Hada, M. Ehara, K. Toyota, R. Fukuda, J. Hasegawa, M. Ishida, T. Nakajima, Y. Honda, O. Kitao, H. Nakai, M. Klene, X. Li, J. E. Knox, H. P. Hratchian, J. B. Cross, V. Bakken, C. Adamo, J. Jaramillo, R. Gomperts, R. E. Stratmann, O. Yazyev, A. J. Austin, R. Cammi, C. Pomelli, J. W. Ochterski, P. Y. Ayala, K. Morokuma, G. A. Voth, P. Salvador, J. J. Dannenberg, V. G. Zakrzewski, S. Dapprich, A. D. Daniels, M. C. Strain, O. Farkas, D. K. Malick, A. D. Rabuck, K. Raghavachari, J. B. Foresman, J. V. Ortiz, Q. Cui, A. G. Baboul, S. Clifford, J. Cioslowski, B. B. Stefanov, G. Liu, A. Liashenko, P. Piskorz, I. Komaromi, R. L. Martin, D. J. Fox, T. Keith, M. A. Al-Laham, C. Y. Peng, A. Nanayakkara, M. Challacombe, P. M. W. Gill, B. Johnson, W. Chen, M. W. Wong, C. Gonzalez and J. A. Pople, in: *Gaussian03 Revision W*, Wallingford, CT (2004).

57. R. S. Mulliken, *J. Chem. Phys.* **23**, 1833 (1955).

58. C. M. Breneman and K. B. Wiberg, *J. Computational Chem.* **11**, 361 (1990).

59. M. Dienwiebel, G. S. Verhoeven, N. Pradeep, J. W. M. Frenken, J. A. Heimberg and H. W. Zandbergen, *Phys. Rev. Lett.* **92**, 126101 (2004).

60. T. Frauenheim, U. Stephan, P. Blaudeck, D. Porezag, H. G. Busmann, W. Zimmermannedling and S. Lauer, *Phys. Rev. B* **48**, 18189 (1993).

61. R. E. Stallcup, L. M. Villarreal, S. C. Lim, I. Akwani, A. F. Aviles and J. M. Perez, *J. Vac. Sci. Technol. B* **14**, 929 (1996).

62. R. J. Cannara, M. J. Brukman and R. W. Carpick, *Rev. Sci. Instrum.* **76**, 053706 (2005).

63. B. V. Derjaguin, V. M. Muller and Y. P. Toporov, *J. Colloid Interface Sci.* **53**, 314 (1975).

64. K. L. Johnson, K. Kendall and A. D. Roberts, *Proc. R. Soc. Lond. A* **324**, 301 (1971).

65. B. N. Taylor and C. E. Kuyatt, http://physics.nist.gov/Pubs/guidelines/TN1297/tn1297s.pdf (1994).

66. J. A. Greenwood, *Proc. R. Soc. Lond. A* **453**, 1277 (1997).

67. C. A. Klein, *Mater. Res. Soc. Bull* **27**, 1407 (1992).

68. J. Turley and G. Sines, *J. Phys. D: Appl. Phys.* **4**, 264 (1971).

69. E. S. Zouboulis, M. Grimsditch, A. K. Ramdas and S. Rodriguez, *Phys. Rev. B* **57**, 2889 (1998).

70. M. Chandross, C. D. Lorenz, M. J. Stevens and G. S. Grest, *Langmuir* **24**, 1240 (2008).

71. K. S. Kim and J. A. Hurtado, *Key Eng. Mater.* **183-1**, 1 (2000).

72. V. P. Godbole, A. V. Sumant, R. B. Kshirsagar and C. V. Dharmadhikari, *Appl. Phys. Lett.* **71**, 2626 (1997).

73. A. V. Sumant, C. V. Dharmadhikari and V. P. Godbole, *Mater. Sci. Eng. B* **41**, 267 (1996).

Part 3
Adhesion and Friction Measurements

Theoretical and Experimental Study of the Influence of AFM Tip Geometry and Orientation on Capillary Force

Alexandre Chau [a], **Stéphane Régnier** [b], **Alain Delchambre** [a] **and Pierre Lambert** [a,*]

[a] BEAMS Department, Université libre de Bruxelles, CP 165/56, 1050 Bruxelles, Belgium
[b] ISIR, 4 Place Jussieu BP 173, 75252 Paris Cedex 05, France

Abstract
Adhesion issues are present in many disciplines such as, for example, surface science, microrobotics or MEMS design. Within this framework, this paper presents a study on capillary forces due to capillary condensation. A simulation tool had already been presented using Surface Evolver and Matlab to compute the shape of a meniscus in accordance with the Kelvin equation and contact angles. The numerical results of this simulation complied well with literature results. One very important result is the ability to compute the evolution of the capillary force depending on the tilt angle of the gripper with respect to the object. The main contribution of this new paper is a test bench and the related experimental results which validate these numerical results. We present here new experimental results illustrating the role of humidity and tilt angle in capillary forces at the nanoscale.

Keywords
Condensation, capillary forces, adhesion, AFM, tilt angle, Kelvin equation

1. Introduction

When considering applications at small scale, adhesion cannot be ignored. For example, capillary adhesion is important in micromanufacturing [1] or in assembly of small components [2]. Contrary to marco-world ruled by gravity, small scale applications are governed by surface forces. Indeed, when size diminishes and the objects are scaled down, surface forces become more important and the major opposing force to picking up and releasing micro- and nanocomponents becomes the force of adhesion [3]. The force needed to separate two objects is also known as pull-off force. Adhesion can also prevent structures like RF-MEMS or any high aspect ratio structures from normal functioning [4].

The adhesion force is actually composed of different components: electrostatic force, van der Waals force, chemical forces and capillary force. Electrostatic

---

[*] To whom correspondence should be addressed. E-mail: pierre.lambert@ulb.ac.be

Adhesion Aspects in MEMS/NEMS
© Koninklijke Brill NV, Leiden, 2010

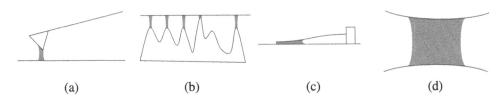

(a)   (b)   (c)   (d)

Figure 1. Some case studies in which capillary condensation can occur: (a) adhesion between an AFM tip and a substrate; (b) capillary bridges between roughness asperities and a — rough or not — substrate; (c) stiction of a cantilevered beam, and (d) scheme of the problem tackled in this work (two solids S_1 and S_2 are linked by a capillary bridge. Besides geometry, the main parameters of this problem are the contact angles made by the liquid on S_1 and S_2, surface tension γ of the liquid and the surrounding humidity).

force can be avoided by choosing materials properly (conductive materials to avoid charge accumulation and similar junction potentials to avoid a capacitive effect between two objects brought close to one another). Van der Waals force arises from the intrinsic constitution of matter: it is due to the presence of instantaneous dipoles. It becomes non-negligible at the nanometer scale. Chemical forces are due to the bondings between objects. It is active when objects are in contact (i.e., the distance between them is about an intermolecular distance).

Capillary force between two objects is due to the presence of liquid between them (see Fig. 1). It has already been shown [5, 6] that it can be used to manipulate submillimetric objects (with 300–500 µm characteristic dimension) by manually placing a liquid droplet on the object (a fraction of µl). At nanoscale, the liquid comes from capillary condensation of ambient moisture.

This nanomanipulation will require force modulation, in order to pick up and release components: the picking force should be larger than the other forces while the force applied on the component during release should be lower than its weight or any other adhesion force. It is proposed here to control the force by tilting the tip (an application could be an AFM tip) with respect to the object. Models which assume objects and meniscus to be axially symmetrical [7, 8], therefore, become inapplicable. A more general model thus has been developed to compute the capillary force without this constraint of axisymmetry.

We already published in [9] a theoretical study aiming at assessing the use of capillary force as a picking principle. The main result of this theoretical study is reproduced with permission in this new paper (Fig. 4), but the main contribution of this new paper consists in experimental results on capillary forces at the nanoscale.

This paper is organized as follows: first, the basic equations will be recalled (Section 2), and their validity discussed. Our simulation tool will then be presented (Section 3.1) together with some theoretical results (Section 3.3). The section on experimental study presents the experimental details and setup used in this work (Section 4.1) and the results are presented in Section 4.2. After discussion in Section 5, conclusions are drawn in Section 6.

2. Model

2.1. Introduction

This section details the different equations involved in the problem description: (i) the Kelvin equation [3, 10], which governs the curvature of the liquid meniscus according to environmental parameters (humidity and temperature), and (ii) the approaches encountered in the literature to compute the capillary force, i.e., the Laplace approach and the energetical approach (based on the derivation of the total surface energy U_S). Both methods are presented in our previous work [9] but the first method is not convenient for 3D-geometries (i.e., non-axially symmetric geometries) and will, therefore, not be presented in this work anymore.

In all cases, gravity is neglected, i.e., the meniscus height h is so small that the hydrostatic pressure difference $\rho g h$ is much smaller than the capillary pressure: this is the case for $h \ll 1$ mm.

2.2. Kelvin Equation

In any method used to compute the capillary force, the meniscus shape appears explicitly or implicitly. In the Kelvin equation approach, the meniscus geometry is involved through the total mean curvature H of the meniscus, or its inverse r, the mean curvature radius:

$$H = \frac{1}{r} = \frac{1}{2r_K} = \frac{1}{2}\left(\frac{1}{r_1} + \frac{1}{r_2}\right), \tag{1}$$

where r_K is, in the case of capillary condensation, the so-called Kelvin radius, and r_1 and r_2 are the two principal curvature radii. The Kelvin radius is governed by the Kelvin equation [3, 10] which is the fundamental equation for capillary condensation. It links the curvature of the meniscus with environmental and materials properties:

$$r_K = \frac{\gamma V_m}{RT \log_e(p/p_0)}, \tag{2}$$

where V_m is the molar volume of the liquid, R is the gas constant (8.31 J/mol K), T the temperature (in Kelvin) and p/p_0 is the relative humidity (RH), between 0 and 1. Typically, for water, this gives $r_K = 0.54$ nm$/\log_e$(RH), which gives for RH = 90%, a Kelvin radius of about 5 nm at 20°C.

2.3. Energetical View

The total energy of the meniscus can be expressed as:

$$U_S = \gamma_{LV} A_{LV} + \gamma_{LO} A_{LO} + \gamma_{LT} A_{LT} + \gamma_{OV} A_{OV} + \gamma_{TV} A_{TV}, \tag{3}$$

where γ_{ij} is the interfacial tension of the $i-j$ interface: Liquid, Vapor, Object and Tip, and A_{ij} are the areas of these interfaces.

The capillary force in the z-direction can then be computed using a classical derivative of the energy with respect to the distance between the objects. If the problem is axisymmetrical, the z-direction is the axis of symmetry, otherwise it can be any direction along which the force is calculated. Note that the energy is derivatized with respect to the separation distance z to obtain the capillary force in the direction of this separation, but according to [11], it must be emphasized whether the derivative is done assuming constant volume of liquid or constant pressure. In the case of constant volume of liquid, de Boer and de Boer [11] give for the force (positive is taken to be attractive):

$$F(z) = \frac{dU_S}{dz}, \tag{4}$$

while in the case of constant pressure:

$$F(z) = \frac{dU_S}{dz} - \frac{\gamma}{r_K} \frac{dV}{dz}. \tag{5}$$

In this work, according to de Boer and de Boer [11], before separation, the initial condition for the capillary volume will be established by the Kelvin radius. Then, the constant volume assumption will be used since it is reported by de Boer and de Boer that the force–distance curve is rate-independent for withdrawal times from 0.01 to 10 s, which is typically the case in this work. In other words, withdrawal times are short enough to guarantee that the liquid does not evaporate. This point will be discussed later on in the next subsection.

2.4. Validity of the Equations

Both methods presented here are based on macroscopic assumption on the nature of the liquid and solids: matter is continuous and so are its properties such as surface tension. It has been shown experimentally that the Kelvin equation is valid down to menisci of radii in the range 4–20 nm for cyclohexane condensed between mica surfaces [12] and in the range 5–65 nm for water [13]. For smaller radii, the discrete nature of matter should be taken into account, *via* molecular dynamics or Monte-Carlo calculations [14]. The results will thus have to be interpreted keeping in mind that their validity is not proven for very small sizes of meniscus (i.e., for menisci with radii <4 nm).

Another question raised by the literature is the parameter which is kept constant during derivatization: should the latter be computed keeping the volume or the curvature constant [15]? Different mechanisms play a role here, mainly the condensation/evaporation rate. It seems that the condensation takes place on the millisecond (<5 ms [15]) scale while the evaporation needs more time. It has been measured that meniscus stretching at tip-object distance is much larger than r_K. Capillary condensation also has a long term (up to tens of days) component, which will not be considered here [16].

If experimental investigation could provide an estimation of the characteristic times involved in the processes, it would seem natural to compute the volume condensed — fulfilling the Kelvin equation — at the smallest tip-object distance and then compute the evolution of the force with constant volume when retracting the tip (equation (4)).

3. Numerical Study

3.1. Simulation Tool

The numerical model already presented in [9] is able to cope with non-axisymmetrical menisci, based on the energetical approach described previously. The solver makes use of the software Surface Evolver (SE) [17] to compute the meniscus shape and henceforth to compute equation (3). The base of Surface Evolver is the evolution of a surface toward a local minimum of energy, representing the surface of the meniscus with triangularized mesh. If needed, the mesh can be evolved and refined until satisfactory result is found. The meniscus shape is consequently found by energy minimization, fulfilling both the Kelvin equation on the shape of the meniscus and the contact angles. This solver can be freely downloaded (http://www.susqu.edu/brakke/evolver/evolver.html) and is well documented. In equation (4) we replace the derivative by a finite difference:

$$\frac{dU_S}{dz} = \frac{U_S(z_i + \Delta z) - U_S(z_i)}{\Delta z}, \tag{6}$$

where $U_S(z_i + \Delta z)$ and $U_S(z_i)$ are computed with Surface Evolver. All computations have been first tested on a personal computer before being batched on the computing cluster Hydra of the Université libre de Bruxelles. (Hydra — the HPC cluster at the VUB/ULB Computing Centre.)

3.2. Available Shapes

To develop complex shapes without having to define each point of the tip (or the object), analytical shapes have been used. In the xz plane (see Fig. 2), usual profiles can be chosen: circular, conical or parabolical, while in the (x, y) plane, the section of the tip can be described as a polar function.

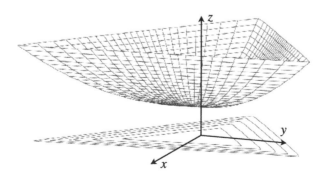

Figure 2. Example of a tip and its projection on the (x, y) plane. Here, the section is a triangle and the profile is a parabola. (Reproduced with permission from [9] © 2007 IEEE.)

Elementary sections are:

- Circle: $r(\theta) = R$;
- Triangle: $r(\theta) = c/(2\sqrt{3}\cos\theta)$ for $-\pi/3 < \theta \leqslant \pi/3$;
- Square: $r(\theta) = c/(2\cos\theta)$ for $-\pi/4 < \theta \leqslant \pi/4$.

In a similar way, any regular polygon of side length c can be very easily implemented. Actually, virtually any section can be represented as it is developed in Fourier series (in polar coordinates) in order to obtain an analytical shape in the domain $\theta = [0; 2\pi]$.

The profile and section are then coupled to obtain the complete tip that has to be used in the equations. An example of a tip with parabolic profile and triangular section is shown in Fig. 2. Such a geometrical description can also be applied to the object.

3.3. Theoretical Results

3.3.1. Introduction
The results presented in this section are twofold. First, validity of the code is shown in Section 3.3.2. As literature provides results only for axisymmetrical shapes, those results will be the only ones that can be used as a proof for the model, since no results are available for other shapes.

Then, in Section 3.3.3, more complex shapes will be investigated (effects of tip tilting).

3.3.2. Validation
In [8], analytical solutions are derived from the base equations. They provide results for the sphere–plane and sphere–sphere cases separated by a liquid bridge. The results (see Fig. 3) are presented with respect to the so-called filling angle [3], which is the angle defining the position of the triple line on the sphere (measured from the vertical axis). One can see that the correspondence is very good with the model from [8] and that the approximation of [3] ($F = 4\pi R\gamma\cos\theta$) is valid for very small filling angles.

The discrepancies between the model results and [8] can be explained by the meshing of the meniscus. The mesh refinement has to be limited to keep the computation time acceptable. In general, the total computation time for a configuration is in the 1–5 min range.

In [7], the shape of the tip is a parabola. The force values are also in good agreement as shown in [9]. Results in [18] and [19] have also been used as benchmarks.

3.3.3. Tilted Tips
One main feature of our model is its ability to compute non-axisymmetrical tips. Therefore, we presented in [9] the influence of the tilt angle τ on the force for a conical tip (see Fig. 4).

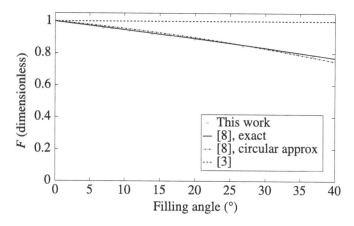

Figure 3. Comparison of normalized capillary force F (dimensionless) at the contact between a sphere and a plane for several models ($\theta_1 = \theta_2 = 40°$). The positive value of force means an attractive force. (F is the force divided by the quantity $2\pi R(\cos\theta_1 + \cos\theta_2)$, the filling angle is the angle defined by the arc of the sphere which joins the triple line to the contact point.)

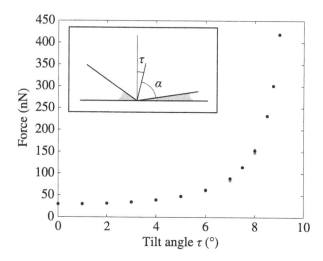

Figure 4. Force between a tilted tip (the tilt angle is τ) and a plane. The aperture angle of the cone (α) is 80°, temperature is 298 K, the relative humidity is 90%. Both contact angles are 30° and the surface tension is 72 mN/m. The positive value of force means an attractive force. The different points at the same tilt angle are for different mesh refinements. They give an idea of the numerical uncertainty in the results. The enclosed sketch defines the aperture angle of the cone α and the tilt angle τ. (Reproduced with permission from [9] © 2007 IEEE.)

The conical tip for which the results are presented has an aperture angle α of 80° (the tilt angle is thus limited to a maximum of 90–80 = 10°). One can see that the force can vary from about 30 nN to over 400 nN, simply by tilting the tip over the plane. This principle is expected to serve as a basis to modulate the capillary force and hence to allow micro/nanomanipulation of components.

4. Experimental Study

4.1. Experimental Test Bench

In order to validate the model, a test bench has been developed, following an AFM type design (see Fig. 5): a Thorlabs CPS198 laser (detail 1) is reflected by an AFM tip (detail 4) onto a Pacific Silicon Sensor QP50-6SD2 photodiode (detail 7), using lenses (details 3–6) and mirrors (details 2–5). When a force is applied on the AFM tip, the tip is deflected, changing the direction of the reflected beam. This modification can be measured using the photodiode. The laser spot displacement on the photodiode can then be converted to a force since the tip has a known stiffness (about 0.8 N/m).

Currently, the force can be measured between an AFM tip and a substrate placed in an environmental box (Fig. 6, detail 9). The substrate can be moved vertically over a 25 mm range with a 200 nm resolution (with a PI M-126 translation stage), and over a further 200 µm range with a 1 nm resolution (with a PI P-528 nanopositioning system, detail 8).

The tip and substrate (mica, used 30 min after cleavage, is assumed to be atomically flat) are approached until contact, then the pull-off force is measured when retracting the substrate. (Silicon tips were supplied by Nanoandmore. They have a nominal stiffness $k_m = 0.8$ N/m, and two different curvature radii of 90 nm and 150 nm: they have been modelled as elliptical tips.) As the tip and the substrate can be enclosed in a small environmental box, humidity can be controlled and the

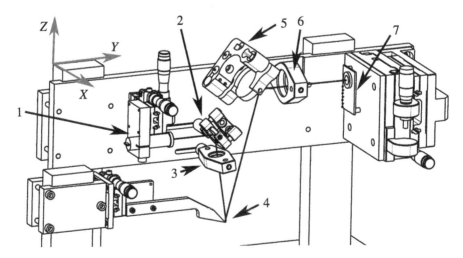

Figure 5. Test bench developed: the solid line indicates the optical path of the laser beam emitted from (1), reflected by a mirror (2) and passing through a lens (3) before being reflected by a cantilever (4). The reflected beam is sent towards a second mirror (5) before passing through a second lens (6) and illuminating a photodiode (7) whose voltage output is a measure of the beam deflection. Knowing the stiffness of the cantilever, this voltage output can be converted into a force with a resolution of the order of 10 nN.

variations of the force with respect to humidity can be measured. A typical pull-off measurement has already been presented in [9].

The sources of error are the quality of substrate and tip surfaces, the stiffness uncertainty (estimated from the cantilever thickness a 10% uncertainty for the range $(0.75–1.40) \times k_m$), electrical noise, positioning stages precision, and contact detection variability. The temperature variation effects can be ignored since only a variation of $0.03°$ has been registered during a 30-min experiment. All these errors contribute to the experimental scatter depicted in Fig. 7.

4.2. Experimental Results

The model presented in this article computes only the capillary component of the adhesion force, while experimental results include all other effects (van der Waals and electrostatic forces). In order to compare model results with experimental ones, we chose to vary a parameter that had an effect only on capillary force.

The most relevant parameter meeting this criterion is relative humidity. Indeed, the assumption that relative humidity has only minor impact on other forces than capillary force is acceptable. The variation of the capillary force with respect to relative humidity can thus be computed and experimentally measured.

In Fig. 7, different batches of pull-off measurements have been made with varying RH. A least squares fit is then made on the points. It is shown as a dashed line. For the same geometry, computations have been made, and the results are given by

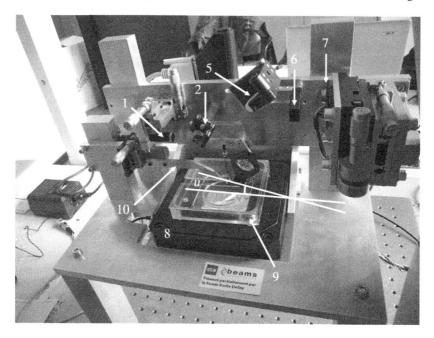

Figure 6. In addition to Fig. 5, we can see in this figure the nano-positioning stage (8), the environmental box (9) and the part on which the cantilever is glued (10). The angle u is the angle between the substrate to be placed on the bottom of the environmental box and the cantilever (4, not shown here).

the solid line. (As already said, only the slope of these curves is to be observed, i.e., the dependence on humidity.) The simulation gives a slope of 0.23 nN/% while the best fit leads to a slope equal to 0.27 nN/% with a standard deviation equal to 0.035 nN/%. In these simulations the contact angles have been assumed close to zero since they cannot be measured at this scale.

Similarly, the measurements and computations of Fig. 7 have been done for different tip to substrate tilt angles. The results are shown in Fig. 8. Results quali-

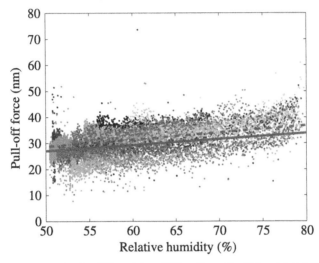

Figure 7. Force measurements: for different humidities, the pull-off force has been measured. The dashed line is the linear regression of the measurement points. The solid line is the result of the model.

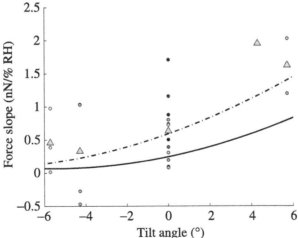

Figure 8. Slope of the pull-off force *vs* humidity with respect to the tilt angle. For each tilt angle, the mean is given by the triangles. The 4th triangle must be treated with care as it is based on a single batch. The solid line is the result of the model; the dot-dashed line is the result of the model for a 4° tilt bias because this 4° tilt deviation is assumed to be a measure for the uncertainty in the initial value of the angle u. It is shown for sensitivity demonstration purpose.

tatively fit but a significant deviation can be observed. One assumption is that there is an error in the angle u between the tangent to the mica substrate placed on the bottom of the environmental box (Fig. 6, detail 9) and the cantilever which is glued on the holding part (Fig. 6, detail 10). This error can come from different sources: glueing of the cantilever on the holding part, manufacturing and assembly errors in the chain linking the holding part to the environmental box, alignment error between the bottom of the box and the mica substrate. This error cannot be measured easily in our setup, consequently we decided to study the sensitivity of the capillary force to this angle parameter u. To do so, the environmental box was tilted with respect to the reference horizontal orientation. The capillary force as a function of this tilt angle (i.e., variation of u) is plotted in Fig. 8.

5. Discussion

In the previous paragraphs, it was shown that our model could be used to compute the capillary force between two objects for usual shapes: spheres, cones, planes... and reproduce existing results. In addition, the three-dimensional capabilities of the model allow the user to compute capillary force for complex configurations with simple shapes or even with complex shapes (e.g., pyramids, rounded pyramids that can model the Berkovich AFM tips).

Results presented in Figs 7 and 8 show fair agreement between experimental and theoretical measurements. Different sources of measurement noise were identified out of which the main one was the surface modification. To overcome the scatter, we repeated the measurements to be able to have statistical results. Another method to limit the scatter could be to work with atmospheres of controlled composition, not just humidity control.

A very interesting result presented is that the force can be varied using tip tilting. This should allow the user to pick up and release a part by controlling the tilt angle of the tip with respect to the object.

To give the order of magnitude, a force of about 50 nN is sufficient to lift a cube of about 280 µm for a density of 2300 kg/m^3 (approximately the density of silicon). For objects with such masses, a tip should be able to pick up, manipulate, and then release them.

The manipulable part weights can also be extended using different tip shapes or using multiple tips.

6. Conclusions

This paper presents a three-dimensional model for the computation of the capillary force, allowing to compute the effects of capillary condensation, with configurations that are not mandatorily axisymmetrical. The model has been validated by comparing it with existing theoretical results. The model was also compared to experimental results using a dedicated test bench. The comparison has shown a fair correspondence even if the quantitative correlation is arguable. Experiments under controlled atmosphere could improve measurements.

It was shown here that the tilt angle of a tip with respect to a flat substrate is an important parameter to vary the capillary force between them. This result is promising for a new application in micromanipulation of components. Nevertheless, experiments are going on to determine how to make use of this effect with better repeatability.

Acknowledgement

This work has been funded by a grant from the *F.R.I.A. — Fonds pour la Formation à la recherche dans l'industrie et l'agriculture.*

References

1. D. Wu, N. Fang, C. Sun and X. Zhang, *Sensors Actuators A* **128**, 109–115 (2006).
2. C. H. Mastrangelo, *J. Microelectromechanical Syst.* **2**, 33–43 (1993).
3. J. N. Israelachvili, *Intermolecular and Surface Forces*, 2nd edn. Academic Press (1992).
4. T. Kondo, S. Juodkazis and H. Misawa, *Appl. Phys. A* **81**, 1583–1586 (2005).
5. P. Lambert and A. Delchambre, *Langmuir* **21**, 9537–9543 (2005).
6. P. Lambert, F. Seigneur, S. Koelemeijer and J. Jacot, *J. Micromech. Microeng.* **16**, 1267–1276 (2006).
7. T. Stifter, O. Marti and B. Bhushan, *Phys. Rev. B* **62**, 13667–13673 (2000).
8. F. M. Orr, L. E. Scriven and A. P. Rivas, *J. Fluid Mech.* **67**, 723–742 (1975).
9. A. Chau, S. Régnier, A. Delchambre and P. Lambert, in: *Proceedings of IEEE International Symposium on Assembly and Manufacturing*, pp. 215–220 (2007).
10. A. W. Adamson and A. P. Gast, *Physical Chemistry of Surfaces*, 6th edn. Wiley (1997).
11. M. P. de Boer and P. C. T. de Boer, *J. Colloid Interface Sci.* **311**, 171–185 (2007).
12. L. R. Fisher and J. N. Israelachvili, *Colloids Surfaces* **3**, 303–319 (1981).
13. M. M. Kohonen and H. K. Christenson, *Langmuir* **16**, 7285–7288 (2000).
14. J. Jang, M. A. Ratner and G. C. Schatz, *J. Phys. Chem. B* **110**, 659–662 (2005).
15. L. Sirghi, R. Szoszkiewicz and E. Riedo, *Langmuir* **22**, 1093–1098 (2005).
16. F. Restagno, L. Bocquet and T. Biben, *Phys. Rev. Lett.* **84**, 2433–2436 (2000).
17. K. Brakke, *Expl. Math.* **1**, 141–165 (1992).
18. A. de Lazzer, M. Dreyer and H. J. Rath, *Langmuir* **15**, 4551–4559 (1999).
19. O. H. Pakarinen, A. S. Foster, M. Paajanen, T. Kalinainen, J. Katainen, I. Makkonen, J. Lahtinen and R. M. Nieminen, *Model. Simul. Mater. Sci. Eng.* **13**, 1175–1186 (2005).

Odd–Even Effects in the Friction of Self-Assembled Monolayers of Phenyl-Terminated Alkanethiols in Contacts of Different Adhesion Strengths

Yutao Yang [a], Andrew C. Jamison [b], David Barriet [b], T. Randall Lee [b] and Marina Ruths [a,*]

[a] Department of Chemistry, University of Massachusetts Lowell, 1 University Avenue, Lowell, MA 01854, USA

[b] Department of Chemistry, University of Houston, 4800 Calhoun Road, Houston, TX 77204, USA

Abstract

We have studied the frictional properties of self-assembled monolayers (SAMs) of phenyl-terminated alkanethiols, $C_6H_5(CH_2)_nSH$ ($n = 13$–16) on template-stripped gold. The friction force was measured with atomic force microscopy (AFM), and the magnitude of the adhesion was controlled by immersing the sliding contact in ethanol (giving low adhesion) or dry N_2 gas (giving enhanced adhesion relative to ethanol). We observed a linear friction force as a function of load ($F = \mu L$) in the systems with low adhesion and a non-linear friction force when the adhesion was higher. The non-linear behavior in the adhesive systems appeared to be area-dependent ($F = S_c A$) and was compared to contact areas calculated using the extended Thin-Coating Contact Mechanics (TCCM) model. In ethanol, the coefficient of friction μ was found to be systematically higher for odd values of n (i.e., for the monolayers in which the terminal phenyl group was oriented closer to the surface normal).

Keywords

Phenyl-terminated alkanethiols, self-assembled monolayers (SAMs), friction, adhesion, odd–even, atomic force microscopy (AFM), Thin-Coating Contact Mechanics (TCCM) model

1. Introduction

Self-assembled monolayers (SAMs) are commonly used as model boundary lubricants that reduce friction and protect against wear of surfaces that are in close proximity (in contact or at separations of a few molecular diameters) and at high pressures. Among the properties that affect the frictional response of self-assembled systems are packing density [1–4], molecular chain length [5, 6] and rigidity [7], strength of anchoring to the underlying substrate [8], and end-group functionality

---

* To whom correspondence should be addressed. Tel.: (978) 934-3692; e-mail: marina_ruths@uml.edu

Adhesion Aspects in MEMS/NEMS

[6, 9], all of which have some effect on the lateral cohesion of the monolayer [3, 5, 6, 10, 11] and on the ease with which defects are formed during sliding [1, 3–6].

Monolayers containing aromatic moieties are of interest from a fundamental perspective because of their stronger and more complex intermolecular interactions compared to the better-known alkanethiol and alkylsilane monolayers [10, 11]. Aromatic compounds exhibit potentially useful electronic and optical properties [10], and their stiffness is of practical use for forming end-functionalized monolayers where the orientation of the end-group is unaffected by the gauche defects found in alkane-based systems [10, 11]. The orientation and close-packing in the aromatic systems are affected by the stiffness of the molecules, and it has been shown that even the introduction of a single $-CH_2-$ group between the aromatic moiety and the group anchored to the surface enables a better packing of the resulting monolayer [12, 13], and correspondingly diminished friction [8, 14–16].

The friction of monolayers containing aromatic groups is also of interest from a practical point of view. It is known that nitrogen-, oxygen-, and sulfur-containing aromatic and heteroaromatic molecules contribute to the natural lubricity of mineral-oil-based fuels [17–19], importantly, the hydrogenation process used to lower the aromatic content leads to an increase in friction. Furthermore, biodiesel is commonly blended with standard diesel fuel to improve its properties, and there is also a variety of aromatic and heteroaromatic friction-reducing additives [17–19]. Despite these important applications, there is remarkably limited information available on the molecular-level lubricating properties of aromatic compounds [7, 20]. To this end, we have examined simple aromatic monolayers [8, 14, 15] and a series of polyaromatic thiol-based monolayers [16] that are highly rigid and give a relatively high friction. In this work, we use atomic force microscopy (AFM) to study a series of phenyl-terminated alkanethiol self-assembled monolayers [21] with the same end-group functionality but higher molecular packing than the simple aromatic thiols studied previously.

Friction in single-asperity contacts is commonly observed to depend on the strength of adhesion [8, 16, 22–27], therefore the measurements were done in ethanol, where the adhesion (attraction due to van der Waals forces) is low, and in dry N_2 gas, where the adhesion is higher. As in several of these previous studies [16, 23], we observe here also different functional forms of the friction force F *versus* load (normal force) L in adhesive *versus* non-adhesive systems. In ethanol (low adhesion), the friction force is a linear function of load ($F = \mu L$, where μ is the coefficient of friction), whereas in dry N_2, it shows a non-linear load dependence that is generally associated with a dependence on the contact area ($F = S_c A$, where S_c is the critical shear stress and A the contact area). These non-linear data were compared to contact areas calculated with the extended Thin-Coating Contact Mechanics model recently developed by Reedy [28, 29]. We find an odd–even effect in the coefficient of friction μ as a function of the number of methylene units in the alkane chain, which correlates with the different orientations of the terminal phenyl group of the monolayers.

2. Materials and Methods

2.1. Self-Assembled Monolayers (SAMs)

Phenyl-terminated alkanethiols, $C_6H_5(CH_2)_nSH$, were synthesized as described in Ref. [21], where details on the characterization of the compounds with $n = 13$–15 are given. These adsorbates have been shown to form well-ordered SAMs on gold with a packing density similar to that of n-alkanethiols and a herringbone packing of the terminal groups [21]. The analytical data for the new adsorbate ($n = 16$) are provided here.

16-Phenylhexadecanethiol ($C_6H_5(CH_2)_{16}SH$). ^1H NMR (300 MHz, CDCl$_3$): δ 7.17–7.28 (m, 5 H), 2.60 (t, $J = 7.8$ Hz, 2 H), 2.52 (q, $J = 7.5$ Hz, 2 H), 1.56–1.63 (m, 4 H), 1.25–1.36 (m, 24 H). ^{13}C NMR (75.6 MHz, CDCl$_3$): δ 143.11, 128.54 (2 C), 128.35 (2 C), 125.68, 36.15, 34.22, 31.68, 29.82 (4 C), 29.79 (2 C), 29.75 (2 C), 29.68, 29.50, 29.24, 28.54, 24.82. Mass: $m/z = 334$ (molecular weight 334.6).

Ellipsometry was used to measure the thickness of the $C_6H_5(CH_2)_{16}SH$ monolayers formed on two slides made from Si(100) wafers, on which 10 nm chromium and 100 nm gold had been deposited. Measurements were done at three distinct points on each slide, and the average value, obtained using a refractive index of 1.45, is given in Table 1. Further technical details and the thicknesses of monolayers with $n = 13$–15 are given in Ref. [21] (cf. Table 1).

For the measurements of friction, SAMs were formed on flat substrates by immersing template-stripped gold on polystyrene backing [14, 30] in 0.4–1 mM solu-

Table 1.
Film thicknesses (T), contact angles (θ) and surface energies (γ_{YD}). Tip radii (R), and coefficients of friction (μ) measured in ethanol

n	T (nm)	θ_{adv} (°)	θ_{rec} (°)	γ_{YD} (mJ/m^2)[b]	Bare Si tip		SAM-covered tip	
					R (nm)	μ^c	R (nm)	μ^c
13	2.09[a]	93 ± 2	83 ± 3	35	132	0.25 ± 0.02	66	0.31 ± 0.01
					132	0.26 ± 0.03	88	0.33 ± 0.02
							88	0.30 ± 0.02
14	2.19[a]	92 ± 2	78 ± 2	36	132	0.14 ± 0.01	143	0.21 ± 0.01
					132	0.19 ± 0.01	143	0.23 ± 0.01
15	2.27[a]	93 ± 2	80 ± 2	34	132	0.30 ± 0.01	130	0.38 ± 0.01
					132	0.32 ± 0.01	130	0.42 ± 0.01
					137	0.28 ± 0.02	103	0.37 ± 0.04
					137	0.31 ± 0.06	103	0.34 ± 0.04
16	2.45	94 ± 1	81 ± 2	35	137	0.17 ± 0.01	81	0.32 ± 0.01
					137	0.16 ± 0.01	81	0.29 ± 0.01

[a] Reference [21], $\Delta T = 0.1$ nm.
[b] Equation (1), $\Delta\gamma_{YD} = 1$ mJ/m^2.
[c] Standard deviations from linear fits as in Fig. 2.

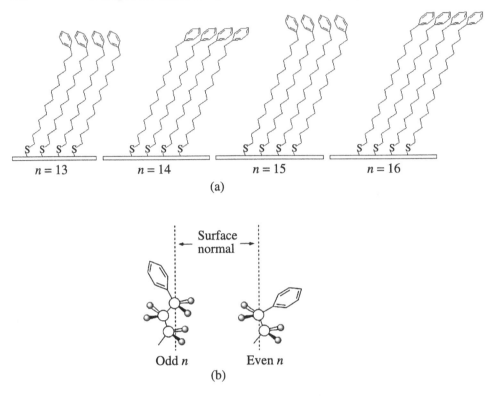

Figure 1. (a) Schematic illustration of the structure of SAMs on gold derived from phenyl-terminated alkanethiols, $C_6H_5(CH_2)_n SH$, $n = 13$–16. (b) Expected orientation of the terminal phenyl group at ca. 30° and 60° from the surface normal for odd and even values of n, respectively.

tion of the thiol in ethanol (Sigma-Aldrich, $\geqslant$99.5%) for 24–48 h. The samples were then removed from the solution, rinsed with ethanol, and blown dry with a stream of dry N_2 gas. The rms roughness of the template-stripped gold was 0.2–0.4 nm, measured over 1 µm². Monolayers were also formed on gold-covered AFM tips (see below). Polarization modulation infrared reflection-absorption spectroscopy of phenyl-terminated SAMs on gold has shown that there is a larger chain twist in monolayers with even n [21], and has confirmed the expected orientation of the terminal phenyl group [21, 31]. Schematic drawings of the self-assembled monolayer structures, including the orientation of the terminal phenyl group for odd and even values of n, are shown in Fig. 1.

2.2. Surface and Interfacial Energy

Advancing and receding contact angles of water were measured with a Krüss Drop Shape Analysis System 100. Typical drop volumes were 5–10 µl. The values in Table 1 are averages of 5–8 measurements on different positions on 3–5 different samples and are given with their standard error (standard deviation of the mean). The advancing contact angles agree well with previous data for $n = 13$–15 [21], where no discernible difference was seen between the different chain lengths. Using the Young–Dupré (YD) equation [32], $W = \gamma_{LV}(1+\cos\theta)$, where γ_{LV} is the surface tension of the liquid, and assuming that the work of adhesion, W, is composed of the dispersion and polar components of the surface energy of the solid and surface tension of the liquid (the Owens–Wendt approach) [33], $W = 2(\sqrt{\gamma_S^d \gamma_L^d} + \sqrt{\gamma_S^p \gamma_L^p})$, then one obtains

$$2(\sqrt{\gamma_S^d \gamma_L^d} + \sqrt{\gamma_S^p \gamma_L^p}) = \gamma_{LV}(1+\cos\theta). \tag{1}$$

Using equation (1), the surface energy of the solid surface (i.e., the self-assembled monolayer), $\gamma_{YD} = \gamma_S = \gamma_S^d + \gamma_S^p$, can be obtained from the advancing contact angles in Table 1 and Ref. [21] for liquids with different surface tensions $\gamma_{LV} = \gamma_L^d + \gamma_L^p$ (water: $\gamma_L^d = 21.8$ mJ/m^2, $\gamma_L^p = 51$ mJ/m^2; methylene iodide [33]: $\gamma_L^d = 48.5$ mJ/m^2, $\gamma_L^p = 2.3$ mJ/m^2; nitrobenzene [34]: $\gamma_L^d = 38.7$ mJ/m^2, $\gamma_L^p = 5.1$ mJ/m^2). The resulting values of γ_{YD} are listed in Table 1 ($\Delta\gamma_{YD} = 1$ mJ/m^2). In all four SAM systems, the polar contribution to the surface energy of the solid is quite low, $\gamma_S^p \approx 1$ mJ/m^2, i.e., the surface energy of the monolayers arises mainly from dispersion intermolecular interactions.

For comparison with the surface energies from the contact angle measurements and with interfacial energies obtained from our AFM experiments below, values can also be calculated using van der Waals–Lifshitz theory [32]. Some of our measurements of friction were performed by scanning the monolayers with bare (unfunctionalized) Si tips carrying a native oxide layer, and others with monolayer-functionalized gold-covered tips. The van der Waals interactions in the first system were approximated by an asymmetrical three-layer system, $\gamma_{vdW} = A_{132}/(24\pi D_0^2)$, where A_{132} is the Hamaker constant ($1 = $ monolayer, $3 = $ medium (N$_2$ or ethanol), $2 = $ amorphous SiO$_2$), and $D_0 = 0.165$ nm is the cut-off separation at contact [32]. The interfacial energies in systems with monolayer-functionalized tips were calculated for a symmetrical 5-layer system (gold/monolayer/medium/monolayer/gold, materials 1/2/3/2/1 in equation (2)), as $\gamma_{vdW} = -F_{vdW}(D_0)/(4\pi R)$, where F_{vdW} is the non-retarded van der Waals force between a sphere and a flat surface [32],

$$\frac{F_{vdW}}{R} = -\frac{1}{6}\left(\frac{A_{232}}{D^2} - 2\frac{\sqrt{A_{232} \times A_{121}}}{(D+T)^2} + \frac{A_{121}}{(D+2T)^2}\right), \tag{2}$$

where R is the radius of curvature (AFM tip radius), D is the separation distance (at contact, $D = D_0 = 0.165$ nm), T is the monolayer thickness (Table 1), A_{121} is the Hamaker constant for material 1 interacting across material 2, and A_{232} the Hamaker constant for material 2 interacting across material 3. The Hamaker constants [32] were calculated using a bulk refractive index and dielectric constant of the aromatic compounds estimated from data for compounds with a structure similar to the molecules in our SAMs, 11-phenylheneicosane (1-decylundecylbenzene) [35] and 1-phenylpentadecane [36] (r.i. $= 1.48$ [35, 36] and $\varepsilon = 2.2$), and bulk values for N_2 and ethanol. The Hamaker constant for gold interacting across a vacuum or in air ($A = 45.5 \times 10^{-20}$ J) [37] was used in the combining relation [32] $A_{121} = (\sqrt{A_{11}} - \sqrt{A_{22}})^2$. In the asymmetric systems (bare Si tip probing a monolayer), $\gamma_{vdW} = 31$ mJ/m^2 in dry N_2 gas and 2 mJ/m^2 in ethanol. In the symmetric systems (monolayer-covered tip probing monolayer), $\gamma_{vdW} = 32$ mJ/m^2 in N_2 and 3 mJ/m^2 in ethanol.

2.3. Friction Force Microscopy

The friction force, F, was measured over a scan length of 1 µm with atomic force microscopy (AFM) in lateral or friction mode, using a Multimode AFM with Nanoscope IIIa controller (Veeco). Raw data were collected as "friction loops" in Scope mode at different loads (normal force) L, and the measured voltages were converted to force as described in Refs [38] and [39]. The dependence of the friction force on scan rate was found to be weak (in the rate range 0.6–122 µm/s; not shown) and a scan rate of 2 µm/s was chosen for these experiments. The statistical error in F (standard deviation of the mean, from averaging the sliding portion of the friction loop) was ca. 0.2 nN at $F < 50$ nN, and 0.5 nN at higher F, and is not shown in Fig. 2 (and later in Fig. 4) since it is similar to the height of the symbols. The experiments were conducted with bare Si tips (CSC17, MikroMasch) carrying a native silicon oxide layer, and with gold-covered tips (CSC38/Cr-Au) functionalized with phenyl-terminated alkanethiol (the same compound as on the flat substrate). The normal and lateral spring constants of the rectangular cantilevers were determined from their resonance frequency [40, 41] and dimensions measured with scanning electron microscopy (JEOL-7401F), as described previously [16, 38, 39]. In these experiments, the normal spring constants were in the range 0.20–0.55 N/m and the lateral spring constants in the range 25–92 N/m. The tip radii, R, were determined by reverse imaging of a calibration sample (TGT01, MikroMasch) in two orthogonal directions. In this particular batch of tips, $R = 65$–200 nm, which was larger than the manufacturer's specification. This larger radius was advantageous for friction measurements on our self-assembled monolayers, since these typically showed a reversible transition (cf. Results) at pressures of ca. 1 GPa. With smaller tip radii this transition would be reached at very low loads, giving a more limited range of data. The uncertainty in the radius was $\Delta R = 3$ nm ($R < 100$ nm) or 5 nm ($R \geqslant 100$ nm) [16]. Experiments in ethanol were conducted in a fluid cell. For experiments in dry N_2, the AFM was enclosed in a home-made plastic chamber that was continuously purged with a slow stream of N_2 gas (Airgas, 99.5%). The humidity was monitored with a Vaisala DM70 dewpoint meter. A relative humidity of $\leqslant 0.7\%$ was reached after purging for 2 h.

2.4. Contact Mechanics

At low loads, the radius of the contact area between an AFM tip and a monolayer-functionalized substrate can be similar to the thickness of the monolayer. If the elastic modulus of the monolayer is significantly lower than that of the substrates,

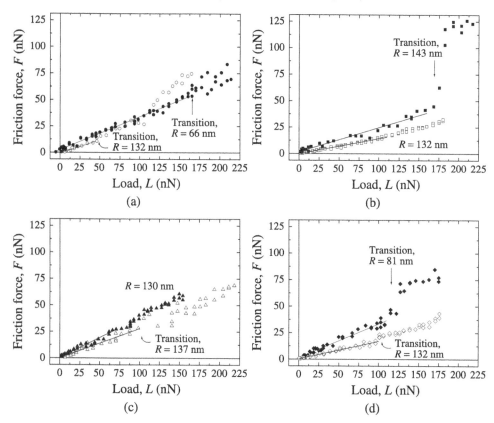

Figure 2. Friction force, F, as a function of applied load, L, measured in ethanol. Open symbols represent data obtained by scanning a monolayer-functionalized flat gold surface with a bare Si tip. Data obtained with monolayer-functionalized gold-covered tips are shown as filled symbols. (a) $n = 13$, $R_{Si} = 132$ nm, $R_{Au} = 66$ nm. (b) $n = 14$, $R_{Si} = 132$ nm, $R_{Au} = 143$ nm. (c) $n = 15$, $R_{Si} = 137$ nm, $R_{Au} = 130$ nm. (d) $n = 16$, $R_{Si} = 132$ nm, $R_{Au} = 81$ nm. The coefficients of friction μ obtained from linear fits to the low-load data are given in Table 1 and Fig. 3.

deformations at low loads will occur in the monolayer. At increased loads, the effective stiffness of such a layered system is still affected by the presence of the monolayer, and the analysis is more complicated than assumed in contact mechanics models for homogeneous elastic bodies, such as the Johnson–Kendall–Roberts (JKR) model [42], the Derjaguin–Muller–Toporov (DMT) model [43], or extensions of these [44]. Furthermore, in nanoscopic contacts, the pressure distribution should be affected by the atomic level structure of the substrate [45]. It is not yet established how well macroscopic models apply, although it has been shown that the influence of the substrate structure is reduced when a molecularly thin film is present in the contact [46].

In this work, we use the extended Thin-Coating Contact Mechanics (TCCM) model for the relationship between contact area and load. The details of this model [28, 29] and examples of its application to monolayer systems are shown elsewhere

[16, 47, 48], and only selected information needed for the discussion of the present work is shown here. In this model, the probe (spherical indenter) and the flat substrate are assumed to be rigid, which is a reasonable approximation at low load in our systems, since the Young's modulus of self-assembled monolayers is expected to be only a few GPa (cf. Discussion), but 78.5 GPa for gold [49], 70–80 GPa for SiO_2 [50], and 170 GPa for Si [38, 39].

The F vs. L data in our adhesive systems (Fig. 4) were compared to $F = S_c A$, where S_c is a constant, the critical shear stress, and A is the contact area at a given load L. The relationship between L, the radius of the contact area, a, and the work of adhesion ($W = 2\gamma$) in the extended TCCM model is given in non-dimensional form by [29]

$$\overline{L} = \frac{\pi}{4}\bar{a}^4 - \zeta^{1/2}\pi\bar{a}^2(2\overline{W})^{1/2} - 2\pi\overline{W}(1 - \zeta), \tag{3}$$

where $\overline{L} = L/(E_u Rh)$, $\bar{a} = a/(\sqrt{Rh})$, and $\overline{W} = W/(E_u h)$. The uniaxial strain modulus is $E_u = E(1 - v)/[(1 + v)(1 - 2v)]$, where E is Young's modulus (in this case, 0.5 GPa, see Discussion) and v is Poisson's ratio (0.4). The film thickness, h, is the thickness of one monolayer, $h = T$, in the case of a bare Si tip scanning a monolayer on the flat substrate, or two monolayers in contact, $h = 2T$, in experiments with monolayer-functionalized tips (cf. Tables 1 and 2). $\zeta = 2Wh/E_u\delta_c^2$ is a transition parameter ($0 \leqslant \zeta \leqslant 1$), a measure of the ratio of elastic deformation to the effective range of the surface forces. $\zeta = 0$ and $\zeta = 1$ correspond to limits where the range of adhesion is large and small compared to the elastic deformations (i.e., the DMT and JKR-like limits, respectively). The critical separation, δ_c, was assumed [29] to be 1 nm, as in a previous application of this model to monolayer systems [16].

In these experiments, the uncertainty in S_c mainly arises from propagation of the uncertainties in R, h, E_u, and W in the calculation of A [16]. The uncertainty in S_c can be calculated by differentiating equation (3) and using $\Delta R = 3$ nm ($R < 100$ nm) or 5 nm ($R \geqslant 100$ nm), $\Delta h = 0.1$ nm (monolayer on only the flat gold substrate, $h = T$) or 0.2 nm (monolayers on both surfaces, $h = 2T$), $\Delta E_u = 0.2$ GPa (ca. 20%), and $\Delta W = 0.002$ J/m². Following the procedure described in Ref. [16], we estimate that the relative uncertainty in S_c in the current experiments is 20%. Similarly, by differentiating the equation for ζ, its uncertainty is found to be 20%.

3. Results

3.1. Friction in Ethanol

The friction force F between monolayers on flat substrates and bare Si tips or monolayer-functionalized gold-covered tips was measured in ethanol. Representative results are shown as a function of load, L, in Fig. 2. In all systems, F increased linearly with L until a transition or plateau regime was reached, after which F was more scattered and also less reproducible from experiment to experiment. The on-

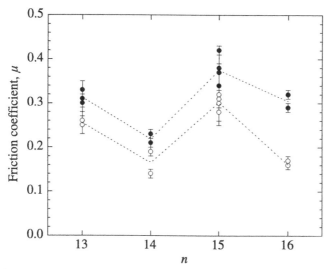

Figure 3. Coefficients of friction measured in ethanol for different n (cf. Fig. 2). Open and filled symbols indicate values obtained with bare Si tips and monolayer-functionalized tips, respectively. The error bars are the standard deviations from linear fits like those in Fig. 2. The dotted lines are drawn as guides.

set of this transition or plateau, which will be discussed in detail below, is indicated with an arrow in the cases where it could be clearly identified. In our analysis, we focus on the reproducible friction at low loads, below the transition. Experiments in addition to the ones in Fig. 2 were performed with tips having different radii and on monolayers prepared under identical conditions but on different occasions. The coefficients of friction μ obtained from linear fits ($F = \mu L$) to the low-load data in all these experiments are given in Table 1 and Fig. 3. The coefficients of friction were found to vary with the number of methylene units in the alkane chain, n, as shown in Fig. 3, and were independent of the tip radius, R, in the cases where tips with different R were used. The values of μ were significantly lower than those of simple aromatic and polyaromatic monolayers (thiophenol, $\mu = 1.2$–1.4; phenyl-thiophenol, $\mu = 0.9$–1.1 and terphenylthiol, $\mu = 0.6$) [16], but higher than those of a close-packed CH_3-terminated alkanethiol monolayer under similar conditions ($\mu = 0.02$–0.1, cf. Ref. [5], and discussion in Refs [14] and [15]). The coefficient of friction, μ, was higher with the monolayer-covered tips, which is different from what is observed for CH_3-terminated monolayers, where two confined monolayers generally give lower friction [1]. Our experiments in N_2 (see below) gave the expected result: higher S_c with a monolayer on only the flat surface, probed by a bare Si tip. Measurements of the friction of octadecanethiol SAMs in ethanol and N_2 (not shown) indicate that in that system, μ and S_c are higher when measured with a bare Si tip than with a monolayer-covered tip. As a whole, these observations are consistent with a model in which the enhanced friction for the monolayer-covered tips compared to that of the bare Si tips arises from attractive interfilm π–π interactions across the interface.

3.2. Friction in Dry N₂ Gas

Measurements of friction were also conducted under dry conditions (r.h. $\leqslant 0.7\%$), with the sliding contact immersed in dry N_2 gas. Figure 4 shows experimental data obtained with bare Si tips and monolayer-functionalized gold-covered tips. In several cases, the same tips were used as in the experiments in ethanol. No tip wear or monolayer damage was observed (cf. Discussion). In the experiments in N_2, where the adhesion was larger than in ethanol (the pull-off forces were larger, cf. Fig. 4, as expected from higher F_{vdw}/R, cf. Materials and Methods), F was not a linear function of L at the lowest loads. At higher loads, we observed transition regimes as

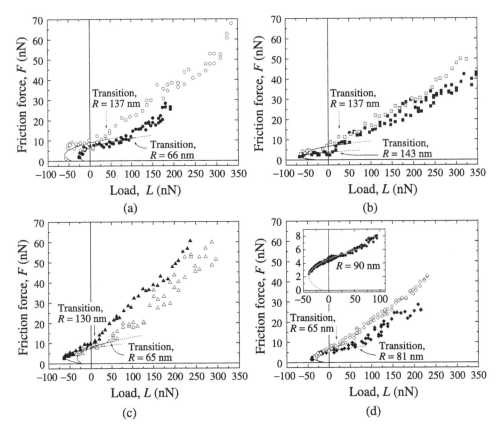

Figure 4. Friction force F vs. applied load L, measured in dry N_2 gas. Open symbols show data obtained using bare Si tips, and filled symbols show measurements with monolayer-functionalized gold-covered tips. (a) $n = 13, R_{Si} = 137$ nm, $R_{Au} = 66$ nm. (b) $n = 14, R_{Si} = 137$ nm, $R_{Au} = 143$ nm. (c) $n = 15, R_{Si} = 65$ nm, $R_{Au} = 130$ nm. (d) $n = 16, R_{Si} = 65$ nm, $R_{Au} = 81$ nm. Insert in panel (d): $n = 16$, monolayer on tip with $R_{Au} = 90$ nm, where L was maintained below the threshold for the monolayer transition. The solid curves represent $S_c A$, where the contact area A was calculated using the extended TCCM model with $E = 0.5$ GPa and $\nu = 0.4$. The curves are intended as comparisons to the low-load data only, below the transition regime. The values of S_c are provided in Table 2 and Fig. 5.

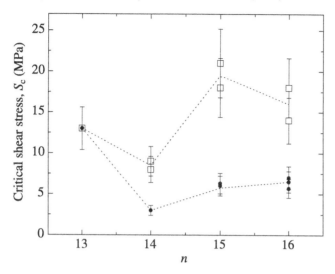

Figure 5. Critical shear stresses S_c obtained in dry N_2 gas for different n (cf. Fig. 4). Open and filled symbols show values obtained with bare Si tips and monolayer-functionalized tips, respectively. The error bars indicate a relative uncertainty of 20% (see text). The dotted lines are drawn as guides.

in the ethanol experiments, and above this regime F was again more scattered, less reproducible, and its functional form varied from experiment to experiment. The reproducible, non-linear data at low L are generally interpreted as area-dependent friction, since it is observed to depend on R.

The solid curves in Fig. 4 are $F = S_c A$, where $A = \pi a^2$ as a function of L was calculated using the extended TCCM model (equation (3)) with the parameters given in Table 2 and $E = 0.5$ GPa ($E_u = 1.1$ GPa). In the TCCM calculation, $h = T$ in the case of a single monolayer in contact with a bare Si tip and $h = 2T$ for two contacting monolayers. It should be noted that in cases where similar F values were measured with a bare Si tip and a monolayer-covered tip (cf. Fig. 4(b) and 4(d)), a larger S_c is, as expected, obtained for the single monolayer (Si tip), since the contact area in such a system is smaller (for given L, R and W). In Fig. 5, S_c is seen to vary with n for $n = 13$–15, whereas the result for $n = 16$ was similar to that of $n = 15$. In each system, the interfacial energy γ_{TCCM} ($= W/2$, $\Delta W = 0.002$ J/m^2) in Table 2 was in good agreement with γ_{vdW} calculated from bulk dielectric properties. The value of γ_{TCCM} for monolayer–monolayer contact (Table 2) can also be compared to the value of γ_{YD} from the contact angle measurements (Table 1), where a good agreement is found.

4. Discussion

In the following paragraphs, we discuss various aspects of the experimental data and details of the data analysis used to extract surface and interfacial energies and parameters describing the friction in different systems. The results are contrasted with those obtained in previous experiments on alkanethiols [9, 51–54] and poly-

Table 2.
Tip radii (R), critical shear stresses (S_c) at $E = 0.5$ GPa, transition parameter (ζ), and interfacial energy (γ_{TCCM}) measured in dry N_2

n	Bare Si tip ($h = T$)				SAM-covered tip ($h = 2T$)			
	R (nm)	S_c (MPa)[a]	ζ[a]	γ_{TCCM} (mJ/m^2)[b]	R (nm)	S_c (MPa)[a]	ζ[a]	γ_{TCCM} (mJ/m^2)[b]
13	137	13	0.27	35	66	13	0.50	32
14	137	8.0	0.29	35	143	3.0	0.61	38
	137	9.0	0.31	38				
15	65	21	0.34	40	130	6.3	0.66	39
	65	18	0.34	40	130	6.0	0.59	35
16	65	18	0.34	38	90	5.7	0.64	35
	65	14	0.34	38	81	7.0	0.73	40
					200	6.5	0.69	38

[a] Uncertainty 20%.
[b] $\Delta\gamma_{TCCM} = 1$ mJ/m^2.

aromatic thiol [16] self-assembled monolayers having different packing densities and rigidities.

4.1. Contact Angles and Interfacial Energy

The advancing contact angle of water (Table 1) showed no dependence on n, which was consistent with a previous study where slight differences were seen only with contacting liquids that were less polar than water [21]. The contact angles observed here were lower than those on monolayers derived from methyl-terminated alkanethiols, which is expected for the more polarizable phenyl groups, but higher than those on simple aromatic and polyaromatic thiols. Thiophenol, phenylthiolphenol and terphenyl thiol show advancing contact angles of 86–88°, and also have larger contact angle hysteresis with receding contact angles of 72–75° [16]. The values for the phenyl-terminated alkanethiols are consistent with their higher packing density (similar to the packing density of alkanethiols on gold) [21] compared to the abovementioned aromatic monolayers (where thiophenol, phenylthiolphenol and terphenyl thiol have molecular areas [16] of 0.4–0.7, 0.33 and 0.22 nm^2, respectively).

On close-packed aromatic structures (anthracene and naphthalene crystals), the highest contact angles of water, 94–95°, are found on crystal planes where the aromatic rings are oriented edge-on [55]. Our values are slightly lower, which is consistent with the structure in Fig. 1, where both odd and even n phenyl groups are oriented so that the face of the aromatic ring is, to some extent, available to the water. Systematic differences in contact angle observed with other contacting liquids [21] suggested that there was a slight difference in the exposure of the phenyl group for odd *vs.* even n.

The values of γ_{YD} (Table 1) are in good agreement with γ_{vdW} for the symmetric (monolayer-covered tip) system (32 mJ/m^2, cf. Materials and Methods), and with literature data on phenyldodecanethiol ($n = 12$), which showed that the monolayer surface energy was due purely to dispersion interactions with $\gamma^d = 34 \pm 2$ mJ/m^2 [31]. The data in Table 2 show that there is also a good agreement with the values of γ_{TCCM} ($= W/2$). Any differences in surface energy for odd and even values of n were too small to be detected with these methods and are thus probably not responsible for the differences in friction. In particular, it is unlikely that small differences in interfacial energy give rise to the distinct dependence on n of the friction in ethanol, where the adhesion was intentionally reduced. Instead, the odd–even effect on the friction might arise from the different orientations of the phenyl groups at the surface (shown schematically in Fig. 1), which will be discussed in detail below.

4.2. Monolayer Transition

In most of the data in Figs 2 and 4, a plateau, dip, or step in the F vs. L data was observed at high load (marked with an arrow in the figures). Similar transition regimes, ascribed to a reversible displacement of the molecules, have been observed for n-alkanethiol monolayers at pressures around 1 GPa [51]. They have also been seen for aromatic thiols and silanes [8, 14–16], and physisorbed fatty acids [23]. For a given monolayer, they occur at similar pressures in adhesive and non-adhesive contacts [15, 23]. The transition is reversible, i.e., as the load is decreased, the lower load regime is recovered. The data above the transition are typically more scattered and their functional form typically varies from experiment to experiment, which was not the case for the low-load data. The friction above the transition was not identical to that of pure gold (not shown), suggesting that the thiol molecules remained in the contact. In the current systems, the onset of the plateau was often gradual, making an accurate determination of a transition pressure difficult, but it was around 1 GPa as in the abovementioned systems. In a few cases, the data did not show a clear plateau (open symbols in Fig. 2(b), and filled symbols in Fig. 2(c), where the fit was limited to pressures below ca. 1 GPa), or the critical pressure was not reached (insert in Fig. 4(d)).

4.3. Friction in Non-adhesive and Adhesive Contacts

The frictional responses in ethanol and dry N$_2$ were clearly different. In ethanol (Fig. 2), we observed a linear dependence on L, with F approaching zero as L was reduced to zero, and with no dependence on R in the cases where several radii were investigated ($n = 13$ and 16, Table 1). The pull-off forces measured in normal force vs. separation curves (not shown) were no more than a few nN, which corresponds to an interfacial energy slightly lower than the calculated value of γ_{vdW} in ethanol (2–3 mJ/m^2, cf. Materials and Methods), but somewhat higher than the force expected at a separation of the average cross-sectional diameter [56] of an ethanol molecule ($D = 0.44$ nm). This observation, which also has been made for polyaromatic SAMs [16], suggests that at the closest separation, a full monolayer of ethanol is not present between the tip and the sample. No layering of the ethanol was observed in the force *vs.* distance curves, which is consistent with the known reduction or absence of layering of solvents between amorphous or slightly rough monolayers [22]. The negative load of only a few nN at pull-off was not always discernible in the *F vs. L* data, since the surfaces tended to separate during scanning at the very lowest load so that no friction data could be obtained. This type of frictional response is commonly interpreted as load-dependent, and similar behavior has been observed in other systems: A linear trend in friction has been observed in single-asperity contact between mica surfaces due to repulsive hydration forces in aqueous electrolyte solution [57] and in self-assembled monolayer and polymer systems under conditions where the adhesion between these surfaces was low [8, 14–16, 22, 23, 26, 27].

In dry N_2, where the adhesion was higher, F was non-linear and depended on R (in the cases where tips with different R were used), which is commonly interpreted as an area-dependence. The stronger adhesion of the surfaces across N_2 arises from the stronger van der Waals attraction across N_2 gas than across ethanol, as shown by the calculation of γ_{vdW} in Materials and Methods. To evaluate the adhesion in terms of pull-off forces (as done above for the case of ethanol), we can consider the lowest load value in each data set in Fig. 4. For a given R, the pull-off forces agree well with those expected based on the calculated van der Waals interactions, which are 10–15 times stronger in N_2 than in ethanol. The different results in ethanol and N_2 do not arise from tip or monolayer damage since the different frictional responses (linear *vs.* non-linear) can be obtained with the same tip when switching from one environment (ethanol *vs.* N_2) and back. A non-linear increase in F with increasing L has been observed previously for adhering, unfunctionalized surfaces [22, 25, 58], and for adhering, self-assembled monolayers [16, 22, 23].

Many models for the dependence of the friction force on load and on contact area have been suggested based on empirical observations. It has been proposed that a dilation of the surfaces is necessary for sliding to occur, and that the external load and the adhesion forces contribute separately to the friction in such systems [22, 57]. In the absence of adhesion, the surfaces need to separate only against the external load (with no influence from the size of the contact area), whereas in adhesive systems there is an additional contribution from the interfacial energy (which acts over the real contact area). Depending on the strength of adhesion and the load regime, one of these contributions might dominate over the other. This situation is commonly expressed as $F = \mu L + S_c A$ (where A is a sublinear function of L) [22]. In one very simple model, the adhesion (interfacial energy) is incorporated into S_c only, which implies that if the interfacial energy were reduced, the second term would be strongly reduced or vanish, and only the linear load term ($F = \mu L$) would remain. In such cases, the friction would not depend on R, and data with different probe sizes could be directly compared with one another. The data in Fig. 2 (cf. values of μ in Table 1) are consistent with this model, which has been demonstrated in other systems with probe sizes differing by 5–6 orders of magnitude [8]. Other models suggest that the friction force always depends on the contact area, and that a linear dependence on L arises from a non-constant (pressure-dependent) shear stress. Changes in the shear stress with pressure are certainly possible, especially considering the monolayer transition observed at higher loads. However, the linear F vs. L behavior has been observed in a wide variety of non-adhesive systems over wide ranges of loads and pressures, and it is unlikely that all of these systems can be rationalized by a pressure-dependent shear stress. In our adhesive systems, S_c appeared to be constant in the investigated range of loads (see Table 2), which has also been observed in other systems [16].

Differences between adhesive and non-adhesive systems have also been demonstrated in computer simulations, although not as two separate, additive terms. A linear dependence of F on L has been seen in molecular dynamics simulations of lubricated contacts (n-hexadecane between slightly rough gold surfaces), with a different slope and a shift along the L axis toward lower L as adhesion was introduced [24]. Recent molecular dynamics simulations of dry (unlubricated) contacts with atomic scale roughness showed sublinear and linear F vs. L-curves with and without adhesion, respectively [25].

We observed systematic differences in the friction with one *versus* two monolayers confined in the contact (i.e., with bare Si tips *vs.* monolayer-functionalized tips). In N_2, the systems with one monolayer showed a higher S_c (i.e., a higher friction), which might be expected if the presence of a monolayer on only one of the surfaces leads to a less well lubricated contact than when there is a monolayer on both surfaces. However, the opposite trend was observed in μ obtained from the measurements in ethanol. This different response in ethanol does not appear to arise from a contact area dependence, since the same proportionality between μ from experiments using bare or functionalized tips was observed irrespective of whether the radius of the monolayer-covered tip was smaller, larger, or approximately equal to the radius of the bare Si tip (cf. Table 2).

4.4. Monolayer Modulus

In experiments on polyaromatic thiol monolayers [16], the friction data in adhesive contact in N_2 could only be replicated (as $F = S_c A$) with areas from the extended TCCM model if a high Young's modulus was chosen, $E \geqslant 7$ GPa, ($\nu = 0.4, \zeta = 0.01$–0.02). Similar data on fatty acid monolayers required a much lower modulus, $E \leqslant 0.7$ GPa ($\nu = 0.4, \zeta > 0.1$) [16]. Moduli chosen outside these ranges gave curves with a rise too high or low compared with the experimental data, or did not reproduce the pull-off region (data at lowest L) well. Following a similar approach, the current data (Fig. 4) were best approximated with a Young's modulus in a narrow range of $E = 0.5 \pm 0.1$ GPa ($E_u = 1.1 \pm 0.2$ GPa).

A wide range of monolayer moduli can be found in the literature. Experiments using AFM to measure local compliance have suggested Young's moduli of $E = 0.2$–0.4 GPa for close-packed Langmuir–Blodgett fatty acid monolayers [59], and measurements of thickness changes during compression gave 1–5 GPa [60].

AFM experiments on the viscoelastic properties of close-packed alkanethiol mono-layers have suggested a Young's modulus of 2 GPa [61]. Computer simulations of alkanethiol monolayers have indicated moduli around 20 GPa [62] and 36 GPa [49, 63], possibly representing ideal systems with few defects. A uniaxial strain modulus of $E_u = 3$ GPa ($E \approx 1.4$ GPa) has been found in molecular dynamics simulations of an alkylsilane monolayer compressed by a flat plate [64] and used successfully to compare calculated contact areas to molecular dynamics simulations of the contact between an AFM tip and the alkylsilane monolayer [48].

A Young's modulus of $E = 0.5$ GPa in our systems is quite similar to the exper-imental values reported for fatty acids, alkanethiols and alkylsilanes [16, 59–61], but differs clearly from the values for polyaromatic monolayers without methylene groups. These observations suggest that the deformations in the current, phenyl-terminated alkanethiol monolayers occur mainly in the alkane portion of the films. Furthermore, the overall lower friction compared to simple aromatic and polyaro-matic thiol monolayers [14–16] is likely due not only to the higher packing in the current systems, but also to the possibility of facile reorientation of the terminal groups that are attached to flexible alkane chains.

4.5. Odd–Even Effects on Friction

The number of methylene units in alkanethiol monolayers or in spacers between functional end-groups and substrates has been shown to affect a number of prop-erties such as the orientation of the terminal methyl group, the advancing contact angle of water and other liquids [52, 65–67], and the nanoscopic friction [9, 21, 52–54, 68, 69]. Interesting odd–even effects on packing density and molecular ori-entation have also been seen in biphenylalkanethiol [70, 71] and terphenylalkane-thiol monolayers [13, 72], but their friction has not been studied.

Experimental work on alkanethiol SAMs on gold has suggested a higher fric-tion with an even number of methylene units [53, 54]. Molecular dynamics sim-ulations [68] of monolayers consisting of close-packed, methyl-terminated alkane chains have shown that compression and sliding induced significant conformational changes in the uppermost part of monolayers where the terminal CH_3- group was initially oriented close to the surface normal. In monolayers obtained by adding a methylene unit to this alkane chain, so that the terminal group was oriented away from the surface normal, compression and sliding induced mainly a slightly larger tilt of the terminal CH_3- group, without extensive defect formation, and this system consistently showed lower friction [68].

A possible rationalization of the odd–even effect that we observe for μ in ethanol is consistent with the computer simulation. Specifically, for even n the phenyl groups are more tilted from the surface normal than for odd n; consequently, less defect formation and energy dissipation can plausibly occur for even n during com-pression and sliding.

5. Summary

We have studied the friction of a series of self-assembled, phenyl-terminated alkanethiol monolayers. The adhesion between the monolayer-covered flat sample and the AFM tip was controlled by immersing the contact in ethanol or in dry N_2 gas. In ethanol, we observed linear friction ($F = \mu L$) that showed no dependence on the tip radius, R. In dry N_2, where the adhesion was stronger, F depended on R. The apparent area-dependence of F in this case was analyzed by comparing the experimental data to $F = S_c A$, where A was the contact area calculated according to the extended TCCM model and S_c was the critical shear stress, which was a constant for each system within the experimental uncertainty. The functional form of the friction data was well described by areas calculated using a Young's modulus of $E = 0.5$ GPa, which is lower than the modulus needed to describe similar data obtained on polyaromatic thiol monolayers. A systematic dependence of the coefficient of friction μ in ethanol on the number of methylene units in the alkane chain was found, with higher values obtained for odd n, where the terminal phenyl group is oriented closer to the surface normal than in films with even n.

Acknowledgements

We thank S. Lee for helpful discussions, J. Mead for access to the contact angle goniometer, T. Petterson for software for analysis of AFM data, and J. Whitten for advice on gold evaporation. Acknowledgment is made to the Donors of the American Chemical Society Petroleum Research Fund for support of this research through Grant #45101-G5 (MR). This work was also supported through NSF CAREER Award #NSF-CMMI 0645065 and by start-up funding (MR) through the NSF-funded Center for High-Rate Nanomanufacturing (CHN) (Award #NSF-0425826). The Robert A. Welch Foundation (grant E-1320) provided generous support for the work at the University of Houston.

References

1. E. E. Flater, W. R. Ashurst and R. W. Carpick, *Langmuir* **23**, 9242 (2007).
2. P. T. Mikulski and J. A. Harrison, *J. Am. Chem. Soc.* **123**, 6873 (2001).
3. M. Salmeron, *Tribology Lett.* **10**, 69 (2001).
4. S. Lee, Y.-S. Shon, R. Colorado Jr, R. L. Guenard, T. R. Lee and S. S. Perry, *Langmuir* **16**, 2220 (2000).
5. A. Lio, D. H. Charych and M. Salmeron, *J. Phys. Chem. B* **101**, 3800 (1997).
6. N. J. Brewer, B. D. Beake and G. J. Leggett, *Langmuir* **17**, 1970 (2001).
7. B. Bhushan and H. Liu, *Phys. Rev. B* **63**, 245412 (2001).
8. M. Ruths, N. A. Alcantar and J. N. Israelachvili, *J. Phys. Chem. B* **107**, 11149 (2003).
9. H. I. Kim and J. E. Houston, *J. Am. Chem. Soc.* **122**, 12045 (2000).
10. A. Ulman, *Acc. Chem. Res.* **34**, 855 (2001).
11. J. F. Kang, A. Ulman, S. Liao, R. Jordan, G. Yang and G.-y. Liu, *Langmuir* **17**, 95 (2001).
12. Y.-T. Tao, C.-C. Wu, J.-Y. Eu, W.-L. Lin, K.-C. Wu and C.-h. Chen, *Langmuir* **13**, 4018 (1977).

13. A. Shaporenko, M. Brunnbauer, A. Terfort, M. Grunze and M. Zharnikov, *J. Phys. Chem. B* **108**, 14462 (2004).
14. M. Ruths, *Langmuir* **19**, 6788 (2003).
15. M. Ruths, *J. Phys. Chem. B* **110**, 2209 (2006).
16. Y. Yang and M. Ruths, *Langmuir* **25**, 12151 (2009).
17. E. S. Forbes, *Wear* **15**, 87 (1970).
18. D. Wei and H. A. Spikes, *Wear* **111**, 217 (1986).
19. D. F. Heenan, K. R. Januszkiewicz and H. H. Sulek, *Wear* **123**, 257 (1988).
20. M. Nakano, T. Ishida, T. Numata, Y. Ando and S. Sasaki, *Jpn. J. Appl. Phys.* **43**, 4619 (2004).
21. S. Lee, A. Puck, M. Graupe, R. Colorado Jr, Y.-S. Shon, T. R. Lee and S. S. Perry, *Langmuir* **17**, 7364 (2001).
22. M. Ruths and J. N. Israelachvili, in: *Springer Handbook of Nanotechnology*, 2nd edn, B. Bhushan (Ed.), Chapter 30, pp. 859–924. Springer-Verlag, Berlin & Heidelberg (2007).
23. M. Ruths, S. Lundgren, K. Danerlöv and K. Persson, *Langmuir* **24**, 1509 (2008).
24. J. Gao, W. D. Luedtke, D. Gourdon, M. Ruths, J. N. Israelachvili and U. Landman, *J. Phys. Chem. B* **108**, 3410 (2004).
25. Y. Mo, K. T. Turner and I. Szlufarska, *Nature* **457**, 1116 (2009).
26. C. R. Hurley and G. J. Leggett, *Langmuir* **22**, 4179 (2006).
27. T. J. Colburn and G. J. Leggett, *Langmuir* **23**, 4959 (2009).
28. E. D. Reedy Jr, *J. Mater. Res.* **21**, 2660 (2006).
29. E. D. Reedy Jr, *J. Mater. Res.* **22**, 2617 (2007).
30. D. Valtakari, *MSc Thesis*, Department of Physical Chemistry, Åbo Akademi University, Åbo (Turku), Finland (2002).
31. F. Cavadas and M. R. Anderson, *J. Colloid Interface Sci.* **274**, 365 (2004).
32. J. N. Israelachvili, *Intermolecular and Surface Forces*. Academic Press, London (1991).
33. H. Y. Erbil, *Surface Chemistry of Solid and Liquid Interfaces*. Blackwell, Oxford (2006).
34. F. M. Fowkes, F. L. Riddle Jr, W. E. Pastore and A. A. Weber, *Colloids Surfaces* **43**, 367 (1990).
35. F. C. Whitmore, J. N. Cosby, W. S. Sloatman and D. G. Clarke, *J. Am. Chem. Soc.* **64**, 1801 (1942).
36. D. R. Lide (Ed.), *CRC Handbook of Chemistry and Physics*, 89th edn. CRC Press, Boca Raton, FL (2008–2009).
37. J. Visser, *Adv. Colloid Interface Sci.* **3**, 331 (1972).
38. Y. Liu, T. Wu and D. F. Evans, *Langmuir* **10**, 2241 (1994).
39. Y. Liu, D. F. Evans, Q. Song and D. W. Grainger, *Langmuir* **12**, 1235 (1996).
40. J. E. Sader, J. W. M. Chon and P. Mulvaney, *Rev. Sci. Instrum.* **70**, 3967 (1999).
41. C. P. Green, H. Lioe, J. P. Cleveland, R. Proksch, P. Mulvaney and J. E. Sader, *Rev. Sci. Instrum.* **75**, 1988 (2004).
42. K. L. Johnson, K. Kendall and A. D. Roberts, *Proc. R. Soc. London Ser. A* **324**, 301 (1971).
43. B. V. Derjaguin, V. M. Muller and Y. P. Toporov, *J. Colloid Interface Sci.* **53**, 314 (1975).
44. R. W. Carpick, D. F. Ogletree and M. Salmeron, *J. Colloid Interface Sci.* **211**, 395 (1999).
45. B. Luan and M. O. Robbins, *Nature* **435**, 929 (2005).
46. S. Cheng, B. Luan and M. O. Robbins, *Phys. Rev. E* **81**, 016102 (2010).
47. E. D. Reedy Jr, M. J. Starr, R. E. Jones, E. E. Flater and R. W. Carpick, in: *Proceedings of the 28th Annual Meeting of the Adhesion Society*, Mobile, AL, pp. 366–368 (2005).
48. M. Chandross, C. D. Lorenz, M. J. Stevens and G. S. Grest, *Langmuir* **24**, 1240 (2008).
49. A. Lio, C. Morant, D. F. Ogletree and M. Salmeron, *J. Phys. Chem. B* **101**, 4767 (1997).
50. R. G. Munro, *Elastic Moduli Data for Polycrystalline Oxide Ceramics*. NISTIR 6853, National Institute of Standards and Technology, Gaithersburg, MD (2002).

51. G.-y. Liu and M. Salmeron, *Langmuir* **10**, 367 (1994).
52. F. Tao and S. L. Bernasek, *Chem. Rev.* **107**, 1408 (2007).
53. S. S. Wong, H. Takano and M. D. Porter, *Anal. Chem.* **70**, 5209 (1998).
54. S. S. Perry, S. Lee, T. R. Lee, M. Graupe, A. Puck, R. Colorado Jr and I. Wenzl, *Polymer Preprints* **41**, 1456 (2000).
55. H. W. Fox, E. F. Hare and W. A. Zisman, *J. Colloid Sci.* **8**, 194 (1953).
56. S. Löning, C. Horst and U. Hoffmann, *Chem. Eng. Technol.* **24**, 242 (2001).
57. A. Berman, C. Drummond and J. Israelachvili, *Tribology Lett.* **4**, 95 (1998).
58. R. W. Carpick, N. Agrait, D. F. Ogletree and M. Salmeron, *Langmuir* **12**, 3334 (1996).
59. R. M. Overney, E. Meyer, J. Frommer, H.-J. Güntherodt, M. Fujihira, H. Takano and Y. Gotoh, *Langmuir* **10**, 1281 (1994).
60. V. V. Tsukruk, V. N. Bliznyuk, J. Hazel, D. Visser and M. P. Everson, *Langmuir* **12**, 4840 (1996).
61. M. Salmeron, G. Neubauer, A. Folch, M. Tomitori, D. F. Ogletree and P. Sautet, *Langmuir* **9**, 3600 (1993).
62. Y. Leng and S. Jiang, *J. Chem. Phys.* **113**, 8800 (2000).
63. R. Henda, M. Grunze and A. J. Pertsin, *Tribology Lett.* **5**, 191 (1998).
64. M. Chandross, personal communication (2008).
65. M. M. Walczak, C. Chung, S. M. Stole, C. A. Widrig and M. D. Porter, *J. Am. Chem. Soc.* **113**, 2370 (1991).
66. P. E. Laibinis, G. M. Whitesides, D. L. Allara, Y. T. Tao, A. N. Parikh and R. G. Nuzzo, *J. Am. Chem. Soc.* **113**, 7152 (1991).
67. N. Nishi, D. Hobara, M. Yakamoto and T. Kakiuchi, *J. Chem. Phys.* **118**, 1904 (2003).
68. P. T. Mikulski, L. A. Herman and J. A. Harrison, *Langmuir* **21**, 12197 (2005).
69. S. Lee, Y.-S. Shon, T. R. Lee and S. S. Perry, *Thin Solid Films* **358**, 152 (2000).
70. H.-T. Rong, S. Frey, Y.-J. Yang, M. Zharnikov, M. Buck, M. Wühn, Ch. Wöll and G. Helmchen, *Langmuir* **17**, 1582 (2001).
71. G. Heimel, L. Romaner, J.-L. Brédas and E. Zojer, *Langmuir* **24**, 474 (2008).
72. W. Azzam, A. Bashir, A. Terfort, T. Strunskus and Ch. Wöll, *Langmuir* **22**, 3647 (2006).

The Pull-Off Force and the Work of Adhesion: New Challenges at the Nanoscale

Nathan W. Moore* and **J. E. Houston**

Surface and Interface Sciences, Sandia National Laboratories, P.O. Box 5800, MS-1415, Albuquerque, NM 87185, USA

Abstract

The pull-off force required to separate two surfaces has become a convenient metric for characterizing adhesion at the micro- and nanoscales using cantilever-based force sensors, such as an atomic force microscope (AFM), e.g., as a way to predict adhesion between materials used in MEMS/NEMS. Interfacial Force Microscopy (IFM) provides unique insight into this method, because its self-balancing force-feedback sensor avoids the snap-out instability of compliant sensors, and can estimate both the work of adhesion and the pull-off force that would be measured using a compliant cantilever. Here, IFM is used to illustrate the challenges of determining the work of adhesion from the pull-off force in a nanoscale geometry. Specifically, adhesion is evaluated between a conical, diamond indenter and three surfaces relevant to MEMS/NEMS adhesion problems: silicon, a model insulator and a compliant polymer surface.

Keywords

Nanomechanics, contact mechanics, pull-off force, work of adhesion, Interfacial Force Microscopy

1. Introduction

Adhesion at the nanoscale and between MEMS/NEMS components can be unique both physically and in the manner in which it is measured [1]. Unpredictable topographies, size-dependent deformation mechanisms, compliance of surface coatings, and the statistical nature of device/surface fabrication are just a few complexities that lead to a rich diversity of adhesion behavior at the nanoscale [1–8]. An equal challenge is that nanoscale adhesion is often most conveniently measured with scanning probe microscopies, which can measure very small forces, but provide only limited information about the contact geometry [9, 10], thus making difficult the validation of suitable contact mechanics models.

Given the above, it is not surprising that classical theories, such as the JKR [11] or DMT [12] theories for single-asperity contacts, sometimes fail to describe ad-

---

* To whom correspondence should be addressed. Tel.: (505) 844-0278; Fax: (505) 844-5470; e-mail: nwmoore@sandia.gov

Adhesion Aspects in MEMS/NEMS

hesion at the nanoscale [3, 6]. Nonetheless, this framework is frequently used to relate the pull-off force required to separate surfaces to the work of adhesion of the contacting materials, using either the original JKR or DMT models or their many variants [13]. This method seems practical, first because the work of adhesion can be reasonably estimated with minimal calculation for many types of adhesion inter-actions, particularly at the macroscale [6, 13–19]. Second, in measurements with a cantilever-based force sensor (e.g., atomic force microscopy or MEMS cantilever), the pull-off force is often the only adhesion metric available [1, 9]. It is well recog-nized that cantilever force sensors are inherently unstable, as the cantilever will snap into or out of contact with a surface whenever the gradient of the force between the tip and the sample exceeds the cantilever spring constant [1, 16]. Replacing a weak cantilever with a stiffer one can avoid this problem, but the reduction in force sen-sitivity may be unacceptable.

Growing recognition of these challenges makes timely a perspective based on measurements with Interfacial Force Microscopy (IFM) [20–22] — a scanning probe technique that is perhaps best known for its ability to avoid jump-in and jump-out instabilities, and thus provide a more complete view of the surface forces that drive nanoscale adhesion. The IFM's unique sensor design maintains stability in a force gradient by tracking the tip's displacement in response to a force, as a differential change in the capacitance of two capacitors mounted on each side of a teeter-totter to which the tip is attached (Fig. 1). The DC voltage on the capacitor pads is continually adjusted using a feedback controller to eliminate the tip dis-placement, with the applied voltage allowing a simultaneous measure of the force encountered by the tip. With this scheme, the IFM sensor is noncompliant, yet can measure attractive and repulsive forces at all tip–sample separations.

In this work, we utilize the IFM's ability to measure the work required to separate two surfaces, as the integral of the force–displacement curve. Then, we approximate the corresponding pull-off force that would be measured by a weak (infinitely com-pliant) cantilever based on the maximum adhesion force value. Estimating the work of adhesion from each method provides a framework for comparing the strengths and weaknesses of each approach to characterizing nanoscale adhesion. Specif-ically, we compare the work of adhesion estimated from each method for three

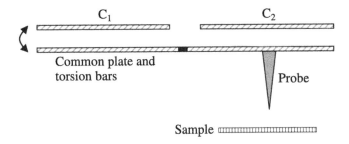

Figure 1. Schematic of the IFM's differential-capacitor force-feedback sensor, showing the capacitor gaps on each end of the common plate, with capacitances C_1 and C_2.

model surfaces with properties relevant to MEMS/NEMS adhesion problems: a silicon surface, an insulator (BaF_2) and a polymer surface [poly(vinyl acetate)].

This work follows an earlier IFM study of adhesion between surfaces decorated with self-assembled alkanethiol monolayers terminated in either CH_3 or CF_3 end groups [23]. There it was shown that the pull-off force was the same within ~10%, but that the work of adhesion differed by nearly a factor of 2. This discrepancy was attributed to the complex mechanics of the adsorbed films under the influence of the long-range Debye component of the van der Waals attraction. We expand on this observation by showing that a convolution of van der Waals and electrostatic forces, capillary attraction, elastic deformations, and/or viscoelastic deformation can lead to a rich variety of adhesion behavior at the nanoscale that can be masked when only the pull-off force is measured. These interactions are particularly relevant to MEMS/NEMS, as there has been considerable effort to develop thin-film coatings to eliminate capillary adhesion between components, and to reduce corrosion and wear [24].

2. Methods

The idea of relating adhesion properties to a pull-off force originates from the theory developed by Johnson, Kendall and Roberts (JKR) [11], which has the key assumptions: (1) the two interacting solids are isotropic, linear elastic materials; (2) attractive forces act over an infinitesimal distance; and (3) the contact is frictionless. Because only a few materials meet these requirements, many refinements have been made, such as the DMT model [12], which addresses assumption (2), and various 'transition' models that are valid over larger parameter spaces (for reviews, see [1, 13, 16, 25]). Of the extensive work in this area, the JKR and DMT models are perhaps the most widely used [26], owing to the simple result that the pull-off force (F_{PO}) is linearly related to the work of adhesion (W) by:

$$F_{PO} = -\xi \pi R W, \qquad (1)$$

where R is the asperity radius and $\xi = 3/2$ and 2 in the JKR and DMT theories, respectively [11, 12]. Which model is more appropriate depends on the magnitude of the adhesion relative to the contact's composite modulus (K). This is usually decided by evaluating the Tabor parameter, $\mu_T = (16RW^2/9K^2z_0^3)^{1/3}$ [27], where z_0 is the range of the interatomic potential, for which 3 Å may be a good estimate [16]. The choice of models is JKR for $\mu_T > 5$, DMT for $\mu_T < 0.1$, and a transition model for intermediate values [16].

More recently, Yang developed a model appropriate for adhesion between a conical indenter on a flat sample of the same material, which, apart from geometry, follows the same assumptions as the JKR model [28]. For frictionless contact, the Yang model gives the work of adhesion as:

$$W_{PO} = [F_{PO}\pi K \cos^2(\theta) \tan^3(\theta)/(54(1 - \nu^2))]^{1/2}, \qquad (2)$$

where ν and θ and are the Poisson's ratio and indenter half-angle, respectively. The elegance of equations (1) and (2) is that these predict a simple relationship between the work of adhesion and the pull-off force ($W_{PO} \propto F_{PO}$ or $W_{PO} \propto F_{PO}^{1/2}$). More difficult is the frequent need to relate the work of adhesion, which is defined on an areal basis, to the work that would be required to separate two real asperities. In general, this requires accurate values for the energies required to create two solid surfaces, their interaction with the intervening medium, and their interaction with each other [16]. For the materials we consider, this is further complicated by the fact that the intervening medium may not have uniform properties throughout the contact, as is true for nanoscale liquid capillaries, which may not occupy the entire region over which attractive forces are important [29, 30]. Further, one may need to consider surface roughness, chemical inhomogeneities at or near the opposing surfaces, viscoelastic deformation, and surface contamination — complications that can play exaggerated roles at the nanoscale [1]. As we show here, the numerical result can also be quite sensitive to the probe shape.

Rather than address all possible scenarios, we restrict our analysis to the minimum required to show key trends. Specifically, in the pull-off method, we use equation (2) to estimate the work of adhesion that would be required to separate two parallel surfaces. Also, because the IFM does not experience a pull-off point, we take F_{PO} to be the minimum of the force-well in the force–displacement curve measured with IFM. This corresponds to the pull-off force that would be measured had the measurement been done with an infinitely complaint cantilever. Although no cantilever is infinitely compliant, this method approximates the pull-off force that would be measured in experiments wherein soft cantilevers are chosen to enhance force sensitivity [9].

To illustrate the difficulties of relating a pull-off force to the work required to separate two real asperities, we also estimate the work of adhesion using a second method. The 'integral method' capitalizes on the IFM's unique ability to measure forces at all tip–sample distances. The integral of the attractive portion of the force–displacement curve divided by the contact area gives the work of adhesion, W_I, required to separate the real contact. While this method appears simple, the challenge is in estimating the true area of a material contact that is only a few nanometers wide. Few techniques can measure a tip's shape to such accuracy, so that the shape of the indentation can only be estimated using contact mechanics models. Another challenge is that sharp conical or spherical tips may adhere laterally to the contact's periphery, creating additional contact area, so that the work of adhesion is overestimated. Again, these shortcomings introduce systematic uncertainty, but do not affect our conclusions, which are concerned only with trends.

To compare W_I to the W_{PO} measured from the pull-off method, we used the same Yang model [28] to calculate the contact area (A) at the equivalent pull-off, or separation, point between the conical indenter and flat substrates, viz.,

$$A = \pi[18(1 - \nu^2)W_{PO}/(\pi K \cos(\theta)\tan^2(\theta))]^2. \tag{3}$$

This area estimate requires an estimate of the work of adhesion, for which the value from the pull-off method (equation (2)) was used to allow a comparison between the pull-off and integral methods. Both the pull-off and integral methods estimate the work of adhesion on an areal basis; however, their definitions are geometry-specific. As stated, W_{PO} represents the work to separate two flat surfaces, while W_I represents the true work to separate the conical and planar surfaces, divided by the estimated area of that particular contact. Because these geometries are not equivalent, one should not expect the two methods to yield identical values. For example, this can occur when the true contact area is enlarged by lateral adhesion between the tip and substrate, or by the presence of a liquid capillary [29]. Nonetheless, because these differences are systematic, one should expect W_{PO} and W_I to be strongly correlated if the pull-off force is, in fact, a good indicator of the work required to separate two surfaces.

It is important to recognize that the values for W_I and W_{PO} are highly sensitive to the half-angle of the indenter used. To show this quantitatively, we note that the numerical calculation of W_I, W_{PO} and their ratio, have the following proportionalities:

$$W_I \propto \sin^2(\theta) \tan^2(\theta), \tag{4}$$

$$W_{PO} \propto \cos(\theta) \tan^{3/2}(\theta), \tag{5}$$

$$W_I / W_{PO} \propto \sin(\theta) \tan^{3/2}(\theta), \tag{6}$$

from equations (1)–(3). A physical difference in angle will also correspond to different contact areas and forces, so that the proportionalities of equations (4)–(6) are only meaningful as an assessment of systematic error introduced by uncertainty in θ. Using equations (4)–(6) with a nominal value of $\theta = 60°$, a 5% uncertainty in θ produces uncertainty in W_I, W_{PO} and the ratio, W_I / W_{PO}, of 18%, 62% and 43%, respectively. In comparison, uncertainty in the work of adhesion determined with a spherical tip scales linearly with uncertainty in the tip radius (equation (1)). An additional systematic error in the calculation of W_I and W_{PO} may arise from the faceted shape of the diamond indenter, for which the cone geometry of the Yang model is only an approximation. A smaller systematic error (<5%) arises from assuming a Poisson's ratio of 0.3 for both tip and substrate, which are assumed identical in equations (2)–(3). For these reasons, we do not attempt to compare the absolute values of W_I and W_{PO} on a point-by-point basis. Instead, we focus only on the relative trends between W_I and W_{PO}. This approach is reasonable because the shape of the diamond indenter can be expected to remain constant under the short times and small elastic loads applied in these experiments.

A Sandia-built Interfacial Force Microscope and control software were used as described elsewhere [20, 21]. The present configuration can measure forces between ∼30 nN and 50 μN with time resolution ∼1 ms, over a wide range of temperature (−50 to +100°C) and relative humidity (0 < RH < 70%). Accuracy in the measured force was limited by absolute error in calibrating the sensor gain

($\pm 10\%$). Accuracy in the tip displacement is limited by absolute error in calibrating the piezo actuator ($\pm 5\%$).

Polished single-crystal BaF_2 (Alfa Aesar, Ward Hill, MA; Lot#G16S051) was cleaned with ethanol and dried in the measurement chamber ($<1\%$ relative humidity, RH) at 75°C for 2.5 h before use. AFM imaging showed that the BaF_2 surface used here was atomically flat over the areas of interest. Si(111) substrates were rinsed in ethanol and water and dried. Poly(vinyl acetate) (Scientific Polymer Products; provided by courtesy of Hongbing Lu, University of North Texas) was in the form of ~4 mm hemispherical pucks, cleaned with ethanol, and dried. The IFM tip was a conical diamond indenter (macroscopic half-angle $\theta = 30°$) and cleaned by rinsing successively in 3:1 concentrated H_2SO_4/30% H_2O_2 (piranha), water, and ethanol, and dried, which left the tip mildly hydrophilic (water contact angle ~70°). Caution! Piranha is a strong oxidant and reacts violently with organic substances. The end radius of the diamond tip was estimated by scanning the tip over a calibration grating (MikroMasch, Si grating #TGG01), viz., $\leqslant 10$ nm, which is an upper bound for the grating radius. This value was used to estimate the Tabor parameter. All measurements were performed at room temperature and $<1\%$ RH except where noted. BaF_2 and Si surfaces were grounded to minimize electrostatic charging.

3. Results and Discussion

To compare what can be learned from the pull-off and integral methods, we briefly examine the adhesion between our diamond tip and three materials showing distinct adhesion mechanisms. The first, adhesion between diamond and Si(111), is shown in Fig. 2(a). The force between tip and substrate gradually becomes more adhesive (negative) as the tip approaches the surface, until the two materials touch near $D \approx 0$, defined as the distance where the force is minimum (most negative), i.e., where the force of elastic compression begins to become significant. That only a slight hysteresis is seen between the force on approach and withdrawal is consistent with the presence of only van der Waals or electrostatic forces.

To show that the long-range attraction is not dominated by van der Waals forces, we use the model of Argento and French [31] to estimate the magnitude of the van der Waals interaction between the conical tip and substrate, viz.,

$$
\begin{aligned}
F_{vdW} = &\{\alpha \rho^2 [1 - \sin(\theta)](\rho \sin(\theta) - D \sin(\theta) - \rho - D)\} \\
&\times [6D^2(\rho + D - \rho \sin(\theta))^2]^{-1} \\
&- \alpha \tan(\theta)[D \sin(\theta) + \rho \sin(\theta) + \rho \cos(2\theta)] \\
&\times [6 \cos(\theta)(D + \rho - \rho \sin(\theta))^2]^{-1},
\end{aligned}
\tag{7}
$$

where ρ is the end-radius of a spheroconical tip and α is the Hamaker constant. In principle, values for α could be estimated using dielectric constants for the interacting materials. However, for an order of magnitude estimate, it is sufficient to assume a typical value, α of the order ~10^{-19} J as values for α span a narrow range for a broad range of materials (~1–50×10^{-20} J) [16]. We use the macroscopic

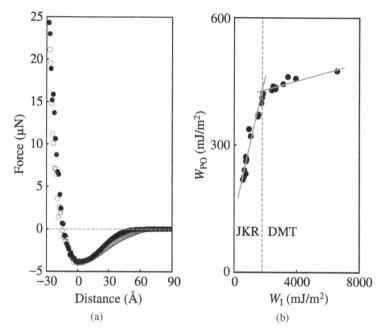

Figure 2. Interaction between a diamond tip and Si(111) at 25°C and <1% *RH*. (a) Example measurement of the normal force during the tip's approach (●) and withdrawal (○). (b) Comparison of the work of adhesion calculated by the pull-off force method (W_{PO}) and the integral method (W_I). Dashed vertical line indicates the approximate transition from the JKR to DMT regimes. Grey lines are best-fit lines in each regime.

half-angle of the diamond indenter ($\theta = 30°$), and values for the unknown ρ ranging from 0 to 1 μm. For these parameters, the magnitude of the attractive van der Waals force (equation (7)) ranges from 10^{-6}–10^{-3} μN for $D = 40$ Å, 10^{-5}–0.2 μN for $D = 3$ Å, and 10^{-4}–1.5 μN for $D = 1$ Å. Similarly, values of $F_{vdW} < 10^{-3}$ μN are estimated for $\rho = 0$ and $1° < \theta < 80°$. Consequently, for the probe geometry here, the van der Waals force is significant over only a few atomic distances. Sumant *et al.* [32], measured the work of adhesion for a diamond-coated tip and a Si(111) substrate, and found $W_{PO} = 59$ mJ/m², which was attributed solely to van der Waals forces. Their value is an order of magnitude lower than what we find in this particular experiment (Fig. 2(b)), suggesting again that here the adhesion is dominated by long-range electrostatic forces, e.g., as can arise from surface charge [32].

Each data point in Fig. 2(b) represents a measure of the work of adhesion every ~4 min in the same location on the sample, as calculated from both the pull-off (W_{PO}) and integral (W_I) methods. As the surface ages over a span of ~100 min, both W_{PO} and W_I decrease commensurately. This decrease in adhesion can be ascribed to chemical changes at the interface, e.g., from triboelectrification or oxidation, rather than from physical changes in the underlying material, which never showed plastic deformation under these small indentation forces (<25 μN).

This example is classic in that at least two of the key requirements of the JKR–DMT framework are fully met: both tip and substrate are elastic, and the adhesion forces (i.e., the van der Waals and electrostatic forces) are conservative forces. That is, the amount of energy dissipated during the formation and rupture of the single asperity contact was insignificant, as evidenced by the small hysteresis between the approach and withdrawal data in Fig. 2(a). For this reason, we find excellent correlation between the works of adhesion calculated using the pull-off force and integral methods. The two methods give values that are correlated in each of the two regions indicated in Fig. 2(b) (dashed line). This division roughly corresponds to the transition between the JKR and DMT regimes ($\mu_T \sim 1.2$, $W_I \sim 1900$ mJ/m^2). This deviation is expected because Yang's model for conical adhesion is strictly valid only for the JKR regime, so that W_{PO} is underestimated for higher values of F_{PO}. That W_I is much larger than W_{PO} may stem from systematic error in estimating the true contact area for the cone-plane geometry (equation (3)), as discussed earlier. For this work, the key detail is the strong correlation between W_I and W_{PO}. This correlation suggests that either method could be used as a general indicator of the adhesion strength between these materials.

The situation changes dramatically when the tip is brought near our model insulator surface. BaF$_2$(111) is a well-defined, ionic surface that exposes a hexagonal array of barium and fluoride ions and is moderately hydrophilic [33]. Figure 3(a) shows a long-range attraction that abruptly changes to a stronger, short-range attraction when the tip is ~ 50 Å from the BaF$_2$ surface. Here the force–distance profile becomes more linear, a strong contrast from the monotonically decreasing attractive force at larger distances. Such short-range, quasi-linear forces can arise from condensation of vapor into a liquid capillary between the tip and substrate [34]. This transition in force might also arise from solid necking in response to short-range, van der Waals forces [26, 34, 35]. That is, at tip–sample separations on the order of a few angstroms, where the van der Waals forces dominate the adhesion forces, the gradient of the attractive force can exceed the effective elastic constants of the tip and substrate materials, causing them to snap into contact [20]. The above analysis using equation (7) can also be used to argue that, here also, the long-range force cannot be explained by van der Waals forces alone, and is therefore likely electrostatic in origin, as would be expected if defects were present in the ionic surface.

The humidity dependence (Fig. 3(a), inset) suggests that both capillary attraction and elastic necking may contribute to the short-range adhesion. Below $\sim 45\%$ *RH*, the integral of the force curve, I, fluctuates seemingly randomly, while above $\sim 45\%$ *RH*, I increases steadily with humidity. This transition occurs near the 45% *RH* required to produce a net monolayer coverage of water on an isolated BaF$_2$(111) surface [33]. We anticipate that capillary formation is 'frustrated' at low humidities, before the surface is wetted fully [36, 37]. We also note that such a liquid capillary would have a very small volume ($< 10^5$ Å^3 as estimated using the method of Sirghi *et al.* [34]), which would preclude more than ~ 1 BaF$_2$ unit being dissolved before the meniscus is saturated with Ba$^+$ and F$^-$ ions. Such small amounts of dissolution

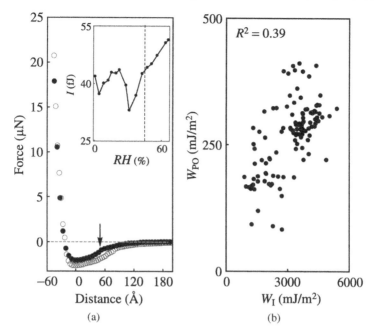

(a)   (b)

Figure 3. Interaction between a diamond tip and a model insulator surface, $BaF_2(111)$. (a) Example measurement of the normal force during the tip's approach (●) and withdrawal (○) (25°C, 1% *RH*). Arrow marks the onset of the short-range attraction. Inset shows how the integral of the force curve, I, changes with relative humidity at 25°C; dashed vertical line indicates the humidity corresponding to monolayer adsorption on isolated $BaF_2(111)$ surfaces [33]. (b) Comparison of the work of adhesion calculated by the pull-off force method (W_{PO}) and the integral method (W_I), from measurements over a range of temperature (23–90°C) and relative humidity (1–66%).

have been shown to allow hillocks of salt to grow inside a nanoscale meniscus formed between a NaCl crystal surface and an AFM tip [38]; however, this process occurs over a timescale of minutes to hours. Thus, for the measurements here, which lasted only a few seconds, we do not anticipate that dissolution played a dominant role in the surface mechanics or the measured adhesion forces.

Here, the possible confluences of long-range electrostatic and short-range van der Waals forces, capillary attraction, and short-range DMT-type necking control the adhesion in a complicated manner, even at low humidities. For this reason, Fig. 3(b) shows no apparent correlation between W_I and W_{PO} ($R^2 = 0.39$). This contrasts with the diamond/Si interaction (Fig. 2), which showed strong correlation between W_I and W_{PO}, even in the DMT regime. Both Si and BaF_2 behave elastically with a relatively high Young's modulus (~150 and 54 GPa, respectively), and produce similar pull-off forces. Yet with BaF_2, the relationship between W_I and W_{PO} is unclear. This result also contrasts with the supposition by several authors (e.g., [14, 27]) that the JKR–DMT type framework is relatively insensitive to the nature of the adhesion forces. The large scatter in the data (Fig. 3(b)) echoes a common challenge in characterizing adhesion at nanoscale dimensions, namely, that the geometries of the contacting surfaces and any capillary between them are generally unknown, and may be highly variable. In particular, that capillary forces can modify the contact geometry has only recently begun to be explored for a sphere-plane contact geometry [39]. How these effects could be manifested in a cone-plane geometry remains an open question.

We now turn to a third example of an adhesion interaction where the pull-off and integral methods can show qualitatively different trends in the adhesion behavior, namely, in adhesion to compliant polymer films. Because MEMS/NEMS are frequently coated with thin films to reduce friction and corrosion, even structural components that behave as elastic bodies at the macroscale can have surfaces that deform plastically or viscoelastically at the nanoscale [23]. For this we have repeated similar IFM measurements, using the same tip as before, on the surface of poly(vinyl acetate), PVAc.

Figure 4(a) shows three examples of force–displacement profiles measured between our diamond tip and the PVAc surface, in different locations on the sample, after 3, 6 and 9 loading cycles (bottom to top curves, respectively). In each of the force curves in Fig. 4(a), the attractive forces extend up to ~600 Å from the surface (not shown), suggesting that the long-range attraction is electrostatic in origin, as might arise from charged impurities introduced during manufacture of the poly-

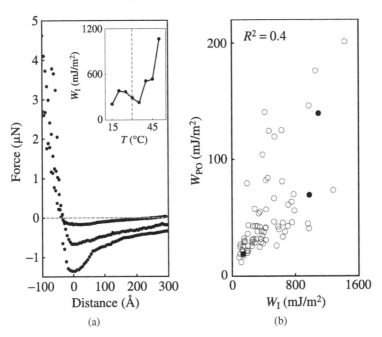

(a)   (b)

Figure 4. Interaction between a diamond tip and the surface of poly(vinyl acetate). (a) Example measurements of the normal force in different sample locations (bottom to top: after 3, 6 and 9 successive compressions, respectively). Inset shows the work of adhesion as a function of temperature (T); each point is an average of ~10 measurements at various speeds and indentation depths; dashed vertical line is the polymer glass-transition temperature. (b) Comparison of the work of adhesion calculated by the pull-off force method (W_{PO}) and the integral method (W_I) for the three curves in part (a) (●) and for all measurements over the range, 15–50°C (○).

mer, for example. This view is supported by the analysis using equation (7), which showed that for the present tip geometry, the van der Waals force is only expected to be significant over a few atomic lengths. In all cases, the forces in Fig. 4 at long range ($D > 100$ Å) are proportional to $1/D^n$, with $n = 1.8 \pm 0.3$. This exponent falls within the range (1–3) determined for a van der Waals force between sphero-conical probes and a flat surface [31]. However, as Argento and French [31] point out, it is not possible to distinguish this scaling from an electrostatic interaction when there is the possibility of plastic deformation, as might be caused by repeated indentations. Here, the magnitude of the long-range adhesion decreases with the cycle number (Fig. 4(a)), implying smaller contact areas and a surface that is less able to conform to the indenting tip.

The near-surface force behavior is more distinct. The bottom curve in Fig. 4(a) shows a slope discontinuity near $D = 60$ Å. This may arise from viscoelastic (time-dependent) deformation of the polymer surface onto the tip. Under the influence of attractive force, a viscoelastic material can be expected to adhere to the tip in a manner that depends on the entire history of the tip–sample contact. Here, repeated compression resulted in a more elastic response (cf. Fig. 2), e.g., as could arise from either a reshaping of the polymer surface or from a local change in mechanical properties as the molecular chains within the polymer are reorganized under the applied forces. The temperature dependence also suggests that the adhesion is influenced by viscoelastic deformation of the polymer onto the tip. Temperature was varied over 15–50°C in a random sequence for the same sample, which may introduce complicated history effects, such as water adsorption/desorption, that we do not attempt to quantify here. Although we do not see evidence of capillary condensation in the force profile (Fig. 4(a)), PVAc is extremely hygroscopic, and even small amounts of absorbed water affect its viscoelastic properties [40]. Nonetheless, it can be seen that above 35°C (near the glass transition temperature, $T_g \approx 30$°C [40]), the work of adhesion increases with temperature to about twice the values near room temperature (Fig. 4(a), inset). This transition is expected, because increased mobility of the polymer chains above T_g allows the polymer to flow onto the tip more rapidly, creating larger contact areas.

These subtle changes at the polymer surface are not predicted by the pull-off force, which, in Fig. 4(a), simply decreases with cycle number. Further, the pull-off force is not strongly correlated with the work of adhesion for this material. Looking at the entire data set, the works of adhesion calculated with each method (W_{PO} and W_I) are not strongly correlated ($R^2 = 0.40$). Considering just the three examples in Fig. 4(a) (dark circles in Fig. 4(b)), the two measurements with the strongest adhesion have W_I within 12% of each other, while the work of adhesion calculated from the pull-off force (W_{PO}) differs by ~100%. That is, in these measurements, the pull-off force does not predict the material's adhesion behavior. In this case, measuring only the pull-off force with a cantilever would show quantitatively different trends, and obscure the effects of viscoelastic deformation. We anticipate that these challenges are not unique to PVAc, as viscoelasticity and load history are known determinants in the adhesion of some thin films used on MEMS/NEMS components, including SAMs and gold films [23, 41–43].

4. Conclusions

We have shown three situations where the pull-off and integral methods estimate different values or trends for the work of adhesion, as measured in a cone-plane geometry. This follows an earlier observation of the same discrepancy, resulting from deformation of self-assembled monolayers adsorbed on a sphere-plane geometry [23]. This prior work, as well as the measurements with PVAc, involve 'soft' material interactions where agreement with the strict requirements of a JKR-type framework is not expected *a priori*. However, neither does one generally know the mechanical properties of an interface *a priori*. That is, when only the pull-off force can be measured, one cannot know if it is an appropriate metric for a given material.

This work also shows that measuring only the pull-off force can mask the physical origins of the adhesion. In contrast, because the integral method accounts for forces at all surface separations, it is extremely sensitive to any material behavior that deviates from the assumptions of simple elastic theories, such as the JKR or DMT models. That is, the integral method can determine the work required to separate surfaces regardless of the adhesion mechanism or contact geometry. The lack of correlation between the pull-off and integral methods in some cases echoes the importance of continuing work to understand these details. In particular, the utility of the integral method is limited by the accuracy of estimating the true contact area. Further, the work of adhesion calculated in this manner may be difficult to relate to other measurement geometries. Toward this end, a promising technique is *in situ* mechanics testing (e.g., [44, 45]), wherein contact geometries can be measured and models justified. A growing field where such breakthroughs are needed is in measuring adhesion between biological materials (e.g., cells, membranes and even individual proteins), for which pull-off forces have been used extensively [10, 17, 46]. Increasingly, MEMS/NEMS technologies are also being used to characterize these materials [1]. The development of displacement-controlled techniques with adequate sensitivity to measure adhesion forces between these materials will be of ongoing interest to both communities.

Acknowledgements

We thank Prof. Hongbing Lu and Direk Cakiroglu for providing the PVAc and for useful discussions, and Maarten de Boer for reviewing this manuscript. This work was supported by the Division of Materials Sciences and Engineering, Office of Basic Energy Sciences, U.S. Department of Energy and by the Laboratory Directed Research and Development program at Sandia National Laboratories. Sandia National Laboratories is a multi-program laboratory operated by Sandia Corporation, a Lockheed Martin Company, for the Department of Energy's National Nuclear Security Administration under Contract DE-AC04-94AL85000.

References

1. B. Bhushan (Ed.), *Nanotribology and Nanomechanics: An Introduction*. Springer, Berlin (2005).
2. J. F. Waters, S. Lee and P. R. Guduru, *Int. J. Solids Struct.* **46**, 1033 (2009).
3. P. Attard and J. L. Parker, *Phys. Rev. A* **46**, 7959 (1992).
4. M. P. de Boer, *Expl. Mech.* **47**, 171 (2007).
5. Y. R. Ding, *J. Adhesion Sci. Technol.* **22**, 457 (2008).
6. D. E. Packham, *Intl J. Adhesion Adhesives* **16**, 121 (1996).
7. P. Attard, *J. Adhesion Sci. Technol.* **16**, 753 (2002).
8. G. Kumar, S. Smith, R. Jaiswal and S. Beaudoin, *J. Adhesion Sci. Technol.* **22**, 407 (2008).
9. V. Craig, *Colloids Surfaces A* **129–130**, 75 (1997).
10. P. M. Claesson, T. Ederth, V. Bergeron and M. W. Rutland, *Adv. Colloid Interface Sci.* **67**, 119 (1996).
11. K. L. Johnson, K. Kendall and A. D. Roberts, *Proc. R. Soc. London Ser. A* **324**, 301 (1971).
12. B. V. Derjaguin, V. M. Muller and Y. P. Toporov, *J. Colloid Interface Sci.* **53**, 314 (1975).
13. X. Shi and Y.-P. Zhao, *J. Adhesion Sci. Technol.* **18**, 55 (2004).
14. E. Barthel, *J. Colloid Interface Sci.* **200**, 7 (1998).
15. H. J. Butt, B. Cappella and M. Kappl, *Surf. Sci. Rep.* **59**, 1 (2005).
16. J. N. Israelachvili, *Intermolecular and Surface Forces*. Academic Press, San Diego, CA (1991).
17. D. Leckband and J. Israelachvili, *Quarterly Rev. Biophys.* **34**, 105 (2001).
18. F. L. Leite and P. S. P. Herrmann, *J. Adhesion Sci. Technol.* **19**, 365 (2005).
19. M. Nosonovsky and B. Bhushan, *Phys. Chem. Chem. Phys.* **10**, 2137 (2008).
20. S. A. Joyce and J. E. Houston, *Rev. Sci. Instrum.* **62**, 710 (1991).
21. J. E. Houston and T. A. Michalske, *Nature* **356**, 266 (1992).
22. R. C. Thomas, J. E. Houston, R. M. Crooks, T. Kim and T. A. Michalske, *J. Am. Chem. Soc.* **117**, 3830 (1995).
23. J. E. Houston, C. M. Doelling, T. K. Vanderlick, Y. Hu, G. Scoles, I. Wenzl and T. R. Lee, *Langmuir* **21**, 3926 (2005).
24. S. H. Kim, D. B. Asay and M. T. Dugger, *Nano Today* **2**, 22 (2007).
25. U. D. Schwarz, *J. Colloid Interface Sci.* **261**, 99 (2003).
26. J. A. Greenwood and K. L. Johnson, *J. Phys. D: Appl. Phys.* **31**, 3279 (1998).
27. D. W. Xu, K. M. Liechti and K. Ravi-Chandar, *J. Colloid Interface Sci.* **315**, 772 (2007).
28. F. Yang, *J. Mater. Res.* **21**, 2683 (2006).
29. P. C. T. de Boer and M. P. de Boer, *Langmuir* **24**, 160 (2008).
30. M. P. de Boer and P. C. T. de Boer, *J. Colloid Interface Sci.* **311**, 171 (2007).
31. C. Argento and R. H. French, *J. Appl. Phys.* **80**, 6081 (1996).
32. A. V. Sumant, D. S. Grierson, J. E. Gerbi, A. Carlisle, O. Auciello and R. W. Carpick, *Phys. Rev. B* **76**, 235429 (2007).
33. G. E. Ewing, *Chem. Rev.* **106**, 1511 (2006).
34. L. Sirghi, R. Szoszkiewicz and E. Riedo, *Langmuir* **22**, 1093 (2006).
35. R. C. Major, J. E. Houston, M. J. McGrath, J. I. Siepmann and X. Y. Zhu, *Phys. Rev. Lett.* **96**, 177803-1 (2006).
36. M. P. Goertz, X. Y. Zhu and J. E. Houston, *Langmuir* **25**, 6905 (2009).
37. D. B. Asay and S. H. Kim, *Langmuir* **23**, 12174 (2007).
38. H. Shindo, M. Ohashi, K. Baba and A. Seo, *Surface Sci.* **358**, 111 (1996).
39. J. Zheng and J. L. Streator, *Trans. ASME, J. Tribology* **129**, 274 (2007).
40. W. G. Knauss and V. H. Kenner, *J. Appl. Phys.* **51**, 5131 (1980).

41. L. Chen, Y. Du, N. E. McGruer and G. G. Adams, in: *Proceedings of the ASME/STLE International Joint Tribology Conference, Parts A and B*, San Diego, CA, p. 803 (2008).
42. L. Chen, N. E. McGruer, G. G. Adams and Y. Du, *Appl. Phys. Lett.* **93**, 053503 (2008).
43. Y. Du, G. G. Adams, N. E. McGruer and I. Etsion, *J. Appl. Phys.* **103**, 064902 (2008).
44. O. Y. Komkov, *J. Friction Wear* **28**, 19 (2007).
45. B. L. Weeks and J. J. DeYoreo, *J. Phys. Chem. B* **110**, 10231 (2009).
46. P. Hinterdorfer and Y. F. Dufrêne, *Nature Methods* **3**, 347 (2006).

Interfacial Adhesion between Rough Surfaces of Polycrystalline Silicon and Its Implications for M/NEMS Technology

Ian Laboriante [a], Brian Bush [a], Donovan Lee [b], Fang Liu [a], Tsu-Jae King Liu [b], Carlo Carraro [a,c] and Roya Maboudian [a,c,*]

[a] Department of Chemical Engineering, University of California, Berkeley, CA 94720, USA
[b] Department of Electrical Engineering and Computer Sciences, University of California, Berkeley, CA 94720, USA
[c] INM — Leibniz Institute for New Materials, Campus D2 2, 66123 Saarbrücken, Germany

Abstract
An electrostatically actuated double-clamped cantilever beam test structure has been designed and fabricated to determine the adhesion forces between co-planar, impacting polycrystalline silicon (polysilicon) surfaces. To examine the effect of apparent contact area, dimples of varying sizes have been included in the test structure. By measuring the cantilever beam profile, through optical interferometric methods, as a function of applied bias, the force of adhesion has been determined for various device geometries. The results reveal a weak dependence of adhesion on apparent contact area, rather than a linear dependence. Fabrication process artifacts, observed and discussed here, contribute to but do not suffice to explain this observed weak scaling. The results strongly suggest that contact on the micrometer scale between rough, rigid materials such as polysilicon involves only a few asperities.

Keywords
Adhesion, apparent work of adhesion, double-clamped cantilever beam, MEMS, NEMS, reliability, rough surfaces, stiction

1. Introduction

Static adhesion, commonly called stiction, remains one of the biggest hurdles preventing a larger number of micro-/nanoelectromechanical systems (M/NEMS) products from entering the mainstream [1–5]. Despite the critical role stiction plays in M/NEMS reliability, a thorough quantitative understanding of this phenomenon is currently lacking. This is due to the fact that the contact mechanics of rough and

* To whom correspondence should be addressed. Tel.: +1-510-643-7957; Fax: +1-510-642-4778; e-mail: maboudia@berkeley.edu

Adhesion Aspects in MEMS/NEMS
© Koninklijke Brill NV, Leiden, 2010

rigid surfaces, such as those of polycrystalline silicon (polysilicon), the most widely used structural material in M/NEMS, is a complicated multiscale phenomenon. As a consequence, when two polysilicon surfaces are brought together under a given load, only a few asperities make physical contact [6–8] and surface forces from the near-contact region may contribute significantly to the net adhesion.

Many earlier attempts to determine adhesion forces for various interfaces have been performed on model surfaces either through single asperity [9–11] and microsphere probe [12, 13] measurements using atomic force microscopy, *a priori* surface roughness characterizations of the structural material [14], multiple asperity contact studies using surface force apparatus [15, 16], or analysis based on electrical contact resistance measurements [17–19]. A fundamental understanding of adhesion phenomenon for representative MEMS devices is of paramount importance for future design considerations such as choice of structural materials, details of the microfabrication processes, operating environmental conditions, and tailoring of surface chemistry. One of the most widely used device-level adhesion microinstruments to date is an array of single-clamped cantilever beams, often referred to as a cantilever beam array (CBA) [6, 20–22]. Here, we provide an alternative method to characterize interfacial adhesion using a microfabricated double-clamped beam (DCB) test structure [17, 22–25]. This test structure provides an excellent model for MEMS switches and for advanced micro-/nano-electromechanical non-volatile memory computer data storage applications [26]. Compared to the more common single-clamped MEMS cantilevers, doubly anchored structures of similar dimensions are less prone to stiction and out-of-plane deflection due to strain gradient effects. In addition, this test structure allows for the determination of the adhesion force, rather than work of adhesion which is obtained from the CBA analysis. From a design point of view, it is often the adhesion force that is of most interest.

There is an extensive body of literature describing theoretical approaches for estimating adhesion in microsystems derived from a fractal description of surface topography [27–29], random process models [30], and numerical and analytical methods [31, 32]. These methods are largely based on the concept that the idealized geometrical information on the microasperities and the expected adhesion can be correlated using theoretical models [33–35] with various simplifying assumptions. However, predictions of adhesion forces using theoretical models are inadequate to describe the nature of contact between two interacting rough surfaces as these approaches have been found to yield adhesion values with a wide range of discrepancies [36–38]. While these studies provide valuable insights into the nature of contact between rough surfaces, they do not include important information such as the dynamic nature of cyclically contacting microasperities. This is relevant for multilayer MEMS fabrication technology which often leads to interacting surfaces with correlated topography [39], since the effective area of interaction is governed largely by surface topography. In addition, interacting MEMS surfaces often involve very few contact points so that interactions between surfaces that are nearly in contact may be comparable or even larger than the interactions at the actual points of contacts [6–8].

This paper presents the results of investigation aimed at elucidating the adhesion forces in polycrystalline silicon MEMS as a function of the apparent contact area. The main objective of this study was to determine how adhesion scaled with apparent area for contacts involving rigid, rough surfaces. The direct, quantitative measurement of apparent work of adhesion for these surfaces is also reported. This approach validates the design and process parameters of the test structure, systematically determines adhesion forces with good precision, and provides new insight into the nature of contact for coplanar, impacting, rigid surfaces.

2. Experimental Section

2.1. Device Structure, Fabrication and Operation

A set of DCB test structures with a wide range of dimensions on a single die were designed and fabricated at the UC Berkeley Microfabrication Laboratory using standard multilayer surface micromachining processes involving a low-stress Si_3N_4 isolation layer, polysilicon bottom electrode and landing pad (poly 0), silicon oxide sacrificial layer and polysilicon structural layer (poly 1). The device is schematically depicted in Fig. 1, showing both the non-contact and contact states. It consists of a 15 μm wide and 2 μm thick polysilicon beam, with length in the range from 160 to 240 μm. The beam is suspended 2 μm above a grounded landing pad (also called the drain) and two symmetrically positioned interconnected electrodes for electrostatic actuation (also called actuation pads). The incorporation of a contact dimple on beam defines the apparent interaction area between the beam and the landing pad. The dimple size is in the range from 1×1 μm^2 to 8×8 μm^2. The separation gap between the dimple and landing pad is 1.1 μm.

The fabrication process is described in greater detail elsewhere [5]. Briefly, polysilicon films are deposited *via* low pressure chemical vapor deposition (LPCVD) at 375 mTorr and ~615°C from SiH_4 gas with *in situ* phosphorus doping. Low-temperature oxide (LTO) forms the sacrificial spacer between the two polysilicon layers. After device fabrication is completed, the substrate is coated with a protective layer of photoresist and diced. A hot piranha solution (2:1 mixture of conc. H_2SO_4 and 35% H_2O_2) then is used to remove the photoresist, after which the die is annealed at 1050°C for 3 hours in a N_2 purged furnace to minimize the residual stress in the structures. The release of the microdevices is conducted as follows. The die is immersed in a 1:1 HF:HCl solution for 20 min with gentle agitation to remove the sacrificial oxide layer, followed by successive water and isopropyl alcohol (IPA) rinses. The final step involves the transfer of the die to an IPA solution and drying in supercritical CO_2 (Tousimis Autosamdri 815B, Rockville, MD) to avoid release-related adhesion [40]. A scanning electron micrograph (NovelX mySEM, Lafayette, CA) of a representative released microstructure is shown in Fig. 2.

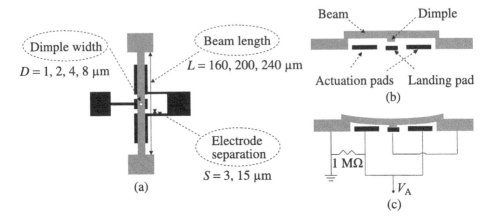

Figure 1. Schematic representations of a double-clamped beam adhesion test structure. The source electrode (beam) is pulled into contact with the underlying drain electrode (landing pad) *via* electrostatic actuation. (a) is a top-view rendition of an as-fabricated device indicating various dimple sizes, lengths and electrode lateral separations; (b) and (c) side-view representations showing non-contact and contact states, respectively. Typical actuation voltage (V_A) required to bring the beam into contact with the landing pad is $\sim$100 V.

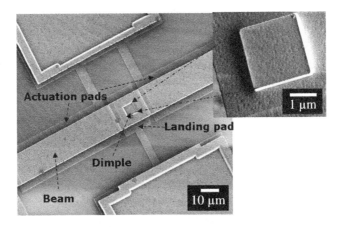

Figure 2. An SEM micrograph of a representative double-clamped beam adhesion test structure that is 200 µm in length. Inset is an SEM micrograph of contacting surface with a 2×2 µm^2 dimple.

The magnitude of residual strain is known to be significantly affected by the film microstructure, deposition process parameters, and annealing temperature. Microstrain gauges fabricated on the same chip are used to determine the average residual strain prior to adhesion measurements. The technique is described elsewhere [41] and summarized here for completeness. A close-up optical view of the aforementioned gauge is shown in Fig. 3. The gauge consists of two test beams connected to an indicator beam through a junction separated by a small distance. Compressive and tensile residual strains in the film cause the test beam to contract or expand, respectively, and thus result in a rotation of the indicator beam either clockwise

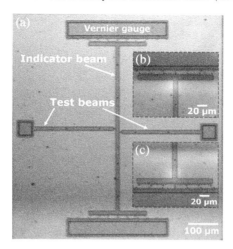

Figure 3. (a) Optical image of a representative microstrain gauge used to characterize the residual strain, ε, of the polycrystalline silicon films. This device has been fabricated together with the double-clamped beam adhesion test structures on the same die. (b) and (c) close-up views of microstrain indicator beams opposite the vernier gauges. Slight rotation of the beam clockwise indicates small amount of compressive stress in the beam.

(compressive) or counterclockwise (tensile) with respect to the vernier gauges positioned at the ends of the indicator beam. The average residual strain is calculated *via*

$$\varepsilon = \frac{\delta_i d_{ti}}{L_t L_i},$$

(1)

where ε is the residual strain, δ_i the displacement of the vernier gauge, d_{ti} is the distance between the two junctions of the test and the indicator beams, and L_t and L_i are the lengths of the test and indicator beams, respectively. The average residual strain for this specific die is found to be $\sim 4.2 \times 10^{-5}$ compressive, given that $\delta_i \sim 2.1$ µm, $L_t = 250$ µm, $L_i = 1000$ µm and $d_{ti} = 5$ µm. Qualitative evaluation of each device using optical interferometry verifies the observed low ε value and minimal curvature as manifested by the absence of any optical fringes in the non-contact state (Fig. 4, left image).

An atomic force microscope (Digital Instruments, Multimode Nanoscope IIIa with extender electronics module, Santa Barbara, CA) operating in tapping mode using a silicon cantilever (MikroMasch, San Jose, CA) is employed to characterize the topography of as-fabricated polycrystalline silicon surfaces.

2.2. Determination of Work of Adhesion

The apparent work of adhesion, defined as the energy required to separate unit areas of two adhering surfaces, is determined *via* double-clamped beam test structures of

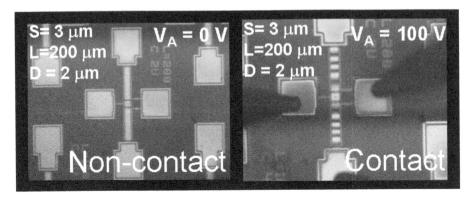

Figure 4. (Left) An optical interferometer image of an as-fabricated 200 µm long double-clamped beam adhesion test structure at initial non-contact state or applied voltage $V_A = 0$ V as depicted in Fig. 1(b). (Right) The source electrode (beam) is pulled into contact with the underlying drain electrode (landing pad) by applying a voltage to the actuation pad, typically in the range of 85–110 V depending on the size of the actuation pad and the electrode separation. The voltage at which contact occurs, as manifested by a change from 3 to 4 fringes, is recorded here as $V_{pull-in}$. The applied voltage is further increased by 10 V above $V_{pull-in}$ and the process is reversed by decreasing the V_A until the beam pulls out of contact from the landing pad, denoted as $V_{pull-off}$ (transition from 4 to 3 fringes).

various lengths. For such an experimental technique, the work of adhesion W_{adh} [1, 23] is obtained according to equation (2):

$$W_{adh} = \frac{\pi^4}{4} \frac{E h^2 t^3}{l_d^4} + \frac{\pi^2}{4} \frac{\varepsilon E h^2 t}{l_d^2}, \tag{2}$$

where E is Young's modulus of polysilicon, h is the separation gap between the beam and the substrate, t the beam thickness, ε the residual strain and l_d is the detachment length. Here, the beam is pulled into contact with the underlying land-ing pad by applying a bias voltage between the actuation and the landing pads. If the restoring force exceeds the adhesion force, the beam will recover from contact back to its free-standing state after removal of the bias voltage. However, when the restoring force is not high enough to overcome adhesion force, the beam remains adhered to the landing pad (substrate). The characteristic length at which the beams exhibit a transition from the adhered to the free-standing state is called the detach-ment length, l_d, and is determined by observing the adhesion behavior of an array of devices of various lengths. The parameters E, h, and t in equation (2) are design specifications where, $E = 170$ GPa, $h = 1.1$ µm, and $t = 2$ µm. The residual strain, ε, is estimated from the strain gauge, as described in Section 2.1.

2.3. Determination of Adhesion Force

Adhesion force measurements are carried out on as-fabricated double-clamped can-tilever devices. Systematic optical interferometric measurements of electrostatically actuated double-clamped beams are performed to determine the beam profile and the actual deflection during testing. Each die is mounted on the stage of a probe station (Signatone S-1160, Gilroy, CA) with vibration isolation features. A Michelson-type interferometer with an incoherent tungsten halogen light source filtered by a 550-nm green interference filter is used to measure the beam profile. When the beam is deflected, alternating bright and dark fringes appear as a consequence of constructive and destructive interference of green light, respectively. The distance between two successive dark (or bright) fringes is equivalent to half the wavelength of green light, or 275 nm. By analyzing the interferometer image, the beam profile under a given actuation condition is obtained. A home-built function generator capable of supplying DC voltages up to 180 V and with a voltage resolution of 0.1 V is used to pre-bias the beam before contacting the substrate. A second power supply (Agilent E3647A, Santa Clara, CA) with a higher voltage resolution of 0.01 V is coupled to the function generator and is used to sequentially bring the dimple into contact with the landing pad. The voltage is applied to the device through a set of micropositioners with fine microprobe tips. A 100 kΩ resistor is connected in series between the power supply and the device to avoid excessive current through the device. Interferometer images are recorded using a CCD camera (Sony SSC-C370) through a Leica DM LM optical microscope.

The adhesion force for coplanar impacting surfaces is determined in this type of electrostatically actuated device by considering relevant forces when the beam disengages from contact. Thus the adhesion force can be correlated to the difference between the restoring force and the electrostatic force at pull-off as given by equation (3):

$$F_{\text{adh}} = k\Delta z - \left(\frac{1}{2}\right) w\varepsilon_0 V_{\text{pull-off}}^2 2 \int_0^x \frac{dx}{g^2(x)}, \tag{3}$$

where the first term on the right hand side represents the restoring force of the spring and the second term is the opposing electrostatic force acting on a deformable capacitor. Here, k is the cantilever spring constant, Δz is the displacement when the beam is in contact with the substrate, w is the width of the actuation electrodes, ε_0 is electrical permittivity of air, $V_{\text{pull-off}}$ is the voltage at which the beam is removed from contact, and g is the distributed gap between the beam and landing pad over the actuation pad length, x. The distributed gap, g, for each device is determined by integrating the measured air gap as a function of actuation pad length, x, while the beam is in contact with the substrate. Equation (3) assumes a constant restoring force and negligible compressive/tensile stress.

Adhesion force calculations require accurate determination of the spring constant. Equation (4) is used to determine the spring constant of the beam from the applied voltage, V, and the corresponding displacement, Δz, before the beam comes into contact with the underlying landing pad.

$$k\Delta z = \left(\frac{1}{2}\right) w\varepsilon_0 V^2 2 \int_0^x \frac{dx}{g^2(x)}. \tag{4}$$

(a)   (b)   (c)   (d)

Figure 5. (a) A representative AFM topographic image (8×8 µm^2 scan size, 10 nm z-range) of the surface of polycrystalline silicon structural material. The average rms roughness measured in a 10×10 µm^2 scan area is 5 nm. Images (b)–(d) are SEM micrographs of the dimples with 2×2, 4×4 and 8×8 µm^2 sizes. Cross sections of these dimples from AFM measurements revealed raised edges highlighted here by the arrows.

3. Results and Discussion

Several beams in the die are removed to characterize dimple profiles and roughness using AFM. A representative AFM topographic image of polysilicon landing pad is shown in Fig. 5(a), indicating a root mean square (rms) roughness of 5 nm over the examined scan size. This value is typical for polysilicon films prepared *via* LPCVD method at $\sim$615°C from SiH$_4$ gas [42]. Topographic imaging of several dimples yields profiles that are consistent with the fabrication specifications and the average rms roughness is comparable to that of the landing pad. However, cross-sectional examination of the dimple profiles *via* AFM and SEM measurements reveals the existence of a raised dimple perimeter as shown in Figs 5(b)–(d). The raised dimple perimeter is an artifact of the reactive ion etching (RIE) definition step during fabrication.

To determine the work of adhesion, DCB devices with different beam lengths are tested. It is found that the DCB devices with the longest beams (240 µm in length) adhere permanently once in contact with the substrate, irrespective of the dimple size. In this case, the restoring force is not high enough to overcome the adhesion force between the two surfaces. On the other hand, devices that are 160 µm long require considerable applied biases in order to actuate the beam to contact due to high stiffness (high spring constant). Similar results are obtained for all dimple sizes. Devices with an intermediate length of 200 µm can be pulled into contact and upon removal of the bias return to their equilibrium state. Therefore, in order to estimate the apparent work of adhesion, a value of 220 ± 20 µm is chosen for the beam detachment length. From this value, the apparent work of adhesion is estimated to be 18 ± 6 mJ/m^2 for native oxide-coated surfaces under 50% RH, in agreement with a previously reported value of 20 mJ/m^2 for polysilicon beams with similar topography [40].

The displacement of the beam is analyzed from the actual interferometric beam profiles as illustrated in Fig. 6. The beam is deflected by slowly increasing the applied bias voltage and the beam profile is deduced for each specific bias voltage.

A computer program is used to convert pixel intensity of the output interferometer image into a vertical deflection of the beam *versus* position along the beam
plot. Deflection-to-contact (equivalent to the effective dimple-landing pad separation of 1.1 μm) is manifested by the appearance of four fringes on each side of the
dimple as shown in Fig. 4. The data in Fig. 6 are summarized in Fig. 7, showing the
centerline gap distance *vs.* applied voltage.

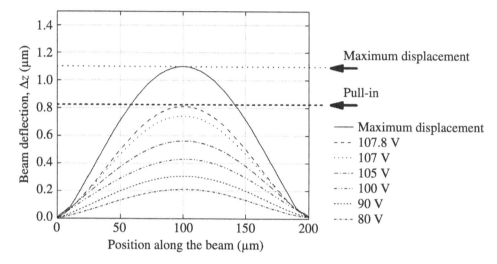

Figure 6. Beam profiles of a 200 μm long DCB taken at different applied voltages (load). $V_{\text{pull-in}}$ is
107.9 V.

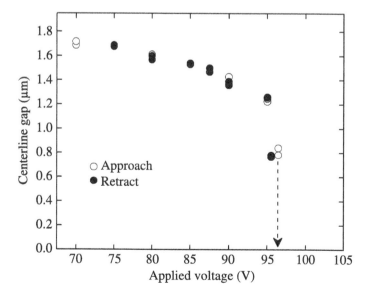

Figure 7. Plot of beam deflection (expressed as centerline gap) as a function of applied voltage. The
broken arrow traces the gap beyond the beam displacement threshold at which $\delta F / \delta z = 0$.

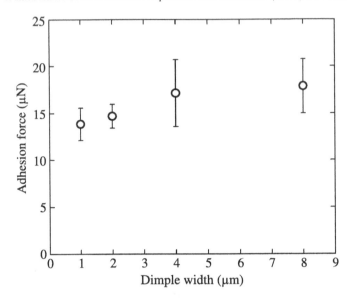

Figure 8. Adhesion force measured as a function of dimple width (square root of the apparent contact area). Vertical line at each data point represents standard deviation of at least 50 different measurements.

From the beam deflection data shown in Fig. 6 and calculations using equation (4), the 200 µm long beam of interest is found to have an average spring constant of 56.0 N/m compared to the calculated value of 40.8 N/m [43]. The measured spring constant values are found to be insensitive to the applied voltage prior to pull-in, indicating no beam stiffening effects [44]. Evaluation of the devices with intermediate lengths of 200 µm demonstrates a pull-off voltage that varies systematically with the dimple size, thus enabling adhesion force measurements. The adhesion force for a given dimple width is calculated from the measured spring constant and the average value of pull-off voltages obtained from different measurements using equation (3). Figure 8 represents the mean adhesion force measured on native oxide-coated surfaces with varied dimple width at 50% relative humidity and ambient room temperature conditions. Each data point represents the average of at least 50 different measurements from three separate dies. Results show that for polysilicon surfaces under these conditions, the adhesion force does not scale with the apparent dimple area.

Assuming two perfectly flat surfaces, a square dependence on dimple width would be expected. If the data were affected by the raised perimeter of the dimples (Fig. 5), as should be, the force would be directly proportional to dimple size. The dependence observed in our experiments is much weaker, and it strongly suggests that the real contact in micrometer scale involving rough, rigid material such as polysilicon consists of only a few asperities. The relatively high standard deviation around the mean value also illustrates the random nature of contacts in real surfaces at the micrometer scale.

4. Conclusions

A new microfabricated double-clamped beam adhesion force test structure is presented, and the range of device parameters which provide measurable adhesion hysteresis is provided. Experimental methodologies to quantify adhesion force as a function of device parameters have been developed and verified through systematic optical interferometry measurements. The apparent work of adhesion for polycrystalline silicon surfaces has also been estimated using the new test structure.

This study provides insight into the nature of contact between impacting rough and rigid surfaces. The results show a weak scaling of adhesion with the apparent contact area. While the observed trend may be affected by the raised perimeter profile revealed by AFM measurements of the apparent contact region, the study clearly demonstrates that mechanical contacts of microfabricated surfaces are intrinsically random and consist of only a few asperities in direct interaction.

Acknowledgements

The authors would like to acknowledge many helpful discussions with Prof. Roger Howe and Dr. J. Provine of Stanford University. All devices were fabricated in the UC Berkeley Microfabrication Laboratory. This work is supported by the Center on Interfacial Engineering for MEMS (CIEMS) under the auspices of DARPA MTO, and by the Center of Integrated Nanomechanical Systems (COINS) under the auspices of the National Science Foundation.

References

1. R. Maboudian and R. T. Howe, *J. Vac. Sci. Technol., B* **15**, 1 (1997).
2. R. Maboudian and C. Carraro, *Annu. Rev. Phys. Chem.* **55**, 35 (2004).
3. W. R. Ashurst, M. P. de Boer, C. Carraro and R. Maboudian, *Appl. Surface Sci.* **212–213**, 735 (2003).
4. R. Maboudian, *Surface Sci. Rep.* **30**, 209 (1998).
5. R. Maboudian and R. T. Howe, *Tribology Lett.* **3**, 215 (1997).
6. M. R. Houston, R. Maboudian and R. T. Howe, *J. Appl. Phys.* **81**, 3474 (1997).
7. B. Bhushan and M. T. Dugger, *Wear* **137**, 41 (1990).
8. F. W. DelRio, M. P. de Boer, J. A. Knapp, E. D. Reedy Jr, P. J. Clews and M. L. Dunn, *Nature Mater.* **4**, 629 (2005).
9. J. A. Greenwood and J. B. P. Williamson, *Proc. R. Soc. London, Ser. A* **295**, 300 (1966).
10. A. Kohno and S. Hyodo, *J. Phys. D: Appl. Phys.* **7**, 1243 (1974).
11. A. L. Weisenhorn, P. K. Hansma, T. R. Albrecht and C. F. Quate, *Appl. Phys. Lett.* **54**, 2651 (1989).
12. W. A. Ducker, T. J. Senden and R. M. Pashley, *Nature* **353**, 239 (1991).
13. M. Kappl and H.-J. Butt, *Particle Particle Syst. Char.* **19**, 129 (2002).
14. D. J. Dickrell III, M. T. Dugger, M. A. Hamilton and W. G. Sawyer, *J. Microelectromech. Syst.* **16**, 1263 (2007).
15. R. A. Quon, R. F. Knarr and T. K. Vanderlick, *J. Phys. Chem. B* **103**, 5320 (1999).
16. M. Benz, K. J. Rosenberg, E. J. Kramer and J. N. Israelachvili, *J. Phys. Chem. B* **110**, 11884 (2006).

17. S. Majumder, N. E. McGruer, G. G. Adams, P. M. Zavracky, R. H. Morrison and J. Krim, *Sensors Actuators A* **93**, 19 (2001).
18. K. Komvopoulos, *Wear* **200**, 305 (1996).
19. A. Mikrajuddin, F. G. Shi, H. K. Kim and K. Okuyama, *Mater. Sci. Semicond. Process.* **2**, 321 (1999).
20. M. P. de Boer and T. A. Michalske, *J. Appl. Phys.* **86**, 817 (1999).
21. C. H. Mastrangelo and C. H. Hsu, in: *Proc. IEEE Solid-State Sensor and Actuator Workshop*, Hilton Head, SC, USA, p. 208 (1992).
22. C. H. Mastrangelo, *Tribology Lett.* **3**, 223 (1997).
23. R. Legtenberg, H. A. C. Tilmans, J. Elders and M. Elwenspoek, *Sensors Actuators A* **43**, 230 (1994).
24. X. Rottenberg, I. de Wolf, B. K. J. C. Nauwelaers, W. de Raedt and H. A. C. Tilmans, *J. Microelectromech. Syst.* **16**, 1243 (2007).
25. K. Najafi and K. Suzuki, in: *Proc. IEEE Micro Electro Mechanical Systems*, Salt Lake City, Utah, USA, p. 96 (1989).
26. W. Y. Choi, T. Osabe and T.-J. K. Liu, *IEEE Trans. Electron Devices* **55**, 3482 (2008).
27. L. Kogut and K. Komvopoulos, *J. Appl. Phys.* **94**, 6386 (2003).
28. T. R. Thomas, B.-G. Rosen and N. Amini, *Wear* **232**, 41 (1999).
29. A. Majumdar and B. Bhushan, *J. Tribology* **113**, 1 (1991).
30. P. R. Nayak, *J. Lubrication Technol.* **93**, 398 (1971).
31. V. Bakolas, *Wear* **254**, 546 (2003).
32. M. Ciavarella, G. Demelio, J. R. Barber and Y. H. Yang, *Proc. R. Soc. London, Ser. A* **456**, 387 (2000).
33. K. L. Johnson, *Contact Mechanics*. Cambridge University Press, London (1998).
34. C. M. Mate, *Tribology on the Small Scale*. Oxford University Press, New York (2008).
35. M. P. de Boer and P. C. T. de Boer, *J. Colloid Interface Sci.* **311**, 171 (2007).
36. J. I. McCool, *Wear* **107**, 37 (1986).
37. A. W. Bush, R. D. Gibson and T. R. Thomas, *Wear* **35**, 87 (1975).
38. D. Bachmann and C. Hierold, *J. Micromech. Microeng.* **17**, 1326 (2007).
39. F. W. DelRio, M. L. Dunn, L. M. Phinney, C. J. Bourdon and M. P. de Boer, *Appl. Phys. Lett.* **90**, 163104 (2007).
40. R. Maboudian, W. R. Ashhurst and C. Carraro, *Tribology Lett.* **12**, 95 (2002).
41. J. Zhang, *PhD Dissertation*, University of California, Berkeley (2007).
42. B. G. Bush, F. W. DelRio, J. O. Opatkiewicz, R. Maboudian and C. Carraro, *J. Phys. Chem. A* **111**, 12339 (2007).
43. V. K. Varadan, K. J. Vinoy and K. A. Jose, *RF MEMS and Their Applications*. John Wiley & Sons, Hoboken, New Jersey (2003).
44. Q. Jing, T. Mukherjee and G. K. Fedder, in: *Proc. IEEE/ACM Int. Conf. on Computer Aided Design*, San Jose, CA, USA, p. 367, November 10–14 (2002).

Effect of Air–Plasma Pre-treatment of Si Substrate on Adhesion Strength and Tribological Properties of a UHMWPE Film

M. Abdul Samad, Nalam Satyanarayana and Sujeet K. Sinha [*]

Department of Mechanical Engineering, National University of Singapore, 9 Engineering Drive 1, Singapore 117576

Abstract

Ultra High Molecular Weight Polyethylene (UHMWPE) film is dip coated onto piranha and air–plasma treated silicon substrates, to study the effect of surface pre-treatment on the adhesion strength and the tribological properties of the film. After the pre-treatment, water contact angle measurements were conducted and the surface free energy of the Si substrate was calculated. It is observed that the air–plasma pre-treatment reduced the water contact angle (4.3°) considerably when compared to that of the piranha pre-treatment (21.3°), which resulted in an increase of the surface free energy of the Si substrate. Scratch tests were conducted for studying the adhesion property of UHMWPE films coated onto Si substrate. It was found that the plasma pre-treatment enhanced adhesion between the UHMWPE film and the Si substrate by more than two times when compared to the piranha pre-treatment. Wear tests were conducted to study the effects of pre-treatments on the tribological properties of the UHMWPE film. UHMWPE film coated onto plasma pre-treated Si showed a wear life of about 50 000 cycles (25 times higher) as compared to that of 2000 cycles when it was coated onto piranha pre-treated Si, tested at a normal load of 1 N and a rotational speed of 200 rpm.

Keywords

Air–plasma treatment, UHMWPE film, wear durability of Si

1. Introduction

Silicon (Si) is one of the widely used materials in microelectromechanical systems (MEMS) and microsystems. However, bare Si, without proper surface modification, exhibits high friction, adhesion and wear [1]. Several types of ultra-thin films have been proposed recently for improving the tribological properties of Si and MEMS made from Si [2–6]. One important category of these films are the polymer coatings, which have found their way in various tribological applications due

---

[*] To whom correspondence should be addressed. Tel.: +65-6516 4825; Fax: +65-6779 1459; e-mail: mpesks@nus.edu.sg

Adhesion Aspects in MEMS/NEMS
© Koninklijke Brill NV, Leiden, 2010

to their excellent self-lubricating properties, low production cost, ease of coating procedures, and corrosion resistance [7–9]. However, despite their self-lubrication property, these polymer coatings usually suffer from poor adhesion to the substrate, thus resulting in low wear life [10, 11]. Various pre-treatment processes, such as piranha treatment, have been used to enhance the adhesion property of the Si substrate and the polymer film. Piranha treatment, besides using harmful chemicals such as hydrogen peroxide and sulphuric acid, is a laborious process involving long reaction time (approximately 2 h [2]).

Lately, the potential of air–plasma, as an effective pre-treatment process for improving the adhesion strength between the polymer films and the substrates, is being explored extensively due to its various advantages such as no use of harmful chemicals, ease of handling, ease of adaptability to industrial applications and ease of treating intricate geometries [12, 13].

UHMWPE was selected as the polymer coating for its excellent wear resistance and self-lubricating properties as well as for exploring a way to improve the tribological properties of polymer films by suitable environmental-friendly pre-treatment of the substrate prior to the polymer coating [11].

Thus, the main objective of this study was to investigate and compare the effects of air–plasma and piranha pre-treatments of the Si substrate on the adhesion strength and the tribological properties of the UHMWPE film coated onto the Si surfaces. Scratch test was conducted for the adhesion study whereas ball-on-disk test was employed for the tribological evaluation of the films. Various surface characterization techniques such as X-ray photoelectron spectroscopy (XPS), atomic force microscopy (AFM) and field-emission scanning electron microscopy (FESEM) were used to study the surfaces.

2. Experimental Procedures

2.1. Materials

Polished single crystal silicon (100) wafers were used as the substrate. The Si wafers were cut into pieces of approximately 2 cm × 2 cm and then used for the surface modification. UHMWPE polymer powder (Grade: GUR X 143) used for coating the specimens was supplied by Ticona Engineering Polymers, Germany and was procured from a local Singapore supplier ("melt index MFR 190/15" = 1.8 ± 0.5 g/10 min; bulk density = 0.33 ± 0.03 g/cm^3; average particle size = 20 ± 5 μm). Decahydronaphthalin (decalin) was used as the solvent to dissolve the polymer powder prior to dip-coating.

2.2. Pre-treatment Procedures

In the present study, the Si substrate was subjected to two different pre-treatments, piranha and air–plasma pre-treatments, prior to the UHMWPE coating. The pre-treatment procedures were as follows.

2.2.1. Piranha Pre-treatment

Si samples were ultrasonically cleaned successively using soapy water and distilled water followed by rinsing with acetone for 10 min. The samples were then hydroxylated by immersing in a piranha solution, a mixture of 7 : 3 (v/v) 98% H_2SO_4 and 30% H_2O_2 at 60–70°C for 50 min. After piranha treatment, the samples were thoroughly rinsed successively with distilled water and acetone. The total time for treating the Si samples by piranha treatment was approximately 90 min. Further details can be found in [8].

2.2.2. Air–Plasma Pre-treatment

Si samples were cleaned with distilled water and acetone successively in an ultrasonic bath prior to drying using nitrogen gas. The samples were then air–plasma treated using a Harrick Plasma Cleaner/Steriliser. The sample surface was exposed to plasma under vacuum for approximately 5 min using an RF power of 30 W. Care was taken not to expose the surface to any further contamination and thus it was immediately processed for the dip coating. The total time for treating the Si samples by air–plasma treatment was approximately 20 min.

2.3. Coating Procedure

UHMWPE polymer in powder form was dissolved in decalin by heating the solution to 80°C for 30 min followed by another heating sequence to 160°C for 30 min. Magnetic stirrers were used for uniform distribution of heat in the solution and for speeding up the dissolution process. The solution was used once it turned from milky to transparent appearance indicating a complete dissolution. The specimens were dip-coated using a custom-built dip-coating machine which could submerge and withdraw the sample at a speed of 2.1 mm/s [8, 11, 14, 15]. The samples were held in the polymer–decalin solution for 30 s in submerged condition prior to withdrawal. The coated samples were dried in air for 60 s and then post-heat-treated in a hot oven at 120°C for about 20 h. After the post-heat-treatment, the samples were cooled slowly to room temperature in an oven and stored carefully in a desiccator before proceeding to tribological testing. Further details can be found in [2].

2.4. Surface Characterization and Analysis

A VCA Optima Contact Angle System was used for the measurement of contact angles with three different liquids: de-ionized water, ethylene glycol, and hexadecane. A 0.5 µl droplet was used for the contact angle measurement. A total of five independent measurements were performed randomly at different locations on the samples and an average value was taken for each sample. The measurement error was within ±3°.

An atomic force microscope (Dimension 3000 AFM, Digital Instruments, USA) was used to study the surface topography of the Si bare surface after appropriate pre-treatment, and after coated with UHMWPE film. A silicon tip was used for scanning and images were collected in air in the tapping mode.

A Kratos Analytical AXIS HSi spectrometer (XPS) was used for the surface analysis. XPS (Al Kα source) imaging was performed with an X-ray source (1486.6 eV photons) at a constant dwell time of 100 ms and a pass energy of 40 eV. The core level signals were obtained at a photoelectron take-off angle of 90° (with respect to the sample surface). All binding energies (BEs) were referenced to the C_{1s} hydrocarbon peak at 284.6 eV. In peak synthesis, the line width (full width at half maximum or FWHM) for the Gaussian peaks was maintained constant for all components in a particular spectrum. The curve de-convolution of the obtained XPS spectra was performed using XPS Peak Fitting Program, XPSPEAK41.

2.5. Thickness Measurement

Thickness of the polymer film on Si surface was measured by observing the cross section of the sample (after film coating) using field emission scanning electron microscopy (FESEM, Hitachi S4300). The coated samples were cut using a diamond scriber and a plier and mounted with their cross sections horizontal. Ten independent measurements were taken on each sample and the average value was reported. The thickness variation was within ± 1 µm. Before observing the samples under FESEM, gold was deposited on the films at 10 mA for 40 s using a JEOL, JFC-1200 Fine Coater.

2.6. Surface Free Energy Calculations

Surface free energy calculations were carried out using the acid–base approach [16]. This method requires measurement of contact angles on the substrate using three different, completely characterized liquids, two of which have to be polar and the third apolar. The dispersion component of the surface tension (γ_l^d), the acid component of the surface tension (γ_l^+) and the base component of the surface tension (γ_l^-) of all the liquids were obtained from literature. The measured contact angle values are substituted in the following equation [16]:

$$0.5(1 + \cos\theta)\gamma_l = \left(\gamma_s^d \gamma_l^d\right)^{1/2} + \left(\gamma_s^- \gamma_l^+\right)^{1/2} + \left(\gamma_s^+ \gamma_l^-\right)^{1/2},$$

where γ_s^d, γ_s^+ and γ_s^- refer to the dispersion, acid, and base components of the surface tension (surface free energy) of the solid substrate, respectively. These unknowns are calculated by solving the three equations resulting from the substitution of the contact angle values of the three liquids. The total surface free energy of the solid substrate is given by [16]:

$$\gamma_s = \gamma_s^d + 2\left(\gamma_s^+ \gamma_s^-\right)^{1/2}.$$

2.7. Scratch Test

Scratch tests were carried out on a custom-built scratch tester using a conical diamond tip of radius 2 µm. The length of the scratch and the traverse velocity of the tip were kept constant for each scratch as 10 mm and 0.1 mm/s, respectively. Normal load was also kept constant for each scratch. However, the normal load was varied from 10 mN to 100 mN with an increment of 10 mN for every successive scratch. After the test, the scratches were characterized using FESEM/EDS (energy dispersive spectroscopy) technique to ascertain the critical load defined as the load at which the polymer film showed signs of failure which was characterized by the peeling-off or ploughing mechanism and the appearance of Si peaks (because of exposed substrate) in the EDS spectrum. Before observing the scratches under FESEM (Field Emission SEM), gold was deposited on the films at 10 mA for 40 s using a JEOL, JFC-1200 Fine Coater.

2.8. Friction and Wear Tests

Ball-on-disk wear tests were carried out on a UMT-2 equipment (Universal Micro Tribometer, CETR Inc., USA) under dry conditions using ball-on-disk mode. A silicon nitride ball of $\Phi 4$ mm with a root mean square (RMS) surface roughness of 5 nm (as provided by the supplier) was used as the counterface material. The ball was thoroughly cleaned with acetone before each test. The wear track radius was fixed at 2 mm for all the tests. In this study, wear life of the thin film is defined as the number of cycles when the coefficient of friction exceeds 0.3 or when continuous large fluctuations are observed in the coefficient of friction data (indicative of film failure), whichever happens first [17]. The wear tests were carried out at a normal load of 1 N and a rotational speed of 200 rpm corresponding to a sliding velocity of 0.042 m/s. At least ten repetitions were carried out and average values were calculated to report the final friction and wear life data. Tests were carried out in a clean booth environment (class 100) at a temperature of $25 \pm 2°C$ and a relative humidity of $55 \pm 5\%$. After every test, the counterface and sample surfaces were examined under an optical microscope for investigating the wear mechanisms. At least 10 samples from 3 different batches were tested and an average value of wear life was reported.

3. Results and Discussion

3.1. Effect of Pre-treatment on the Wettability and the Surface Free Energy of the Si Substrates

Table 1 shows the contact angles measured with DI-water, ethylene glycol and hexadecane, and surface free energy of the untreated Si, piranha treated Si and air–plasma treated Si. It is observed that the Si surface which was subjected to air–plasma pre-treatment exhibited lowest contact angles leading to an increase in the surface free energy when compared to that of the piranha treated and untreated Si samples which exhibited lower surface free energies.

Figure 1 shows AFM topographical images for untreated Si, piranha-treated Si and plasma-treated Si, and the corresponding roughness values are reported in Table 1. Both the air–plasma and the piranha pre-treatments were effective in reducing the roughness of the Si surface which might be because of the removal of the organic and/or inorganic contaminants.

Table 1.
Contact angles measured with DI-water, ethylene glycol and hexadecane, for various surface pre-treatments of Si and corresponding surface free energy values. The table also lists the RMS (root mean square) roughness of untreated Si and after two different treatment methods used

Surface	DI-water (°)	Ethylene glycol (°)	Hexadecane (°)	Surface free energy (mJ/m^2)	RMS roughness (nm)
Si-untreated	38.4	21.6	6.1	44.7	0.53
Si-piranha treated	21.3	17.8	4.8	45.4	0.25
Si-plasma treated	4.3	3.9	4.7	47.3	0.29

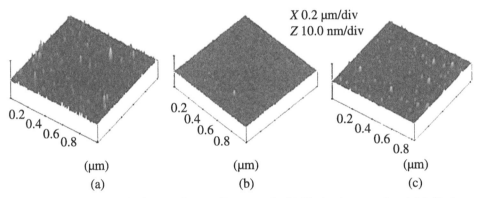

Figure 1. AFM topography images for (a) Si-untreated; (b) Si-piranha treated and (c) Si-plasma treated. The scan area used is 1 µm × 1 µm and the vertical scale is 10 nm.

Figure 2 shows the FESEM/AFM morphology of UHMWPE films on piranha-treated and plasma-treated Si. The morphology of polymer film is almost the same in both cases and we can conclude that the pre-treatment of Si does not have any effect on the morphology of UHMWPE film coated onto Si. It is also observed that the pre-treatment of the Si surface does not affect the thickness of the UHMWPE film. The film thicknesses were calculated as explained earlier and were found to be ~16.6 µm and ~15.2 µm for the air–plasma pre-treatment and piranha pre-treatment, respectively. Further, Si-plasma/UHMWPE showed a water contact angle of 117° whereas Si-piranha/UHMWPE showed a water contact angle of 108°.

3.2. XPS Analysis of the Si Surface

XPS was used to study the chemical state of the Si surfaces after the plasma and piranha treatments. Table 2 shows the atomic percent of C and O on untreated, piranha- and plasma-treated Si surfaces obtained from XPS analysis. It can be observed from Table 2 that after the plasma treatment there is a decrease in the atomic percent of C and an increase in the atomic percent of O when compared to untreated and piranha-treated Si surfaces. The core level spectra (after curve fitting) of C_{1s} and O_{1s} of piranha- and plasma-treated Si surfaces are shown in Figs 3 and 4, respectively. The curve-fit data with peak assignments are shown in Table 3. The C_{1s}

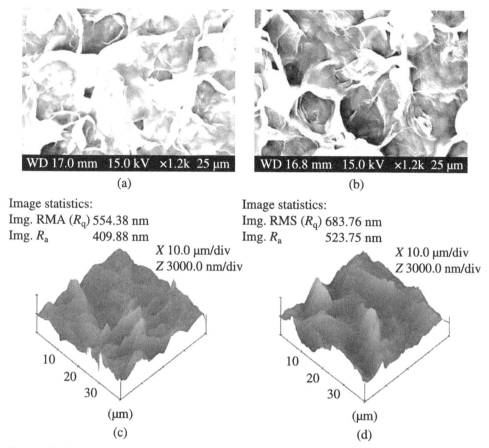

Image statistics:
Img. RMA (R_q) 554.38 nm
Img. R_a 409.88 nm

Image statistics:
Img. RMS (R_q) 683.76 nm
Img. R_a 523.75 nm

X 10.0 µm/div
Z 3000.0 nm/div

X 10.0 µm/div
Z 3000.0 nm/div

(c)   (d)

Figure 2. FESEM/AFM morphology of (a), (c) Si-piranha/UHMWPE and (b), (d) Si-plasma/ UHMWPE films.

Table 2.
Atomic percents of C and O of piranha- and plasma-treated Si surfaces

Surface	C (%)	O (%)
Si-untreated	17.7	37.6
Si-piranha treated	19.7	34.3
Si-plasma treated	16.9	42.9

spectra for both piranha- and plasma-treated Si surfaces contain peak components corresponding to C–C/C–H at 284.6 eV, C–O at 286.2 eV and C=O at 289 eV [18]. However, the peak area is less in the case of plasma-treated Si than that corresponding to the untreated and piranha-treated Si for all the three peaks suggesting that the air–plasma treatment is more effective in removing the organic contaminants from the Si surface.

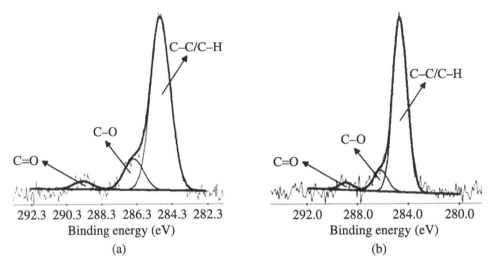

Figure 3. C_{1s} core-level spectra (after curve fitting) of (a) piranha-treated Si and (b) plasma-treated Si.

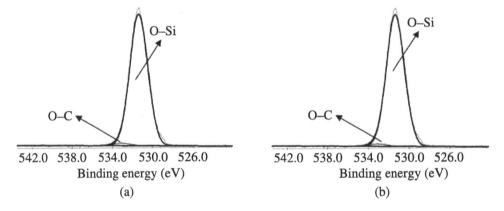

Figure 4. O_{1s} core-level spectra (after curve fitting) of (a) piranha-treated Si and (b) plasma-treated Si.

The O_{1s} spectra of both piranha- and plasma-treated Si surfaces contain peak components corresponding to O–Si and/or OH groups at 532 eV [19] and O–C group at 533.3 eV [18]. However, the peak area of O–Si is higher and that of O–C is lower, in the case of air–plasma-treated Si when compared to those of the untreated and piranha-treated Si. The low peak area of O–C for plasma-treated Si also supports the effective removal of organic contaminants from Si surface through plasma treatment. The higher peak area at a binding energy (BE) of 532 eV for plasma-treated Si suggests that the air–plasma treatment produces more hydroxyl groups on the Si surface when compared to the untreated and piranha-treated Si [19]. The generation of these hydroxyl groups helps in increasing the surface free energy and thus making the surface more hydrophilic which is expected to improve the adhesion between the polymer film and the surface.

Table 3.
Curve-fit data and assignments of the peaks for C_{1s} and O_{1s} spectra of untreated, piranha- and plasma-treated Si surfaces

Surface	Core-level spectrum	Binding energy (eV)	Peak assignment	Area (arb. units)	FWHM[*] (eV)
Si-untreated	C_{1s}	284.6	C–C/C–H	1185	1.30
		286.2	C–O	172	1.30
	O_{1s}	531.9	O–Si	7365	1.60
Si-piranha treated	C_{1s}	284.6	C–C/C–H	1435	1.30
		286.2	C–O	258	1.30
		289.0	C=O	66	1.30
	O_{1s}	532.0	O–Si	6510	1.53
		533.3	O–C	289	1.53
Si-plasma treated	C_{1s}	284.6	C–C/C–H	1031	1.32
		286.2	C–O	131	1.32
		288.9	C=O	47	1.32
	O_{1s}	532.0	O–Si	7414	1.56
		533.3	O–C	133	1.56

[*]FWHM means full width at half maximum.

3.3. Effect of Pre-treatment on the Adhesion Strength between the UHMWPE Film and the Si Substrate

Scratch tests were performed to investigate the effect of the pre-treatment on the adhesion strength between the UHMWPE film and the Si substrate.

Figure 5 shows the FESEM images of scratches made at different loads on Si-untreated/UHMWPE, Si-piranha/UHMWPE and Si-plasma/UHMWPE samples and the corresponding EDS spectra inside the scratches. Results showed that the Si-untreated/UHMWPE and Si-piranha/UHMWPE samples failed at lower critical loads when compared to that of Si-plasma/UHMWPE. UHMWPE film on the untreated and the piranha-treated Si failed at scratch loads of 10 mN and 30 mN, respectively. The EDS spectrum further confirmed the film failure by the presence of Si peak in the peeled off regions. The UHMWPE film on plasma-treated Si failed at a scratch load of ~80 mN, exhibiting a greater scratch resistance. The scratch test results clearly demonstrate that the plasma treatment of Si enhances the adhesion between the UHMWPE film and Si which is much better than that corresponding to piranha-treated Si. The high adhesion strength of UHMWPE film with plasma-treated Si is attributed to the increase in the surface free energy due to the generation of –OH groups on Si as evident from the XPS analysis. As mentioned earlier, the generation of these hydroxyl groups helps in increasing the surface free energy and thus making the surface more hydrophilic which is expected to improve the adhesion between the polymer film and the surface.

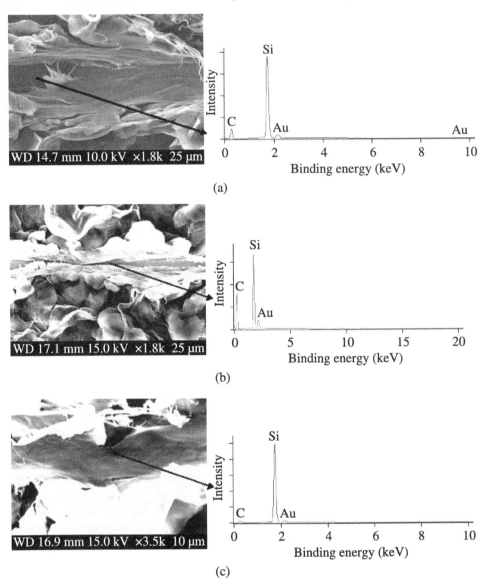

Figure 5. (a) FESEM morphology of a scratch made at 10 mN on Si-untreated/UHMWPE surface and EDS spectrum inside the scratch (right). (b) FESEM morphology of a scratch made at 30 mN on Si-piranha/UHMWPE surface and EDS spectrum inside the scratch (right). (c) FESEM morphology of a scratch made at 80 mN on Si-plasma/UHMWPE surface and EDS spectrum inside the scratch (right).

3.4. Effect of Pre-treatment on the Tribological Properties of the UHMWPE Film

Wear tests were carried out on a ball-on-disk micro-tribometer (CETR Inc., USA) for the Si-piranha/UHMWPE and Si-plasma/UHMWPE samples for comparison

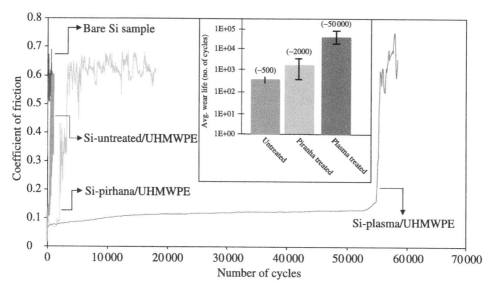

Figure 6. Variation of the coefficient of friction with the number of sliding cycles from typical test runs for bare Si, Si-untreated/UHMWPE, Si-piranha/UHMWPE and Si-plasma/UHMWPE samples. A normal load of 1 N and a rotational speed of 200 rpm (0.042 m/s) were used in the sliding tests. (Inset) Bar graph showing the average wear lives (in terms of number of cycles) for the three different conditions.

purpose. The normal load was kept constant at 1 N and the rotational speed at 200 rpm (0.042 m/s). Ten runs for each condition were conducted.

Figure 6 shows the variation of the coefficient of friction with respect to the number of cycles for bare Si, Si-untreated/UHMWPE, Si-piranha/UHMWPE and Si-plasma/UHMWPE samples (typical data). It can be observed that the UHMWPE film on plasma-treated Si demonstrates a higher average wear life (~50 000 cycles) when compared to that of the UHMWPE film coated onto piranha-treated Si (~2000 cycles). This remarkable improvement (~25 times) in wear durability is attributed to the enhanced adhesion between UHMWPE film and plasma-treated Si and UHMWPE film and as demonstrated by the scratch tests. Thus the two tests (scratch and ball-on-disk) have proved that air–plasma treatment of Si is a very cost-effective and environmental-friendly method of enhancing the adhesion property of a solid lubricant polymer film to Si substrate.

Figure 7 shows the FESEM/EDS images of the wear track for the bare Si (after 500 cycles), Si-untreated/UHMWPE (after 1000 cycles), Si-piranha/UHMWPE (after 2000 cycles) and Si-plasma/UHMWPE (after 10 000 cycles). The respective optical micrographs of the counterface Si_3N_4 balls are also shown. It can be observed that in the cases of Si-untreated/UHMWPE and Si-piranha/UHMWPE, the film failed after 1000 cycles and 2000 cycles respectively, as is clear from the EDS spectra of the wear tracks in which the Si peak represents the exposed surface of the substrate, indicating the film rupture. It can also be observed that there was a considerable polymer transfer onto the sliding ball, indicating poor adhesion of the

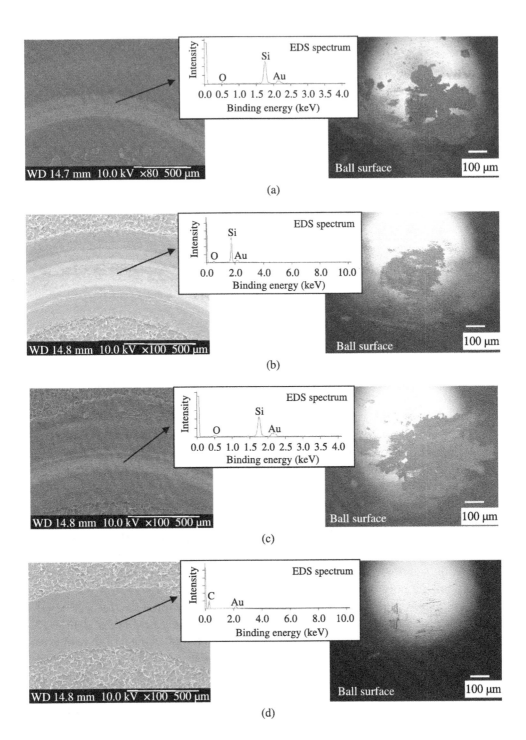

(a)

(b)

(c)

(d)

film to the substrate. On the contrary, the Si-plasma/UHMWPE film did not fail even after 10 000 cycles as evident from the EDS spectrum. No polymer transfer onto the sliding ball is an indication of the strong adhesion of the polymer film to the substrate.

4. Conclusions

In the present work, we studied the effect of plasma pre-treatment of Si surface on the wear durability and adhesion properties of UHMWPE film dip-coated onto Si surface and the results were compared with those corresponding to piranha pre-treatment of Si surface. Following conclusions are drawn from the present study:

1. Plasma pre-treatment of Si improved the hydrophilic nature of the Si surface (with a water contact angle of 4.3°) resulting in an increase in the surface free energy when compared to that of the piranha pre-treatment.

2. The plasma pre-treatment of Si enhanced the adhesion between UHMWPE film (15.2–16.6 µm in thickness) and Si surface when compared to that of the piranha treatment of Si. In scratch tests, the critical load of Si-plasma/UHMWPE (~80 mN) was found to be more than two times of the critical load of Si-piranha/UHMWPE (~30 mN).

3. The UHMWPE film coated onto plasma-treated Si showed very high wear durability (~25 times higher) when compared to the same film coated onto piranha-treated Si which is mainly attributed to the increased adhesion between the UHMWPE film and the plasma-treated Si. In sliding tests against 4 mm diameter silicon nitride ball, Si-plasma/UHMWPE demonstrated an average wear life of ~50 000 cycles as compared to a wear life of only ~2000 cycles demonstrated by the Si-plasma/UHMWPE at a normal load of 1 N and a rotational speed of 200 rpm.

Air–plasma pre-treatment of Si can be used for depositing a lubricating layer in MEMS made from Si for higher wear durability as shown in the case of UHMWPE films.

Figure 7. (a) Wear track and ball surface for untreated Si after 500 cycles, (b) wear track and ball surface from Si-untreated/UHMWPE after 1000 cycles, (c) wear track and ball surface from Si-piranha treated/UHMWPE after 2000 cycles and (d) wear track and ball surface from Si-plasma treated/UHMWPE after 10 000 cycles of sliding. For all cases, left picture shows FESEM/EDS image of wear track and the right picture shows optical micrograph of ball surface. A normal load of 1 N and a rotational speed of 200 rpm (linear speed of 0.042 m/s) were used in the sliding tests. All samples were gold coated for conduction improvement for SEM characterization.

Acknowledgements

This article is based on research work supported by the Singapore National Research Foundation under CRP Award no. NRF-CRP 2-2007-04.

One of the authors (M. Abdul Samad) wishes to thank National University of Singapore (NUS) for providing graduate research scholarship.

References

1. V. DePalma and N. Tillman, *Langmuir* **5**, 868–872 (1989).
2. N. Satyanarayana and S. K. Sinha, *J. Phys D: Appl. Phys.* **38**, 3512–3522 (2005).
3. N. Satyanarayana, N. N. Gosvami, S. K. Sinha and M. P. Srinivasan, *Philosophical Magazine* **87**, 3209–3227 (2007).
4. D. B. Asay, M. T. Dugger and S. H. Kim, *Tribology Letters* **29**, 67–74 (2008).
5. J. J. Nainaparampil, K. C. Eapen, J. H. Sanders and A. A. Voevodin, *J. Microelectromechanical Systems* **16**, 836–843 (2007).
6. J. Zhao, M. Chen, J. Liu and F. Yan, *Thin Solid Films* **517**, 3752–3759 (2009).
7. W. Liu, F. Zhou, L. Yu, M. Chen, B. Li and G. Zhao, *J. Mater. Res.* **17**, 2357–2362 (2002).
8. N. Satyanarayana, S. K. Sinha and B. H. Ong, *Sensors Actuators A* **128**, 98–108 (2006).
9. N. Satyanarayana, S. K. Sinha and L. Shen, *Tribology Letters* **28**, 71–80 (2007).
10. C. Sun, F. Zhou, L. Shi, B. Yu, P. Gao, L. Zhang and W. Liu, *Appl. Surface Sci.* **253**, 1729–1735 (2006).
11. M. Minn and S. K. Sinha, *Surface Coatings Technol.* **202**, 3698–3708 (2008).
12. W. Possart (Ed.), *Adhesion: Current Research and Applications*. Wiley-VCH Verlag GmbH & Co. KGaA, Weinheim, Germany (2005).
13. M. Abdul Samad, N. Satyanarayana and S. K. Sinha, *Surface Coatings Technol.* **204**, 1330–1338 (2009).
14. S. P. Khedkar and S. Radhakrishan, *Thin Solid Films* **303**, 167 (1997).
15. H. J. Butt, K. Graf and M. Kappl, *Physics and Chemistry of Interfaces*, 2nd edn, p. 138. Wiley-VCH, Berlin, Germany (2005).
16. R. J. Good, *J. Adhesion Sci. Technol.* **6**, 1269–1302 (1992).
17. K. Miyoshi, *Solid Lubrication Fundamentals and Applications*, p. 266. Marcel Dekker, New York (2001).
18. G. Beamson and D. Briggs, *High Resolution XPS of Organic Polymers: The Scienta ESCA300 Database*, pp. 277–283. John Wiley and Sons, Chicester, UK (1992).
19. M. Kuemmel, J. Allouche, L. Nicole, C. Boissiere, C. Laberty, H. Amenitsch, C. Sanchez and D. Grosso, *Chem. Mater.* **19**, 3717–3725 (2007).

Part 4

Adhesion in Practical Applications

A Review of Adhesion in an Ohmic Microswitch

George G. Adams [a,*] **and Nicol E. McGruer** [b]

[a] Department of Mechanical and Industrial Engineering, Northeastern University, Boston, MA 02115, USA

[b] Department of Electrical and Computer Engineering, Northeastern University, Boston, MA 02115, USA

Abstract

Due to fundamental scaling laws, the effect of surface forces becomes relatively large for microscale structures, such as MEMS. The operation of an ohmic MEMS microswitch relies on repeated make-and-break contacts, typically between each of two metal tips and a metal drain. Large contact forces are desirable for low contact resistance, but the resulting plastic deformation at this small scale, as well as the breaking through of contaminant films, further increases the adhesion force which can then exceed the restoring force of the actuator. This stuck-closed failure, also known as stiction, is a major hindrance to switch reliability and to the ultimate commercialization of these devices. In this article we review the experimental and modeling efforts which have been directed to achieving an improved understanding of adhesion in an ohmic MEMS microswitch.

Keywords

Adhesion, microswitch, MEMS

1. Introduction

Microswitches have the potential to replace conventional relays and solid-state switches in a number of low-power applications. These microswitches are smaller in size and switch faster than existing mechanical relays, such as reed relays. MEMS switches feature greater isolation as well as lower on-resistance than transistor switches. Electrostatically actuated MEMS switches consume very little steady-state power, unlike reed relays and p-i-n diode switches. The potential uses of MEMS switches include RF switching in applications such as cell phones, phase shifters and smart antennas [1], as well as reed relay replacement in Automatic Test Equipment (ATE) and industrial and medical instrumentation. For many of these applications lifetimes of ten billion cycles or more are expected.

There are two main actuation mechanisms commonly used for MEMS switches — electrostatic and magnetic. The power consumption for electrostatic

---

[*] To whom correspondence should be addressed. E-mail: adams@coe.neu.edu

switches is generally lower but the actuation voltage is higher than in magnetic switches [2]. In both cases the contact force is typically limited to a few hundred micro-Newtons. These low contact forces have caused the contacts to become the greatest barrier to obtaining sufficient reliability for commercial applications.

The low contact force makes it difficult to obtain a low and stable contact resistance, often leading to a resistance which increases steadily with cycling until the switch is no longer useful. It also encourages the choice of gold as a contact material because it is soft (giving it a large real contact area) and relatively inert (producing a clean metal-to-metal contact). However, this combination of a large real contact area and a clean metal contact increases the adhesion force, thereby increasing the possibility of stiction (i.e., a 'stuck-closed' failure). Furthermore, the combination of high adhesion and a soft material can increase failure due to morphology changes as well as material transfer from the contact tip to the drain. Note that at the microscale the effect of adhesion is much greater than at the macroscale due to the scaling of surface forces with area. Furthermore, the contacts in a microswitch occur between the nanoscale asperity peaks; this scaling further increases the importance of surface forces and the possibility of a stuck-closed failure and/or detrimental changes in surface topography.

It is noted that adhesion is a problem in both ohmic contact switches and capacitive switches. However, the physical basis of adhesion differs significantly between these two types of devices. Stiction in capacitive switches is generally due to dielectric charging, whereas in ohmic contacts the adhesion force is due to van der Waals forces, meniscus forces, and, for very clean contacts, metallic bonding. In this review we restrict our attention to adhesion in ohmic switches.

2. Experimental Investigations

A schematic of an electrostatically actuated MEMS switch is shown in Fig. 1. When a voltage difference is applied between the gate and the cantilever structure, the resulting electrostatic force pulls the cantilever downward toward the gate. Contact is then made between the two tips on the end of the cantilever structure and the drain, completing the circuit and allowing current to flow. The dynamic behavior of the switch [3, 4] causes the tips to collide with the drain resulting in an impact force which can be several times greater than the static contact force. Furthermore, as the tips strike the drain, the momentum of the rest of the cantilever structure causes it to flex; when the stored elastic energy is recovered the tip can bounce off the drain one or more times before settling down to the final closed position. The effect of adhesion on tip dynamics, especially on contact bounce, was investigated by Decuzzi *et al.* [5]. It was found that the response could be strongly affected by the adhesion interaction and by the ratio of the tip-to-drain distance and the beam-to-gate separation. This switch dynamics is important for several reasons — the switch is not useable until the bouncing has ceased; each switching event corresponds to two or more contact closing cycles thereby exacerbating any damaging effects due to an

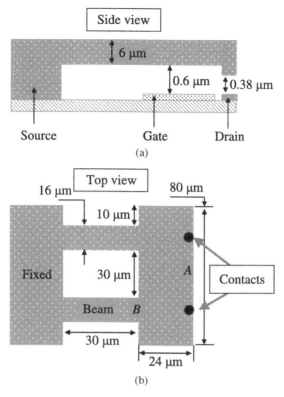

Figure 1. Schematics of (a) side view and (b) top view of an electrostatically actuated microswitch.

impact; and the large impact force can affect the adhesion force due to increased plastic deformation. An SEM image of the switch is shown in Fig. 2. Note that although these tips are nominally flat, other switches have rounded hemispherical tips.

Typical results showing contact resistance *vs* contact force [6] are shown after the first switch closure (Fig. 3(a)) and after 10 000 closing cycles (Fig. 3(b)). The measurements made were of contact resistance *vs* gate voltage, but a simple structural model was used to calculate the corresponding contact force from the gate voltage. Switch dynamics was avoided by slowly ramping the gate voltage so as to close the switch statically. Nonetheless there is some electrostatic snap-in which causes the contact to first occur with a small but finite force. On subsequent loading during the first contact cycle the resistance decreases, probably due to a combination of plastic deformation (which increases the contact area) and a partial mechanical breaking through of any contaminant film. Note that although the tip is nominally flat, there are asperities on the surface and intimate contact occurs between the tips of these asperities and the drain (which is typically smoother than the contact tip). During unloading the resistance remains almost constant and separation occurs with a negative force of about 16 μN.

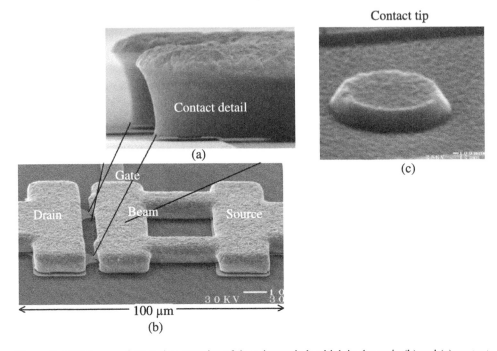

Figure 2. SEM image of (a) region near tips of the microswitch which is shown in (b) and (c) contact tip.

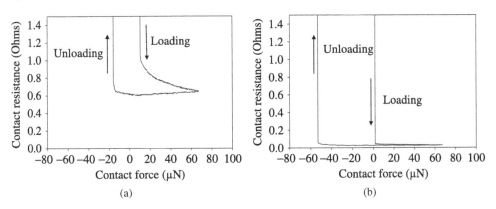

Figure 3. Measured contact resistance *vs* contact force after (a) the first cycle and (b) 10 000 cycles.

On continued cycling the contact resistance decreases (Fig. 3(b)) and becomes relatively insensitive to the contact force. Unfortunately, the decrease in the resistance is inevitably accompanied by an increase in the adhesion force, i.e., to about 54 μN in the case shown. Both the decrease in contact resistance and the increase in the adhesion force are consistent with the notion that the contact area increases due to repeated plastic deformation and/or that the contacts are being cleaned of contaminants by the repeated make-and-break contacts. In fact, the term 'contact scrub' is often used in the switch industry to indicate a cleaning of the contacts with repeated loading/unloading which is accompanied by some tangential sliding. Further cycling generally results in an increase in the adhesion force, often until the restoring force of the structure is insufficient to pull the contacts apart, resulting in a stuck-closed failure.

The results of adhesion testing are sensitive to the materials used and to the surface preparation. From the point-of-view of reducing adhesion, a high hardness and a high elastic modulus are desirable to reduce the contact area under a given force. Unfortunately, the contact resistance increases as a consequence of the reduction in contact area. Thus much of the materials testing has focused on identifying materials with both a high hardness and a high electrical conductivity. Also important are the roles of the contact force and of surface preparation which have also been investigated. The higher the force, the greater is the contact area and the lower is the expected contact resistance. However, the large contact area increases the adhesion force and makes a stuck-closed failure more likely to occur. On the other hand, if the contact force is too small, the less likely is the intimate metal-to-metal contact due to the intervening contaminant film, resulting in an unacceptably high contact resistance. Thus there are various trade-offs in designing the switch contacts.

The role of contact force and surface cleaning was investigated by Schimkat [7] for Au, Au alloy with 5% Ni, and Rh. A cleaning procedure consisting of isopropanol followed by electrical switching cycles (also known as a Schaltreinigung cleaning procedure) was effective in reducing the minimum force needed to attain a stable contact resistance. Contact resistance *vs* force and contact force *vs* position were measured using an experimental setup in dry nitrogen with a piezo-actuator. Forces were in the range of 0.1 mN to 10 mN. Gold was found to be a poor contact material due to high adhesion, whereas both the gold alloy and rhodium were found to be reliable under the conditions tested.

Using interfacial force microscopy, Tringe *et al.* [8] studied the electrical contact between a sputtered gold coated tip and an electroplated gold film, of the type used in a MEMS switch, in a dry nitrogen environment. The contamination layer, consisting mostly of hydrocarbons, was characterized mechanically, electrically, and chemically. The results showed that both the adhesion and contact resistance were dominated by the contaminant film, which can be broken down with high force and voltage. Furthermore, exposure to ozone produced subtle chemical changes in the contamination layer which decreased the resistance significantly and reduced the variability of the contact resistance between contact events.

Studies of hot-switched gold contacts were conducted by Patton and Zabinski using a micro/nanoadhesion apparatus [9, 10] to simulate a switch. This testing was conducted in a well-defined air environment and under precisely controlled operating conditions. The electrical current during hot switching had a significant effect. For low currents (1–10 µA) there was an increase in adhesion after rapid cycling. Adhesion was attributed to the smoothing of the surface and the associated increase in van der Waals forces. Aging of the contacts in air was shown to reduce adhesion. At high current (1–10 mA) there was no measurable adhesion, although it did show necking of the gold on separation. Patton *et al.* [11] applied a self-assembled mono-layer (SAM) of diphenyl disulfide to the Au contacts which were then tested at low and high currents. Testing in the adhesion apparatus was combined with *ex situ* analytical analysis of the contacts consisting of X-ray photoelectron spectroscopy (XPS) and micro-Raman techniques. Hot switching tests were conducted in humid air and in dry nitrogen. At low current these lubricated contacts failed due to an increase in both adhesion and contact resistance at about 10^5 cycles. At high current the contacts failed almost immediately.

The evolutions of the adhesion force and surface morphology in gold switch contacts were measured with repeated cold-switching cycles by Gregori and Clarke [12, 13]. The adhesion force was shown to increase logarithmically with the number of actuation cycles. The increase in the adhesion force measured in air was attributed to mechanical creep of the contacting asperities. Several morphological changes were also observed. Two key parameters were identified, i.e., the plasticity index and the adhesion parameter which provide some insight into the evolution of the contacts.

Lee *et al.* [14] characterized the resistivity and hardness of alloys of platinum, ruthenium, and rhodium with gold. Although the adhesion was not measured, the force of adhesion tends to decrease for harder materials. This behavior of hard materials is because of the smaller contact area due to the reduced plastic deformation. Since alloying of gold increases the hardness, it should decrease the adhesion force. A map which shows where the different alloys fall in the 'hardness *vs* resistivity' plane is given.

Au, a Au–V solid solution, and $Au–V_2O_5$ dispersion films were fabricated and tested as candidates for MEMS contact switches by Bannuru *et al.* [15]. The resistivity and hardness increased with vanadium content, but the ratio of the resistivity increase to the hardness increase was much lower for the $Au–V_2O_5$ films. It also showed a reduction in adhesion which may make the $Au–V_2O_5$ film an attractive candidate for a contact material. Patton *et al.* [16] deposited bimetallic nanoparticles on gold MEMS switch contacts. The nanoparticles consisted of a 10 nm diameter Au core and with smaller Pd particles on the surface. Adhesion and resistance were measured during hot switching. The nanoparticle coated contacts had less adhesion and only slightly greater contact resistance than did the pure gold contacts. Performance and durability were very good at high currents as well.

A recent study by Yang *et al.* [17] compared gold and gold–nickel alloys as contact materials in MEMS switches. The objective was to reduce contact adhesion and wear from what occurs in a gold-on-gold contact. Although low wear is generally associated with high hardness, in contacts in which adhesion is important it may be a combination of a soft material and high adhesion which promotes material transfer and wear. The properties of the Au–Ni alloy were controlled by adjusting the nickel content and thermal processing conditions. A special switching degradation test facility was developed and used to perform testing on the gold–nickel alloy as the lower, and gold as the upper, contact material. Solid solution of Au–Ni samples

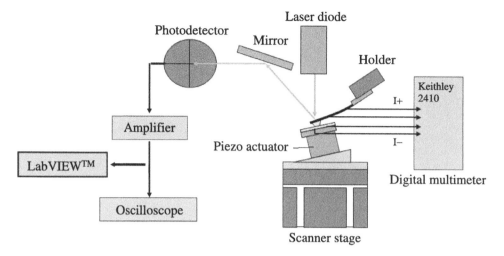

Figure 4. Schematic of the test setup using an AFM along with a separate piezo actuator.

exhibited reduced wear but increased contact resistance, whereas two-phase Au–Ni (20 at% Ni) showed a substantial improvement in reliability with only a modest increase in contact resistance.

Another method of determining the adhesion force is to measure the time required for the switch to open [18]. As in [6] the results for gold-on-gold contacts show that the adhesion force increases as the resistance decreases. It was also found that a small apparent contact area along with a high force reduced the opening time because the additional stored elastic energy in the switch structure helped to separate the contacts. Thus the restoring force of the actuator could be increased by using a larger than needed actuation voltage. They also recommended a small tip area design to reduce adhesion. The experimental model of adhesion developed in [18] was used by Shalaby *et al.* [19] to perform a design optimization.

In order to better study the effects of adhesion in microcontacts, a specially designed contact test station has been designed by Chen *et al.* [20] based on an atomic force microscope (AFM) as shown in Fig. 4. The AFM cantilever is replaced by an extremely stiff ($K \cong 15\,000$ N/m) in-house microfabricated cantilever (Fig. 5(a)) with a hemispherical tip (Fig. 5(b)). The tip is coated with a thin (300 nm) layer of a metal (typically Au or Ru) which makes and breaks contact with a coated chip. The chip is coated with the same or different metal and is mounted on a piezo actuator which sits on a wedge system. The angles of the wedge are chosen so as to reduce the amount of sliding produced by the flexing of the cantilever. The optics of the AFM is used to measure the rotation angle at the end of cantilever which, after calibration, gives the force during loading as well as during unloading, including tensile adhesion forces. The displacement of the piezo actuator is given a ramp function for both loading and unloading. Testing can be halted at any stage and the surfaces observed in an SEM.

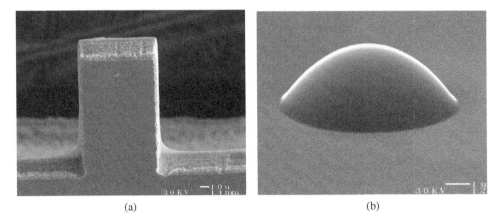

(a)   (b)

Figure 5. Specially designed (a) AFM cantilever with very high stiffness ($K \cong 15\,000$ N/m) and (b) contact bump.

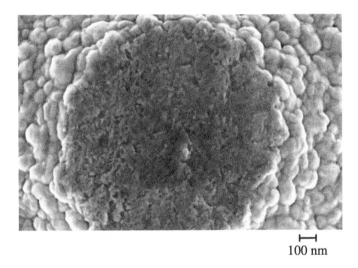

100 nm

Figure 6. SEM image of a flattened contact tip after a brittle separation. Reprinted with permission from Fig. 1(a) in *Applied Physics Letters*, Vol. 93, L. Chen, N. E. McGruer, G. G. Adams and Y. Du, "Separation modes in microcontacts identified by the rate-dependence of the pull-off force", 053503. Copyright (2008), American Institute of Physics.

Two different types of separation modes have been observed from the SEM images of Au-on-Au contacts by Chen *et al.* [21]. In a brittle separation (Fig. 6) the top of the initially hemispherical tip has been flattened by plastic deformation. On separation, however, there is no evidence of adhesion-induced plastic stretching. It is also noted that during continued cycling the area of the flattened region grows as indicated in Fig. 7. The results shown are for different contact tips because after SEM imaging the testing cannot be continued with the same tip. The trend of increased area of the flattened region with cycling remained true for the many tips which were tested.

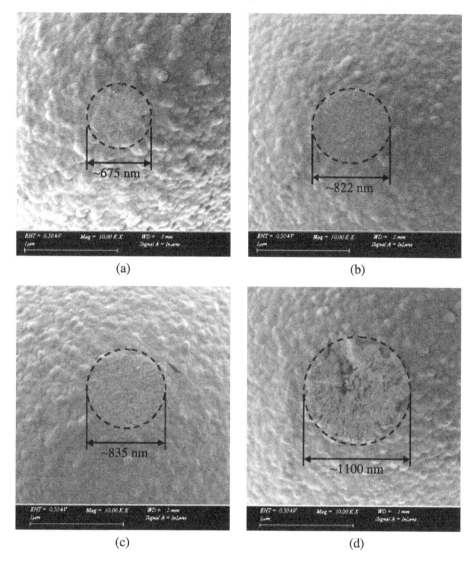

Figure 7. Flattened top of a contact following a brittle separation. The diameter of the flattened region has grown from ~675 nm to ~1100 nm after cycling.

On the other hand, a ductile separation is shown in the high angle SEM image of Fig. 8. Although there is still microscale flattening, it is starkly evident that there are nano-asperities which appear to be stretched during separation, forming the nano-spikes shown. The images in Figs 6 and 8 each show only one type of separation mode, although other images (not shown here) indicate a mixture of brittle and ductile separation regions. Given that testing was stopped at an arbitrary point, it is interesting to speculate as to what would have happened if another contact cycle had been performed. Would those previously stretched nano-asperities have been

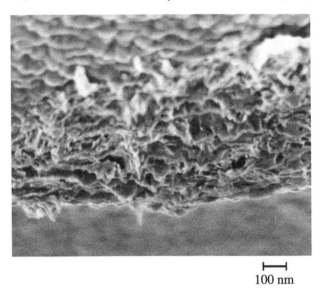

100 nm

Figure 8. High angle SEM image of a contact after a ductile separation.

flattened by the subsequent compressive phase of the next contact? Would then other nano-asperities have been drawn out during the next separation phase? Would material have been transferred from the tip to the drain? These questions have yet to be answered definitively.

A surprising result is that by monitoring the pull-off force it was not possible to reliably predict whether the separation was ductile or brittle [21]. However, if the *rate* of loading and unloading was varied, then the type of separation mode could be determined. Figure 9 shows the pull-off force as a function of the maximum loading force for cycling at 300 Hz and at 0.5 Hz. These results, in which the pull-off force is higher at a lower cycling rate (Higher Force at Lower Rate, HFLR), are typical of a brittle separation. The cycling rate was also swept from 0.5 Hz to 1000 Hz to study the rate-dependency of the pull-off force. During this test, the maximum loading force was set at 200 μN. At each cycling rate, the sample was cycled for 30 s for the pull-off force measurement. In the HFLR mode the pull-off force tended to remain constant for rates below 100 Hz (Fig. 10) after which point it decreased.

This behavior was explained as follows [21]. For brittle separation, the magnitude of the pull-off force depends on the bond strength at the interface. The longer the time in contact and the greater the applied force, the stronger the bond formation is, which, in turn, leads to a higher separation force. Since gold is hydrophilic and the tests were performed in room air, the effect of meniscus forces could be important. Experiments by Szoszkiewicz and Riedo [22] have shown that at 37% relative humidity the mean meniscus nucleation time is 4.2 ms at room temperature. Our rate sweeping experiment shows (Fig. 10) that the pull-off force drops when the cycling rate is greater than 100 Hz. This result suggests that the equilibrium time

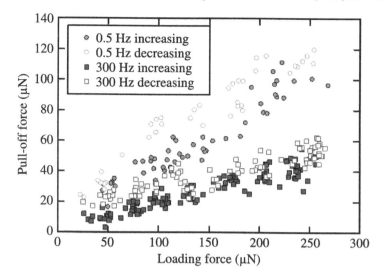

Figure 9. Pull-off force *vs* maximum loading force at a low cycling rate (0.5 Hz in red) and at a high cycling rate (300 Hz in blue) during brittle separation. Reprinted with permission from Fig. 2(a) in *Applied Physics Letters*, Vol. 93, L. Chen, N. E. McGruer, G. G. Adams and Y. Du, "Separation modes in microcontacts identified by the rate-dependence of the pull-off force", 053503. Copyright (2008), American Institute of Physics.

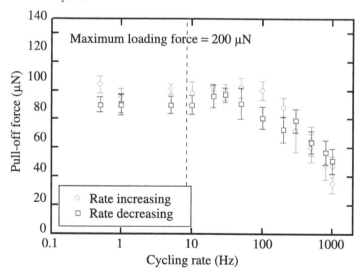

Figure 10. Pull-off force *vs* cycling rate for a maximum loading force of 200 μN. Reprinted with permission from Fig. 3 in *Applied Physics Letters*, Vol. 93, L. Chen, N. E. McGruer, G. G. Adams and Y. Du, "Separation modes in microcontacts identified by the rate-dependence of the pull-off force", 053503. Copyright (2008), American Institute of Physics.

for meniscus condensation is about 5 ms, which is close to the meniscus nucleation time observed in [22]. Furthermore, the measured difference in pull-off force in the HFLR mode is of the same magnitude as the meniscus force effect.

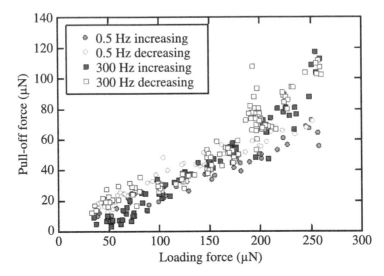

Figure 11. Pull-off force *vs* maximum loading force at a low cycling rate (0.5 Hz in red) and at a high cycling rate (300 Hz in blue) during ductile separation. Reprinted with permission from Fig. 2(b) in *Applied Physics Letters*, Vol. 93, L. Chen, N. E. McGruer, G. G. Adams and Y. Du, "Separation modes in microcontacts identified by the rate-dependence of the pull-off force", 053503. Copyright (2008), American Institute of Physics.

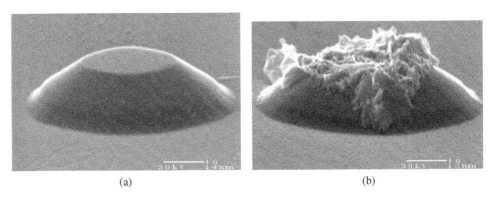

Figure 12. (a) A Au–Ru contact cycled 4×10^6 times and (b) a Au–Au contact cycled 10^6 times. The applied force is 200 μN in each case.

Unlike brittle separation, ductile separation shows an increase in the pull-off force for higher cycling rate, especially at higher maximum contact forces (Fig. 11). Because ductile separation is associated with material stretching and nanostructure neck formation, it could be due to viscous effects in the gold. For nanoscale structures, the dislocation sources are largely reduced resulting in a time-dependent creep flow, which could be responsible for a greater pull-off force at a higher cycling rate.

Figure 12(a) shows a Ru–Au contact cycled 4×10^6 times and Fig. 12(b) shows a Au–Au contact cycled 10^6 times, both at a force of 200 μN. In these cases the tips were flat-topped as shown. As shown, there is considerable ductile separation and/or material transfer on the Au–Au contact tip. Gold-on-gold contact possesses a high work of adhesion and low hardness. This combination is likely the reason for the observed nanoscale ductile separation (Fig. 8) and/or material transfer (Fig. 12(b)).

A test setup using a nanoindenter along with coated silicon cantilevers was used by Gilbert *et al.* [23] to test Au contacts for changes in the adhesion force due to cycling the contacts. Testing was performed under hot-switching conditions. Cycling frequency was limited to 100 Hz with a contact force of 400 μN. Nanoindenter measurements were taken every 10 000 contact cycles. Cycling continued until either a high resistance failure or a stuck-closed condition occurred. The setup also allowed the morphological changes in of the contact bump to be tracked during cycling.

Another means of coping with adhesion forces is through a design which allows for a large opening force. Such a design was proposed by Oberhammer and Stemme [24] using two cantilevers, one of which has a hook at its end to allow for mechanical interlocking. Experimental results are supported by simulations.

3. Modeling and Simulation

The modeling of the adhesion of an elastic sphere with an elastic half-space (or equivalently between two elastic spheres) was first done in the 1970s. In the Johnson, Kendall, Roberts (JKR, [25]) model, the effect of adhesion is included through the influence of surface energy. Thus the adhesion stresses outside of the contact area are not directly included. A stress singularity exists at the contact boundary, much like that at the tip of a crack. The contact problem is solved by minimizing the total potential energy, which includes the elastic strain energy, the work done by the applied force, and the work of adhesion, with respect to the contact radius. The relations among the contact radius (a), the approach (δ), and applied load (P), are given by

$$a^3 = \frac{3P_1 R}{4E^*}, \qquad \delta = \frac{a^2}{R} - \sqrt{\frac{2\pi a \Delta \gamma}{E^*}},$$

$$P_1 = P + 3\pi \Delta \gamma R + \sqrt{6\pi \Delta \gamma R P + (3\pi \Delta \gamma R)^2}, \tag{1}$$

where P_1 is the effective Hertz load due to adhesion, $\Delta \gamma$ is the work of adhesion defined by

$$\Delta \gamma = \gamma_1 + \gamma_2 - \gamma_{12}, \tag{2}$$

where γ_1 and γ_2 are the surface energies of the two contacting bodies and γ_{12} is the interfacial energy. In equation (1) R is the effective radius of curvature given by

$$\frac{1}{R} = \frac{1}{R_1} + \frac{1}{R_2}, \tag{3}$$

where R_1 and R_2 are the radii of curvatures of each of the bodies, and E^* is the composite Young's modulus defined as

$$\frac{1}{E^*} = \frac{1 - v_1^2}{E_1} + \frac{1 - v_2^2}{E_2}, \tag{4}$$

where E_1 and E_2 are the elastic Young's moduli, and v_1 and v_2 are the Poisson's ratios of the contacting bodies. It is noted that the approach (δ) is often referred to as the interference. The pull-off force, i.e., the force needed to separate the adhesion contact is given by

$$P = -\frac{3}{2}\pi\Delta\gamma R, \tag{5}$$

which is the minimum possible value of P, with the negative sign indicating a tensile load. Furthermore, pull-off occurs suddenly at a finite contact radius given by substituting equation (5) into equation (1).

Another theory, by Derjaguin, Muller, and Toporov (DMT, [26]) assumes that the contact stress distribution is the same as in a Hertz contact. Adhesion is included by integrating the intermolecular forces outside of the contact area using the Hertz separation profile. By so doing, the effect of adhesion is simply to increase the contact force by a constant equal to $2\pi\Delta\gamma R$. Thus the relations among the load, contact radius and interference are:

$$\frac{4E^*a^3}{3R} = P + 2\pi\Delta\gamma R, \qquad \delta = \frac{a^2}{R} \tag{6}$$

and the pull-off force is

$$P = -2\pi\Delta\gamma R, \tag{7}$$

with pull-off occurring at a vanishingly small contact radius.

It is noted that for each theory the pull-off force is independent of modulus and thus each could be expected to be valid for a certain range of material properties. However, the results of the two theories differ, most notably in the value of the pull-off force. This apparent inconsistency can be explained in terms of the Tabor parameter (μ, [27]) defined by

$$\mu = \left(\frac{R\Delta\gamma^2}{E^{*2}Z_0^3}\right)^{1/3}, \tag{8}$$

where Z_0 is the equilibrium atomic spacing between the two half-spaces. The Tabor parameter is such that for $\mu \ll 1$ (low work of adhesion, small radius, and high modulus) the DMT theory is valid, whereas for $\mu \gg 1$ (high work of adhesion, large radius, and low modulus) the JKR theory holds.

The model developed by Maugis [28] uses a simple analytical approximation of the adhesion stress in order to arrive at an analytical solution which is valid for a certain range of the Tabor parameter. The results are given by the numerical solution of the following algebraic equations:

$$\frac{\lambda a_M^2}{2}\left[\sqrt{m^2-1}+(m^2-2)\tan^{-1}\sqrt{m^2-1}\right]$$

$$+\frac{4\lambda^2 a_M}{3}\left[\sqrt{m^2-1}\tan^{-1}\sqrt{m^2-1}-(m-1)\right]=1,$$

$$P_M=a_M^3-\lambda a_M^2\left[\sqrt{m^2-1}+m^2\tan^{-1}\sqrt{m^2-1}\right],$$

$$\delta_M=a_M^2-\frac{4}{3}a_M\lambda\sqrt{m^2-1},$$

(9)

where

$$a_M=a\left(\frac{K}{\pi\Delta\gamma R^2}\right)^{1/3},\qquad \lambda=2\sigma_0\left(\frac{9R}{16\pi\Delta\gamma E^{*2}}\right)^{1/3}\cong 1.16\mu,$$

$$m=\frac{c}{a},\qquad P_M=\frac{P}{\pi\Delta\gamma R},\qquad \delta_M=\delta\left(\frac{16E^{*2}}{9\pi^2\Delta\gamma^2 R}\right)^{1/3}.$$

Plots of contact radius vs force and of force vs interference are shown in Figs 13 and 14 respectively for the Hertz, JKR, DMT, and Maugis models, the last for different values of the modified Tabor parameter ($\lambda\cong 1.16\mu$). Note that the pull-off force lies in the range of 1.5–2.0 times $\pi\Delta\gamma R$. Furthermore, each of the three theories reduces to the Hertz contact theory when the work of adhesion can be neglected.

It is important to recognize that all three of these theories are restricted to linear elastic material behavior. The onset of plastic yielding, using the von Mises yield

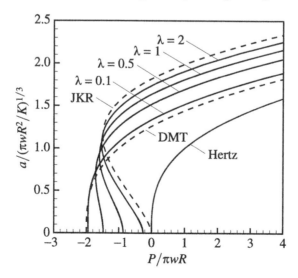

Figure 13. Dimensionless contact radius vs dimensionless force for Hertz, DMT, Maugis (with different values of λ) and JKR theories. Here $w=\Delta\gamma$ and $K=\frac{4}{3}E^*$.

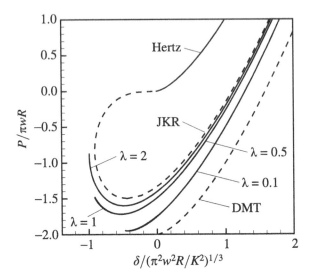

Figure 14. Dimensionless force *vs* dimensionless interference for Hertz, DMT, Maugis (with different values of λ) and JKR theories. Here $w = \Delta\gamma$ and $K = \frac{4}{3}E^*$.

criterion, has been determined and a curve-fit to those results has been presented by Wu and Adams [29]. However, a simpler approximate result can be written for a Poisson's ratio of 0.3 by computing the Hertz stress as if the external force were increased by $2\pi R \Delta\gamma$ leading to the onset of yielding when

$$P \cong 21.6\sigma_Y^3 (R/E^*)^2, \tag{10}$$

where σ_Y is the yield strength. Such a procedure is equivalent to using the DMT theory, but this procedure for predicting the onset of yielding is valid for a broad range of the Tabor parameter. As in non-adhesion contacts, the onset of yielding occurs below the surface of the body with the lower yield strength.

After the onset of plasticity, the behavior of such an elastic–plastic or fully plastic adhesion contact is more difficult to model. In contacts without adhesion, an elastic–plastic contact is said to exist when the subsurface plastic zone is reinforced by surrounding elastic material. After continued loading, the plastic zone expands and a fully plastic contact is said to occur in which the contact stress becomes nearly constant and equal to the hardness ($H \cong 2.8\sigma_Y$) of the softer material. We use the same terminology of elastic–plastic and fully plastic contacts for contacts with adhesion. Simple analytical models of adhesion contacts with plastic deformation do exist [6, 30, 31], as do more recent finite element models [32–36].

Modeling of the combined effects of adhesion and plasticity is more complicated for many reasons. We begin by discussing a simple analytical model by Maugis and Pollock [30] which treats the contact of a rigid sphere with a half-space. The effects of both plasticity and adhesion are accounted for. Some modifications due to Majumder *et al.* [6] are also included. In the elastic regime, the deformation is assumed to be governed by the JKR model. The transition from elastic to elasto-plastic behavior is assumed to occur at a contact radius equal to its value which would initiate yielding without adhesion. With additional loading into the elastic–plastic and plastic regimes, the stress distribution and contact geometry are assumed to be those for the contact between a sphere and a flat without adhesion. The mean contact pressure p_m is assumed to vary with contact radius according to

$$p_m = \sigma_Y \left(1.1 + 0.69 \ln \frac{Ea}{3.9\sigma_Y R} \right). \tag{11}$$

Furthermore, the applied force is related to the contact area by

$$P = \pi a^2 p_m - 2\pi R \Delta\gamma. \tag{12}$$

Eventually, the mean contact pressure becomes equal to the hardness H; this occurs at a critical contact radius. Thereafter, the deformation is purely plastic, and the contact pressure remains constant, with the mean contact pressure in equation (12) equal to the hardness.

Due to plastic deformation, the effective radius of curvature after deformation (R_{eff}) is different than its value (R) before deformation. Since the effect of the adhesion forces on the stress distribution decreases progressively with further loading into the plastic regime, R_{eff} is assumed to be the same as it would be in the absence of adhesion [31]. Therefore,

$$R_{\mathrm{eff}} = \frac{2a_f E}{3\pi p_m (1 - v^2)}, \tag{13}$$

in the elasto-plastic regime and again p_m is replaced by H in the plastic regime.

The asperity penetration (δ) is assumed to be related to the contact radius (a) by equation (1) for elastic contact and by

$$\delta = \frac{a^2}{2R} + \frac{\delta_e}{2} - \frac{1}{2}\sqrt{\frac{2\pi a_e \Delta\gamma}{E^*}}, \tag{14}$$

for the elasto-plastic and plastic regimes, where δ_e and a_e are, respectively, the penetration and contact radius at the onset of plastic deformation.

If the loading is elastic, then the unloading will also be elastic. However, if the loading is either elastic–plastic or plastic, three different modes of unloading are possible. When the contact is gradually unloaded starting from the maximum contact force, the initial unloading is equivalent to the elastic unloading of a flat punch. This initial regime persists until either the average tensile axial stress becomes equal to the hardness, in which case the contacts separate and the separation is called ductile; or the contact force reaches the JKR equilibrium value. If the contact radius in the latter case is less than the contact radius at pull-off, then the contacts separate suddenly in this brittle mode. If the contact radius is greater than the contact radius at pull-off, then subsequent unloading of the contact follows the JKR model until pull-off when the contacts also separate in a brittle manner. Note that although the unloading is assumed elastic, the radius of curvature has changed from R to R_{eff} as a consequence of plastic deformation during the loading phase. This change in the radius affects the pull-off force as well as other quantities. As long as the unloading is in a brittle mode, then subsequent loading/unloading cycles should be purely elastic.

The work by Mesarovic and Johnson [32] has focused on finite element modeling of a smooth hemisphere in contact with either another sphere or a flat, under the influence of adhesion and beyond the elastic limit. It used a dense mesh finite element model of the adhesionless elastic–plastic contact of two spheres. They observed that the contact stress profile was nearly uniform well before the load at which the material hardness is reached. Thus the elastic unloading, which now included adhesion, could be performed analytically from a uniformly loaded state. A decohesion map was presented in terms of two parameters. One parameter χ is the ratio of the adhesion energy to the elastic energy in the recovered crown, whereas the other parameter, S, is the ratio of the theoretical stress to the maximum average contact stress. It was found that under certain combinations of these two parameters, the unloading could be accompanied by plastic deformation; only the onset of plastic deformation was considered.

Chang *et al.* [33] defined the critical interference as that value for which yielding initiates without adhesion. They found that

$$\delta_c = \left(\frac{1.4\pi K_H \sigma_Y}{E^*}\right)^2 R, \quad K_H = 0.454 + 0.41\nu, \tag{15}$$

which generalizes the result given earlier to arbitrary values of Poisson's ratio. The critical contact radius a_c and the critical contact force F_c corresponding to the critical interference are found from the well-known equations for Hertz contact [31], i.e.,

$$a_c = (\delta_c R)^{1/2}, \tag{16}$$

$$F_c = (2/3)K_H 2.8\sigma_Y \pi \delta_c R. \tag{17}$$

Kogut and Etsion [34] performed a finite element analysis to simulate a contact with adhesion between an elastic–plastic hemisphere and a rigid flat. Their analysis was in the spirit of a DMT model, i.e., it neglected the effect of adhesion on the deformation and stress fields of the bodies. Curve-fit equations for the pull-off force were given. The dimensionless interference is defined as the ratio of the interference to the critical interference (without adhesion). The value of the pull-off force is sensitive to the dimensionless interference only in the elastic–plastic range below a dimensionless interference of about 6; in this range it increases. The degree of increase depends on the ratio of the atomic spacing to the critical interference; as this ratio increases so does the pull-off force.

Recently a finite element model has simulated the loading and unloading of an elastic–plastic contact [35]. The effect of adhesion was included through the Lennard–Jones potential using the assumption that the surfaces were locally parallel to each other. Furthermore, the effects of adhesion on the displacements and stress fields were included, making the model applicable for a broad range of material

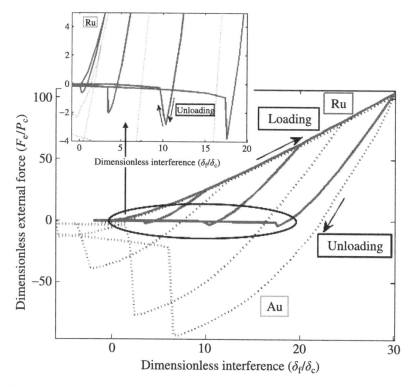

Figure 15. Dimensionless force *vs* dimensionless interference for gold and ruthenium (Table 1) showing brittle separation for ruthenium and ductile separation for gold. Reprinted with permission from Fig. 7 in the *Journal of Colloid and Interface Science*, Vol. 312, Y. Du, L. Chen, N. E. McGruer, G. G. Adams and I. Etsion, "A finite element model of loading and unloading of an asperity contact with adhesion and plasticity", pp. 522–528. Copyright (2007), with permission from Elsevier.

parameters. Two types of separations were predicted for this microscale simulation — brittle and ductile. As has been mentioned, both of these behaviors have been observed experimentally for nanoscale asperities [21].

Typical results are shown in Fig. 15 for force *vs* interference and in Fig. 16 for contact radius *vs* interference for Au and Ru based on data shown in Table 1. There are several distinguishing features of these two separation modes in the microscale simulations. A brittle separation is characterized by predominately elastic unloading, whereas a ductile separation is accompanied by considerable plastic deformation (stretching) on unloading, as well as in some instances the formation of a ductile neck. In a brittle separation the contact radius decreases steadily to a small value before separation, whereas in a ductile separation the contact radius decreases slowly and stepwise before separation occurs suddenly, typically at a significant fraction of the maximum contact radius. Finally the adhesion force for a ductile separation is significantly larger than it is for a brittle separation.

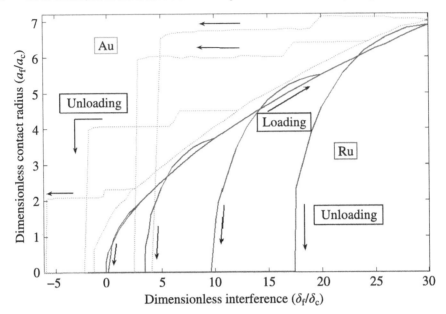

Figure 16. Dimensionless contact radius *vs* dimensionless interference for gold and ruthenium (Table 1) showing brittle separation for ruthenium and ductile separation for gold. Reprinted with permission from Fig. 6 in the *Journal of Colloid and Interface Science*, Vol. 312, Y. Du, L. Chen, N. E. McGruer, G. G. Adams and I. Etsion, "A finite element model of loading and unloading of an asperity contact with adhesion and plasticity", pp. 522–528. Copyright (2007), with permission from Elsevier.

Table 1.
Material properties of Ru and Au used for finite element simulations

Metal	R (μm)	E (GPa)	ν	σ_Y (GPa)	Z_0 (nm)	$\Delta\gamma$ (J/m^2)	δ_c (nm)	E/σ_Y	S	μ
Ru	4	410	0.3	3.42	0.169	1	1.70	120	0.6	1.6
Au	1	80	0.42	0.67	0.184	1	0.41	120	2.7	2.7

Although there are several dimensionless parameters identified which affect the separation mode, the most important of these parameters is given by

$$S' = \frac{\Delta\gamma}{Z_0 H}. \tag{18}$$

Typically brittle separation occurs if S' is less than about 1.2 and ductile separation if S' is greater than approximately 1.2. Because this parameter represents the ratio of the maximum adhesion strength to the material hardness, the fact that this transition from brittle to ductile separation occurs near S' equal to unity is not surprising. Other parameters which play a lesser role are the maximum load level compared to that at which yielding occurs without adhesion, the ratio of the elastic modulus to the yield strength, and the Tabor parameter.

Sabelkin and Mall [36] presented a two-dimensional finite element analysis of a deformable cylinder contacting a deformable flat. The effects of elastic–plastic deformation as well as adhesion were accounted for, with adhesion modeled *via* the Lennard–Jones potential. The model shows hysteresis in the loading/unloading behavior due to the combination of adhesion and plasticity.

3.1. Adhesion in the Contact of Rough Surfaces

A common approach is to model a rough surface by a distribution of asperities with a random distribution of asperity heights in a manner similar to the well-known Greenwood–Williamson (GW, [37]) theory for a non-adhesion rough contact. Such a model which includes adhesion [38] replaces the Hertz contacts of the GW model with JKR adhesion contacts. Fuller and Tabor [38] defined an adhesion parameter for a rough contact which represents the ratio of the standard deviation of asperity heights to the displacement that an individual asperity undergoes before being pulled off the other surface. Alternatively, the adhesion parameter can be viewed as being proportional to the ratio of the force needed to compress an asperity by an amount equal to the standard deviation of asperity heights, to the force required for pull-off. As a consequence, it represents the statistical average of a competition between the higher asperities (tending to separate the surfaces) and the lower asperities (tending to pull the surfaces together). When the adhesion parameter is small, adhesion dominates; when it is large adhesion can be neglected.

Later the Hertz contacts of the GW model were replaced with DMT contacts [39]. Unlike JKR contacts, DMT asperities do not stretch on unloading. Nonetheless tensile loads do exist in the regions surrounding the contacts, leading to a similar overall behavior to the Fuller and Tabor model. The Maugis model of adhesion was also used to represent the asperities in a GW-style model by Adams *et al.* [40]. Curve-fit expressions for the force *vs* penetration, and the contact radius *vs* penetration were given by Morrow *et al.* [41].

Decuzzi and Srolovitz [42] presented scaling laws for partially adhering MEMS contacts. Roughness was represented by a collection of asperities, all of the same height. Using a JKR adhesion model, they compared the force needed to pull the contacts apart with the combination of force and moment needed to pry the contacts apart. As would be expected the separation force decreases when a moment is applied, as the asperity contacts open sequentially. The overall contact then opens in a crack-like manner which requires less force than does the simultaneously pulling apart of all the contacts.

Modeling of the combined effects of adhesion and plasticity in contact tip is more complicated for many reasons. The contact tip is not smooth, but rather contains many nano-scale asperities. Plasticity occurs both at the microscale (in the bulk of the hemisphere) and also at the nanoscale (in and near the asperities). The modeling of plasticity at each of these scales is compounded by the growth of the nanoscale

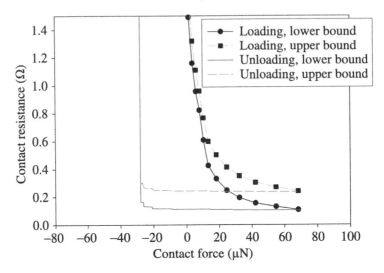

Figure 17. Upper and lower bounds for contact resistance *vs* contact force during loading and unloading.

plasticity into the microscale yield zone. Most of our knowledge of plasticity comes from the macroscale; size effects are known to be very important in plasticity due to the role that dislocations play in plastic deformation. Typically measurements of hardness values increase as the indentation depth decreases from the microscale to the nanoscale.

Majumder *et al.* [6] developed a simplified analytical model of contact resistance and adhesion in a MEMS microswitch. The model includes the single asperity effects of adhesion and plasticity using the approach of Maugis and Pollock [30] as discussed earlier in this article. A finite number of asperities were modeled and the constrictive resistance was calculated as both lower and upper bounds because the spatial distribution of asperities was not known. The results shown in Fig. 17 are in reasonable agreement with the experimental data shown in Fig. 3(b) after 10^4 contact cycles.

MD simulations of individual Au–Au [43] and of Ru–Ru [44] nanoasperity contacts were performed. From the MD simulations, the force *vs* displacement and the contact area *vs* displacement characteristics of these contacts have been determined. In [43] the work of adhesion of gold was artificially varied so as to represent the effect of a contaminant layer. For low work of adhesion the asperity deformed plastically, with only modest elastic rebound. For increasingly larger values of the work of adhesion, there was more and more plastic stretching on separation. For the highest values of the work of adhesion, i.e., for values near the theoretical value, there was material transfer from the asperity to the flat.

The roughness of a MEMS switch consists of nanoasperities on top of a microspherical tip. Unfortunately, the scale of these nanoasperities is so small that a continuum model would fail to capture important molecular phenomena. A molecu-

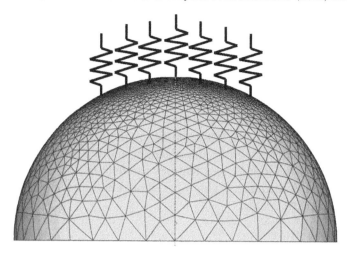

Figure 18. A continuum finite element mesh, with nonlinear spring elements representing nanoscale asperities. Although only 7 springs are shown here the actual model has many more.

lar dynamics (MD) simulation of the entire microsphere would be unrealistic due to the excessive computational time involved. Thus Eid *et al.* [45] presented a multi-scale model of a rough microcontact. Curve-fits of the MD simulations allowed the representation of these nanoasperity contacts by replacing the detailed MD simula-tions with nonlinear spring-like elements (Fig. 18). These spring elements include hysteresis due to adhesion and are, therefore, not true elastic springs, but rather are simply nonlinear relations between force and displacement which differ for loading and unloading. It is noted that there is hysteresis in the loading/unloading curves for a single asperity contact due to the effects of nanoscale plasticity and adhesion. Thus for each loading curve there exists a family of unloading curves, with each un-loading curve depending on the maximum loading level. Due to the computational cost of conducting MD simulations to generate unloading curves for many different maximum loadings, an interpolation scheme was used when the desired unloading curve lies between two existing unloading curves.

In the loading phase, the bottom of the rough sphere is moved upward toward the flat in small increments, causing one asperity after another to come into con-tact. After the maximum displacement is achieved, the system is slowly unloaded. The compressive load in each spring element is reduced progressively. When the maximum tensile force in a particular spring element is attained, then that element will suddenly pull-off the hemisphere. Results of this multi-scale model were ob-tained for loading and unloading of a Ru–Ru contact and a Au–Au contact relevant for a MEMS switch. Load *vs* interference and contact *vs* interference were deter-mined. The Ru–Ru contacts usually show low adhesion and a brittle separation mode, whereas the Au–Au contact shows higher adhesion and a more ductile mode of separation. These trends are qualitatively similar to the smooth bump model.

4. Summary and Conclusions

We have reviewed the state-of-the-knowledge in the adhesion of ohmic microswitch contacts. Gold-on-gold contacts are prone to adhesion and poor durability. Various alloys and the addition of particles both can improve the reliability of gold at the expense of an increase in contact resistance. Separation modes for gold contacts can be either brittle or ductile. More testing of different materials, including alloys and particle additives, is needed. Future work should include testing in a more controlled environment, along with *in situ* cleaning and the design and use of force sensors with better resolution. These features can help to improve our understanding of adhesion in microcontacts. This work should be supplemented by continued multiscale modeling of contacts.

References

1. G. M. Rebeiz and J. B. Muldavin, *IEEE Microwave Magazine* **2**, 59 (2001).
2. P. M. Zavracky, S. Majumder and N. E. McGruer, *J. Microelectromechanical Systems* **6**, 3 (1997).
3. B. McCarthy, G. G. Adams, N. E. McGruer and D. Potter, *J. Microelectromechanical Systems* **11**, 276 (2002).
4. Z. J. Guo, N. E. McGruer and G. G. Adams, *J. Micromech. Microeng.* **17**, 1899 (2007).
5. P. Decuzzi, G. P. Demelio, G. Pascazio and V. Zaza, *J. Appl. Phys.* **100**, 024313 (2006).
6. S. Majumder, N. McGruer and G. G. Adams, in: *Proc. 2003 STLE/ASME International Joint Tribology Conference*, Ponte Vedra Beach, Florida, p. 2003-TRIB-270 (2003).
7. J. Schmikat, *Sensors Actuators* **73**, 138 (1999).
8. J. W. Tringe, T. A. Uhlman, A. C. Oliver and J. E. Houston, *J. Appl. Phys.* **93**, 4661 (2003).
9. S. T. Patton and J. S. Zabinski, *Tribology Intl* **35**, 373 (2002).
10. S. T. Patton and J. S. Zabinski, *Tribology Lett.* **18**, 215 (2005).
11. S. T. Patton, K. C. Eapan, J. S. Zabinski, J. H. Sanders and A. A. Voevodin, *J. Appl. Phys.* **102**, 24903 (2007).
12. G. Gregori and D. R. Clarke, *J. Appl. Phys.* **100**, 094904 (2006).
13. G. Gregori and D. R. Clarke, *Appl. Phys. Lett.* **87**, 154101 (2005).
14. H. Lee, R. A. Coutu Jr, S. Mall and K. D. Leedy, *J. Micromech. Microeng.* **16**, 557 (2006).
15. T. Bannuru, S. Narksitipan, W. L. Brown and R. P. Vinci, *Proc. SPIE, Reliability, Packaging, Testing, and Characterization of MEMS/MOEMS VI*, p. 646306 (2007).
16. S. T. Patton, J. M. Slocik, A. Campbell, J. Hu, R. R. Naik and A. A. Voevodin, *Nanotechnology* **19**, 405705 (2008).
17. Z. Yang, D. J. Lichtenwalner, A. S. Morris III, J. Krim and A. I. Kingon, *J. Microelectromechanical Systems* **18**, 287 (2009).
18. B. D. Jensen, K. Huang, L. L.-W. Chow and K. Kurabayashi, *J. Appl. Phys.* **97**, 103535 (2005).
19. M. M. Shalaby, Z. Wang, L. L.-W. Chow, B. D. Jensen, J. L. Volakis, K. Kurabayashi and K. Saitou, *IEEE Trans. Industrial Electronics* **56**, 1012 (2009).
20. L. Chen, H. Lee, Z. J. Guo, N. E. McGruer, K. W. Gilbert, S. Mall, K. D. Leedy and G. G. Adams, *J. Appl. Phys.* **102**, 074910 (2007).
21. L. Chen, N. E. McGruer, G. G. Adams and Y. Du, *Appl. Phys. Lett.* **93**, 053503 (2008).
22. R. Szoszkiewicz and E. Riedo, *Phys. Rev. Lett.* **95**, 135502 (2005).
23. K. W. Gilbert, S. Mall, K. D. Leedy and B. Crawford, in: *Proc. IEEE Holm Conference on Electrical Contacts*, Orlando, Florida, p. 137 (2008).

24. J. Oberhammer and G. Stemme, *J. Microelectromechanical Systems* **15**, 1235 (2006).
25. K. L. Johnson, K. Kendall and A. D. Roberts, *Proc. Royal Soc. London* **A324**, 301 (1971).
26. B. V. Derjaguin, V. M. Muller and Y. P. Toporov, *J. Colloid Interface Sci.* **53**, 314 (1975).
27. D. Tabor, *J. Colloid Interface Sci.* **58**, 2 (1976).
28. D. Maugis, *J. Colloid Interface Sci.* **150**, 243 (1992).
29. Y.-C. Wu and G. G. Adams, *J. Tribology* **131**, 011403 (2009).
30. D. Maugis and H. M. Pollock, *Acta Metallurgica* **32**, 1323 (1984).
31. K. L. Johnson, *Contact Mechanics*. Cambridge University Press, London, UK (1985).
32. S. Dj. Mesarovic and K. L. Johnson, *J. Mech. Phys. Solids* **48**, 2009 (2000).
33. W. R. Chang, I. Etsion and D. B. Bogy, *J. Tribology* **109**, 257 (1987).
34. L. Kogut and I. Etsion, *J. Colloid Interface Sci.* **261**, 372 (2003).
35. Y. Du, L. Chen, N. E. McGruer, G. G. Adams and I. Etsion, *J. Colloid Interface Sci.* **312**, 522 (2007).
36. V. Sabelkin and S. Mall, *J. Adhesion Sci. Technol.* **23**, 851 (2009).
37. J. A. Greenwood and J. B. P. Williamson, *Proc. Royal Soc. London* **A295**, 300 (1966).
38. K. N. G. Fuller and D. Tabor, *Proc. Royal Soc. London* **A345**, 327 (1975).
39. D. Maugis, *J. Adhesion Sci. Technol.* **10**, 161 (1996).
40. G. G. Adams, S. Müftü and N. Mohd Azhar, *J. Tribology* **125**, 700 (2003).
41. C. Morrow, M. Lovell and X. Ning, *J. Phys. D: Appl. Phys.* **36**, 534 (2003).
42. P. Decuzzi and D. J. Srolovitz, *J. Microelectromechanical Systems* **13**, 377 (2004).
43. P.-R. Cha, D. J. Srolovitz and T. K. Vanderlick, *Acta Materialia* **52**, 3983 (2004).
44. A. Fortini, M. I. Mendelev, S. Buldyrev and D. J. Srolovitz, *J. Appl. Phys.* **104**, 074320 (2008).
45. H. Eid, G. G. Adams, N. E. McGruer, A. Fortini, A. B. de Oliveira, S. Buldyrev and D. J. Srolovitz, *Proc. World Tribology Congress*, p. 376 (2009).

Characterization of Gold–Gold Microcontact Behavior Using a Nanoindenter Based Setup

Kevin W. Gilbert [a], **Shankar Mall** [a,*] **and Kevin D. Leedy** [b]

[a] Department of Aeronautics and Astronautics, Air Force Institute of Technology, Wright-Patterson Air Force Base, OH 45433, USA

[b] Sensors Directorate, Air Force Research Laboratory, Wright-Patterson Air Force Base, OH 45433, USA

Abstract

Gold–gold microcontact behavior of MEMS switches under cycling and hot-switching conditions was characterized experimentally. A nanoindenter based experimental setup was developed where a cantilever beam with contact bump was cycled in and out of physical contact with a flat plate to simulate the action of a MEMS ohmic contact switch. This arrangement offered a simple method to simulate MEMS switches with minimum fabrication effort. Cantilever beam and flat plate were fabricated from silicon, and then sputter coated with 300 nm of gold as the contact material. All contacts failed in adhesion with lifetimes ranging from 10 000 to more than one million cycles. Three failure mechanisms of the contacting surfaces were observed: ductile separation, delamination and brittle separation with short (less than 70 000 cycles), mid (190 000–500 000 cycles) and long (more than one million cycles) life, respectively. Resistance, contact adhesion, threshold force and distance, strain hardening, and plastic deformation were monitored during cycling. Initial contamination of the contact was burnt out quickly during cycling which resulted in a constant threshold force. Contact resistance was practically constant during the cycling in all tests. Time-dependent and plastic deformations of the contact were observed, and these were initially large which then decreased to a constant value with cycling. Thus, elastic–viscoplastic material model(s) with strain hardening capability are needed for the analysis of gold–gold microcontact.

Keywords

Micro-contact, nanoindenter, gold, MEMS switch, contact resistance, adhesion force

1. Introduction

Microelectromechanical Systems (MEMS) based metal-to-metal ohmic contact switches show many advantages in comparison to solid-state switches, such as lower power consumption, smaller size, less weight, lower insertion loss, etc. [1,

---

* To whom correspondence should be addressed. AFIT/ENY, Bldg. 640, 2950 Hobson Way, Air Force Institute of Technology, Wright-Patterson AFB, OH, 45433-7765. E-mail: Shankar.Mall@afit.edu

The views expressed in this article are those of the authors and do not reflect the official policy or position of the United States Air Force, Department of Defense, or the US Government.

Adhesion Aspects in MEMS/NEMS

2]. MEMS ohmic contact switches are commercially available and their reported lifetime is improving each year [3, 4]. However, there are currently no analyses or models which can predict or estimate contact performance as a function of number of switch cycles or which would enable the characterization of switch lifetime performance and behavior. This is due to lack of experimental data which could enable the development of such model(s). Previous studies have investigated the performance of micro-contacts in the pristine state (e.g., [5]), but only a limited number of studies have investigated the mechanics of microcontact under cyclic condition (i.e., during their operation) or have made measurements of microcontact parameters (e.g., [6, 7]). Maugis studied the separation mechanics of contacts in terms of surface forces, deformation, etc. [8, 9]. Chen used silicon cantilevers coated with contact material in a Scanning Probe Microscope setup and studied adhesion and modes of microcontact separation [10, 11]. Gregori *et al.* used a nanoindenter to monitor changes in contact adhesion of an actual microswitch [7]. Dickrell and Dugger used a nanoindenter with a 3.2 mm diameter probe to study resistance degradation under the hot-switching condition [12]. The physical and electrical processes involved over the lifetime of metal-to-metal microcontacts have also been reported [6, 13–15].

However, there is a further need to characterize microcontact behavior under cyclic condition in order to better understand and/or predict MEMS switch performance and reliability. This study is a step in this direction and consisted of two phases. First, an experimental setup was developed to characterize the physics and mechanics of MEMS-scale contacts under cycling condition. This setup was designed using the capabilities of an MTS Nano Indenter® where a cantilever beam fabricated with a contact bump was cycled in and out of physical contact with a flat plate to simulate the action of a MEMS ohmic contact switch. This arrangement offers a simple method to simulate MEMS switch with minimum fabrication effort and accomplish measurements of contact parameters over its lifetime. The second (i.e., main) part of this study involved the characterization of gold–gold microcontact behavior over lifetime utilizing this setup.

2. Experimental Setup

A schematic of the test setup is shown in Fig. 1. The test setup simulated the action of MEMS contact switch by using a silicon cantilever beam with a contact bump on its free end as the upper contact. An example of beam with bump is shown in Fig. 2. The bottom contact was a flat piece of silicon coated with a conductive metal layer (strike plate). Both beam and strike plate were sputter coated with a metallic material of interest, which was gold in the present study. The strike plate was mounted on a piezoelectric transducer (PZT) which mechanically cycled the simulated switch (i.e., cantilever beam and strike plate). The contact was cycled at a frequency of 100 Hz using the PZT which applied the maximum displacement of 2 µm. This resulted in cycling contact force ranging from 0 to 400 µN. Resistance

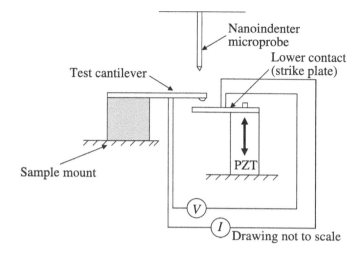

Figure 1. Schematic of experimental setup.

Figure 2. Cantilever beam with contact bump.

of the simulated switch was measured using a four-wire technique before cycling and after application of a certain number of cycles.

At each measurement interval, i.e., after prescribed number of cycles, cycling was paused. The nanoindenter microprobe was then used to apply a contact force of 400 μN on the cantilever beam at a rate of 50   μN/s. The nanoindenter was used in the load controlled mode and had resolution of 50 nN and 0.01 nm. Contact force and resistance were measured during loading and unloading. Detection of contact failure was accomplished by comparing in-contact resistance to out-of-contact resistance. If both measurements showed low resistance, the contact failed in adhesion, i.e., the contact stuck closed (failed closed). If in-contact resistance

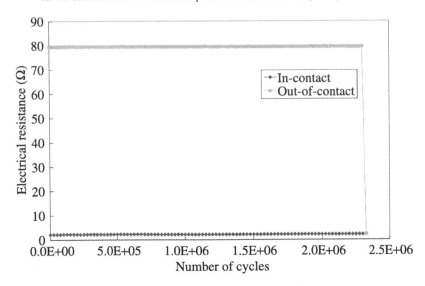

Figure 3. Variation of resistance during cycling showing contact adhesion failure (note out-of-contact and in-contact resistances are equal indicating contact failed to open).

was high, the contact showed a high resistance failure (failed open). An example of resistance measurements indicating an adhesion failure is shown in Fig. 3.

Cantilever beams were fabricated from silicon using the MEMSCAP SOI-MUMPS process followed by a post process to create a contact bump near the end of each cantilever beam. The cantilevers were 450 μm long, 40 μm wide and 22 μm thick with a nominal stiffness of 200 N/m. The cantilevers were sputter coated with 10 nm of chromium as an adhesion layer, followed by gold (contact metal) of 300 nm thickness using a Denton vacuum discovery 18 DC magnetron sputtering system at vacuum of 1.4×10^{-6} Pa. Gold thin film material properties were determined by nanoindentation with a Berkovich tip in an MTS Nano Indenter XP using the continuous stiffness method. Young's modulus of the gold thin film was determined to be 86 GPa, hardness was measured as 1.04 GPa and resistivity was 3.6 μΩ cm [16].

All tests were run in ambient laboratory environment with a current of 0.5 mA, i.e., in the hot-switching condition where hot-switching is defined as when the contact is made and broken with current passing through the contact. The compliance voltage was 40 mV. No evidence of arcing or arc damage was seen based on scanning electron microscopic (SEM) examination of contact surfaces after tests. Contact adhesion force, contact stiffness, and resistance as a function of applied contact cycles were measured in all tests. Further, the force required to obtain stable ohmic contact, referred to as the threshold force in this study, was measured. Energy absorbed during a contact cycle was also measured which is an indication of plastic deformation. Time-dependent deformation of contact was measured as well. The measurement details are provided in [15].

3. Results and Discussion

3.1. Lifetime

A total of seventeen tests were conducted which had life ranging from 10 000 cycles to 2.3 million cycles. These tests can be classified in three categories: short-life failures (Type I) which had contact lifetimes in the range of 10 000–70 000 cycles, mid-life failures (Type II) having contact lifetimes between 190 000–500 000 cycles, and long-life failures (Type III) having contact lifetime greater than 1 million cycles. This classification was based on failure characteristics of contact surfaces. The shorter lifetime (Type I) failures showed ductile separation of contact surfaces, the Type II failures showed delamination of the contact film from the substrate, and the Type III (long-life) failures showed brittle separation of contact surfaces. The classification based on lifetime along with the corresponding failure mechanism is shown in Table 1. The majority of failures were adhesion failure (switch failed closed) while a very few had high resistance failure (failed open). From the SEM examination of contact surfaces after tests, it was observed that even the high resistance (i.e., failed open) failure was due to adhesion of the gold contact film from the cantilever beam contact bump to the lower contact strike plate. The variation in cycles to failure of microcontacts is expected as commercially available MEMS switches show scatter in cycles to failure [3, 4, 17]. Further, this variation in contact lifetime may be due to the fact that the tests in this study were conducted in ambient laboratory air with no environmental control. Environmental and initial surface conditions may play a role in the microcontact's lifetime.

A brief background of the aforementioned failure mechanisms will be provided here for the sake of completeness, and it is based on the previous studies. Adhesion bonds are formed during the microcontact of two solid surfaces. When the surfaces separate and break the adhesion bonds, they either exhibit ductile or brittle separation features. Maugis described these as: "separation between the two solids can occur at the interface ("brittle" or adhesion rupture) or within the softer of the two materials ("ductile" or cohesive rupture)" [9]. In other words, brittle separation occurs at the initial interface between the contacting surfaces, and ductile separation follows a path other than the initial plane of contact. Chen discussed these two separation modes in terms of fracture mechanics [10]. Ductile separation occurs when the energy of separation is absorbed by dislocation nucleation and rupture is

Table 1.

Categorization of gold contact failures

Type	Failure mechanism	Lifetime (cycles)	Number of failures
I	Ductile separation	10 000–70 000	7
II	Contact film delamination	200 000–500 000	7
III	Brittle separation	$>10^6$	3

accompanied by significant plastic deformation. Ductile separation causes surface modification and material transfer. Brittle separation, where the separation occurs along the original interface or along grain boundaries, shows little, if any, plastic deformation [10]. Surfaces which have experienced ductile separation show necking and material transfer, while surfaces which have experienced brittle separation show a surface similar to that resulting from a brittle fracture [10].

3.2. Failure Mechanisms

All short lifetime tests showed evidence of ductile separation of the contact surfaces (Type I). Here adhesion between upper and lower contact surfaces caused failure at locations other than the contact interface when the separation force was applied. Ductile necking and/or material transfer were present during this ductile separation process [10], which is similar to ductile failure of metals under monotonic tensile load. An example of this failure mechanism is shown in Fig. 4, where the gold coated bump on the cantilever beam shows ductile necking and material transfer, while Fig. 5 shows the same bump before cycling (i.e., in its virgin state). This contact failed due to adhesion force between the contact bump and strike plate when it was unable to separate upon the application of 400 μN restoring force after 20 000 cycles.

The mid-lifetime contacts (190 000–500 000 cycles) failed again due to adhesion but exhibited contact film delamination. Here the contact film separated from the substrate (i.e., contact bump) and adhered to the lower strike plate. Further, there were two types of this failure mechanism. The first one occurred when the switch failed closed (i.e., the contact bump remained stuck to the strike plate). However,

Figure 4. Gold coated bump after adhesion failure at 20 000 cycles.

AFIT-SEM 8.0 kV 16.0 mm ×10.0k SE(M) 8/11/2007 16:53 5.00 μm

Figure 5. Gold coated bump before cycling which failed in adhesion in Fig. 4.

when the contact was separated at the end of test, the contact film was found to be attached to the lower strike plate. The second type of this failure mechanism occurred when the 400 μN restoring force was applied by the cantilever beam during cycling; the film was separated from the contact bump. This second type showed a high resistance in the next measurement cycle, i.e., a high resistance failure (failed open). SEM examination of the contact surface showed that the contact film remained adhered to the strike plate, and thus the cause of contact failure was adhesion. An example of the delamination failure is shown in Figs 6 and 7. Figure 6 shows this failure, i.e., separation of gold film from the bump, and Fig. 7 shows adhesion of separated film to the strike plate after 500 000 cycles. Kwon *et al.* reported a similar failure mode during testing of a simulated switch coated with thin film conductive material, which was described as "the surface of several samples was peeled off" [18]. This type of failure mechanism suggests that either subsurface damage developed during cycling or the adhesion force between contact surfaces became larger than the bond strength between the gold film and silicon substrate.

The contact surface of the long-life test (Type III) (more than 1 million cycles) showed much less damage of the gold thin film than the short-life test (Type I). These failures exhibited brittle separation unlike the ductile surface damage seen in Type I failure. Further, there was considerably less material transfer from the bump surface to the strike plate. The contact location on the strike plate also exhibited areas smoothened by the cycling action. This smoothening action over the life of the contact led to an increase in contact adhesion and ultimately failure. A long-life test's contact surface of bump after adhesion failure is shown in Fig. 8. Note the thin lamellar (platelike) features in the gold film on the bump surface in Fig. 8 which can be identified by their lighter color in the SEM image. Figure 9 shows the corresponding contact area on the flat plate. The transferred material from the

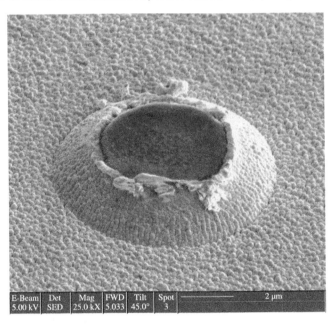

Figure 6. Gold coated contact surface after adhesion failure at 500 000 cycles; contact film separated and adhered to lower contact surface.

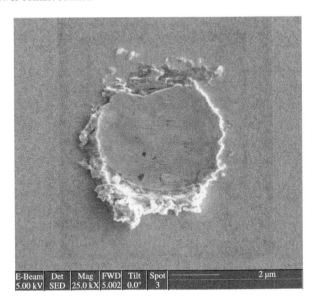

Figure 7. Contact film separated from contact bump and adhered to bottom contact surface after 500 000 cycle adhesion failure.

bump matched with the contact image of the strike plate in Figs 8 and 9, which are indicated by the areas labeled A, B and C in each figure. This suggests a brittle contact separation of the gold film, which can be identified by flattened features on

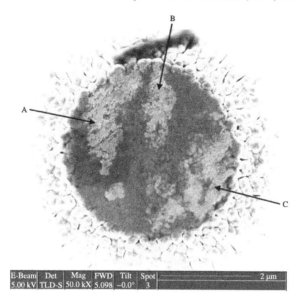

Figure 8. Gold coated contact surface on bump after adhesion failure at 2.35×10^6 cycles.

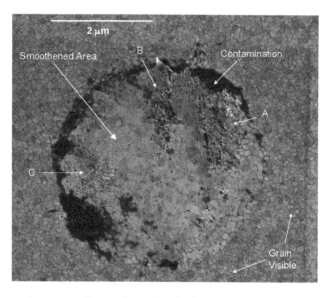

Figure 9. Gold coated contact surface on flat surface (strike plate) after adhesion failure at 2.35×10^6 cycles.

the contact surfaces [10]. Note that brittle fatigue induced fracture surfaces have a similar bright, granular appearance, due to reflection from the flat cleavage surfaces [19]. The bottom contact location on the strike plate also showed a black annular ring around the contact area which indicates contamination of the contact. There also appears to be some grain growth around the edge of the lower contact area indicating possible contact heating. The existence of contamination on the contact surface could explain why some contacts survived longer than others, or why some contacts exhibited ductile and others brittle separation. Note that contamination on contact surfaces reduces the adhesion force [20], i.e., contamination on the surface reduces the surface energy and thus the adhesion force. Contamination will also reduce the likelihood of lattice matching of similar materials when brought into contact. Although contamination on contact surfaces is generally detrimental, small amount of contamination with no electrical resistance may be beneficial in reducing adhesion force [21].

3.3. Microcontact Parameters

3.3.1. Resistance
A widely used measure of MEMS switch performance is the contact resistance. The average resistance of all tests before cycling was 2.2 Ω and the standard deviation was 0.6 Ω. The range of contact resistance was 0.972–3.534 Ω. The resistance varied from test to test due to the difference in parasitic resistance. Parasitic resistances including sheet resistance of the strike plate and the solder/wire-bond joints were included in the resistance measurement. There was about 0.5 Ω parasitic resistance from each of the terminal strips connecting the measurement probe wires to the data acquisition lines. However, change in total resistance during a test was due to change in the contact resistance only because no change in the parasitic resistances occurred in the test. Contact resistance for the gold was estimated to be equal to 0.5 Ω for a contact force of $F_c = 400$ μN using Holm's equation. This equation is given in equation (1), where R is resistance in Ω, ρ is resistivity, H is material hardness, and F_c is contact force [20]; $H = 1$ GPa and $\rho = 3.6$ μΩ cm.

$$R = \frac{\rho}{2}\sqrt{\frac{H\pi}{F_c}}. \tag{1}$$

Thus, contact resistance in each test was higher than the calculated contact resistance due to parasitic resistances and contamination on the contact surfaces. The resistance was practically constant in each test during cycling until failure. Variation of average contact resistance (of all tests) during cycling up to 220 000 cycles is shown in Fig. 10. The bar shows one standard deviation above and below the average value. The contact resistance during a representative long-life (Type III) test is shown in Fig. 11 along with counterparts from short- (Type I) and mid- (Type II) life tests. It is interesting to note that the in-contact resistance was practically constant during the cycling in all tests even with different types of failure mechanisms. This indicates that contamination was not building up between the contact surfaces. Further, the in-contact resistance did not change up to the instant of adhesion failure.

3.3.2. Adhesion
As mentioned earlier, contact cycling was paused after a prescribed number of cycles, whereupon the nanoindenter microprobe was used to actuate the cantilever

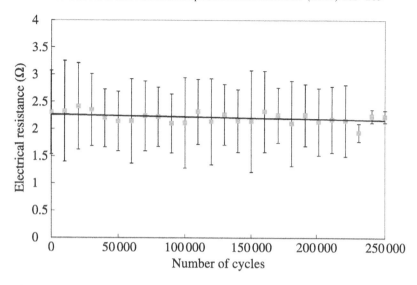

Figure 10. Average resistance *versus* numbers of cycles. Error bar indicates one standard deviation.

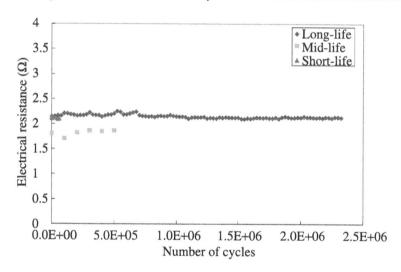

Figure 11. Typical variation of resistance during cycling for three test types.

beam. The load–displacement data consisted of two linear portions during loading up to 400 µN and two linear portions during unloading. Initial linear portion of the loading curve was due to bending action (i.e., deflection) of the cantilever beam. The second linear portion of the loading curve was due to the deformation of the contact. Then, after a five second hold period at the maximum load, the cantilever beam was unloaded until the load was completely removed. As the microprobe retracted, the restoring force of the cantilever beam separated the contacting surfaces. The contact separation was easily determined as the point when the slope of the load–displacement curve changed during unloading due to the significant stiffness

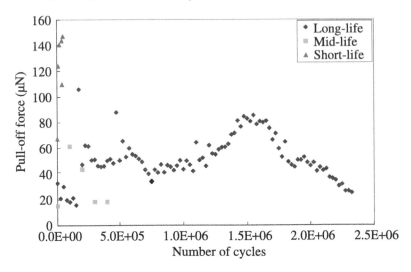

Figure 12. Typical variation of pull-off force during cycling for three test types.

difference before and after the contact. The pull-off force was thus measured at the point of contact separation which was the force required to break the adhesion bond between the contact surfaces.

The present study used the pull-off force as a measure of adhesion between contact surfaces. These terms are used interchangeably because a pull-off force, equal to the adhesion force, is required to separate the contacting surfaces. Pull-off force as a function of number of cycles from representative long-, mid- and short-life tests is shown in Fig. 12. This figure shows an initial increase followed by decrease in the pull-off force in the long-life test. Thereafter similar behavior is repeated. This suggests that contact surfaces were cleaned by mechanical and/or fritting action during initial cycling which increased their adhesion (i.e., pull-off force). Thereafter, continued cycling caused increase in the surface roughness thus reducing contact adhesion. This was followed by smoothening of surfaces which again increased adhesion. This process was repeated with cycling in the long-life tests as shown in Fig. 12.

However, a considerable increase in the adhesion force could cause bonding such that contacting surfaces would not separate upon application of the switch restoring force and thus could cause the adhesion failure. This was the case in the short-life tests where ductile separation occurred. Figure 12 shows that measured pull-off force in a short-life test increased to more than 140 µN which was a considerable increase in the adhesion force. Further, this amount of increase in pull-off force, early in lifetime, matches with several failures before 30 000 cycles. In the case of mid-life test, failure was due to gold film delamination from the substrate even though pull-off force was lower than the short-life tests. This failure was possibly due to subsurface damage in the thin film which developed during cycling. On the other hand, the contact failed due to adhesion after application of large number of cycles (more than 1 million cycles) in the long-life tests, and the failure showed brittle characteristics similar to failure surfaces commonly seen in metallic materials when subjected to high cycle fatigue. It should be mentioned here that a previous nanoindenter based study also observed increase in microswitch pull-off force with cycling [7]. It should further be noted that brittle contact separation is a desirable condition because it could provide longer lifetime and reduce damage to contacts in microswitches [10].

The average pull-off force during the later part of cycling in the long-life tests was about 50 μN. As expected, it was larger than the predicted adhesion force of 21 μN for a clean gold–gold contact based on the Johnson, Kendall and Roberts (JKR) theory [8] of adhesion which is:

$$F_{po} = \frac{3}{2}\pi R \Delta\gamma, \tag{2}$$

where F_{po} is the pull-off force, R is the mean radius of contact, and $\Delta\gamma$ is the energy of adhesion: $\Delta\gamma = \gamma_1 + \gamma_2 - \gamma_{12}$. γ_1 is the surface energy of material 1, γ_2 is the surface energy of material 2, and γ_{12} is the interfacial energy after contact. γ_{12} is zero if both surfaces are of the same material. The surface energy of clean gold is assumed to be equal to 1.12 J/m^2 [22]. There may be several factors affecting adhesion of the contact other than the variability in contact size and surface energy. One important factor is the contamination. It is possible that the fritting effect from the hot-switching condition kept the gold contacts relatively free of contaminants or destroyed contamination buildup during cycling. Lastly, it should be mentioned here that the pull-off force was measured at a constant rate of 50 μN/s. Pull-off force may also be rate dependent as suggested in [10, 11].

3.3.3. Threshold Force and Distance
Contact force required to create a stable electrical contact is referred to as the threshold force in the present study. Similarly, threshold distance is referred to as the deformation beyond initial contact when stable electric contact occurs. The average stable threshold force of all tests in this study was in the range of 10–25 μN. These values are consistent with that reported by Hyman and Mehregany who measured full metallic conduction between gold contacts when 20–60 μN contact load was applied [5]. However, they measured threshold force only up to 60 load/unload cycles. The average threshold forces for three test categories are shown in Fig. 13. In all three test categories, threshold force is initially large, then decreases considerably over the early part of cycling and thereafter it remains practically constant. This trend is similar in all three categories (i.e., long-, mid- and short-life tests). The initial high threshold force indicates contamination of the gold contact in the beginning of the test. Mechanical cycling and fritting most likely removed the contamination from the surface, allowing better ohmic contact later during cycling. The main difference between these three categories was high initial threshold force in the long-life tests which also decreased quickly to a level comparable to the other two categories (i.e., mid- and short-life). This higher initial threshold force sug-

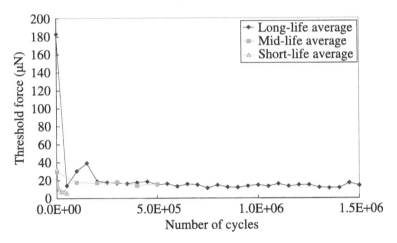

Figure 13. Average threshold force *versus* number of cycles for three test types.

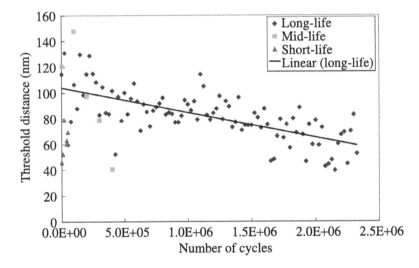

Figure 14. Typical threshold distance *versus* number of cycles for three test types.

gests more contamination of the gold contact before cycling, which was removed quickly during cycling. However, the burn-out period of contamination in all three test categories was similar in spite of the different contamination levels.

Figure 14 shows the comparison of threshold distance *versus* number of cycles relationships among typical short-, mid- and long-life tests. Threshold distance increased up to failure in the short-term tests. It initially increased and then decreased in the mid-life tests. On the other hand, it decreased overall with cycling in the long-life test which is shown by the linear fit to the data as a line in Fig. 14. The decrease in threshold distance with cycling in the long-life tests was possibly due to smoothening of the contact surface, wear and/or damage of the contact surfaces. Hence the decrease in threshold distance with cycling in the long-life test is an-

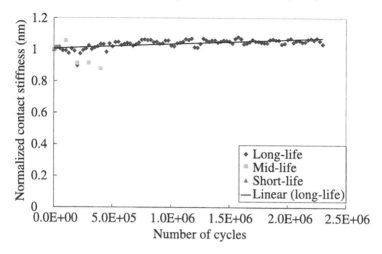

Figure 15. Typical normalized contact stiffness *versus* number of cycles for three test types.

other indicator that the surface was changing, most probably due to smoothening which increased adhesion causing failure. This could also be due to hardening of the contact surfaces with cycling.

3.3.4. Strain Hardening

Contact stiffness was measured from the slope of the unloading portion of the load–displacement curve before start of and during a test. Stiffness at a certain number of cycles was normalized by the initial stiffness, i.e., the slope of the unloading curve before cycling. This was done in order to remove the effects of the experimental frame stiffness, thus it provided the relative change in contact stiffness with cycling. Figure 15 shows the normalized contact stiffness *versus* number of cycles relationship of typical short-, mid- and long-life tests. There was practically no change in contact stiffness in the short-life tests, and a small decrease (∼10%) in the mid-life tests. On the other hand, there was a small increase (<4%) in the long-life tests which suggests that strain hardening occurred in the long-life test. However, there is a possibility that two opposite mechanisms could be present, i.e., plastic deformation causing strain hardening may be occurring simultaneously with an annealing effect from contact heating causing softening. The latter is indicated in the mid-life test which shows relatively more softening than hardening.

3.3.5. Time-Dependent Behavior

During the measurement of load–displacement relationship of the contact pair, a five second hold period was programmed between the end of loading and start of unloading by the nanoindenter. A small deformation of the contact occurred under constant load during this hold period. This was measured and its average value from all tests is shown in Fig. 16. This time-dependent deformation decreased with cycling and stabilized at approximately 2 nm after 150 000 cycles. An example of time-dependent deformation in a long-life test is shown in Fig. 17, which shows

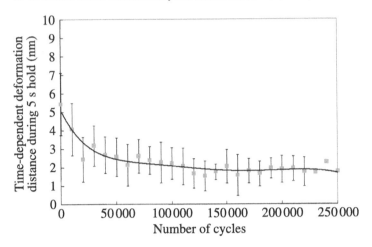

Figure 16. Average time-dependent contact deformation distance for all tests *versus* number of cycles.

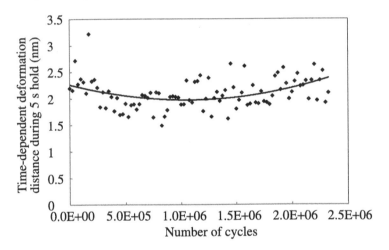

Figure 17. Time-dependent contact deformation distance during 5 s hold in a long-life test.

initially small decrease followed by a slightly increasing trend especially during the later part of cycling. The average of this deformation in the long-life test was about 2 nm. Thus, there was an existence of time-dependent deformation during contact loading, very similar to creep behavior, besides the strain hardening. This time-dependent behavior may be due to heating from the current passing through the contact. The softening temperature of gold is lower at the microscale compared to bulk softening temperature [23]. Previous researchers have calculated temperature in the contact region which showed that the higher temperatures are at the circumference of the contact area [24]. Although the average temperature of the contact region due to contact heating was not close to the melting temperature of gold, the edge of the contact area might have reached a temperature high enough to contribute towards this time-dependent deformation. The time-dependent behavior indicated may also be convoluted with thermal drift of the instrument. However, note that previous researchers had studied room temperature creep in thin-film materials [25]. Bannuru reports that time-dependent plastic deformation in thin films is a cause for concern and that thin films are prone to creep even at room temperature and below the yield stress [26]. Gregori and Clarke suggest that creep in gold microcontacts occurs under load, and that creep is a factor in the development of adhesion force between contact surfaces [27].

3.3.6. Plastic Deformation

Both elastic and plastic deformations were present during each contact load cycle. This was evident from the load–displacement curve. One of the methods to characterize plastic deformation is the energy absorbed during a cycle. An understanding of progression of plastic deformation in microcontacts is important to determine their lifetime. Figure 18 shows typical energy absorbed in short-, mid- and long-life tests. In all three test categories, the energy absorbed, and hence plastic deformation, increased initially. Thereafter, the energy absorbed decreased and hence plastic deformation reduced as shown by both short- and long-life tests. Finally, energy absorbed and hence plastic deformation stabilized to a constant value as seen in the long-life test. This suggests that the majority of the plastic deformation occurred during the initial cycling as commonly seen in many metallic materials under elastic–plastic fatigue loading condition. Thereafter, the gold strain-hardened enough causing reduced or no plastic deformation with further cycling. However, there does not appear to be any indication in the energy absorbed trend (i.e., plastic deformation) which can be used to predict imminent failure of the microcontact.

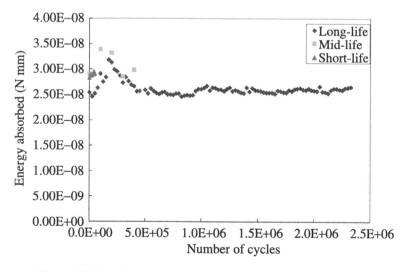

Figure 18. Typical energy absorbed during cycling for three test types.

4. Summary

Understanding microcontact behavior and characterization of its parameters under cycling condition is desirable for the improvement of MEMS switches' performance. Further, gold is a common microcontact material in MEMS switches due to its attractive material properties and ease of fabrication. However, few studies involving measurement of contact parameters of the gold–gold microcontact have been accomplished. The present study, therefore, characterized different parameters of gold–gold microcontact under cycling condition.

Three failure types were observed. Type I had short life, i.e., less than 70 000 cycles which exhibited ductile separation of contact surfaces. Type II had mid (intermediate) life ranging from 190 000 to 500 000 cycles where the gold contact film was separated (i.e., delaminated) from the substrate (i.e., contact bump). Type III had long life, i.e., more than one million cycles. The failure surfaces from Type III tests showed brittle separation characteristics. The primary cause of failure was adhesion, i.e., contacting surfaces remained stuck together (failed closed) in all three test types. This was due to the fact that contact surfaces were initially smoothened out and electrically cleaned from the fritting action with cycling which increased adhesion (i.e., pull-off force), followed by changes to the surfaces with further cycling causing varying levels of adhesion. This process was repeated with continued cycling until the adhesion between contact surfaces became so high that the restoring force of the cantilever was no longer able to pull them apart.

The force to develop stable electrical contact (threshold force) was initially large which decreased considerably very early in cycling, and thereafter remained practically constant. This suggests fritting and mechanical action quickly destroyed initial contamination on gold–gold contacts. Time-dependent and plastic deformations in the gold–gold contact material were also present. These effects were relatively dominant during initial cycling and then settled down to a constant value with further cycling. This suggests that elastic–viscoplastic material model(s) with strain hardening capability are needed for the analysis of gold–gold contact. Lastly, contact resistance was practically constant during cycling in all tests even with different types of contact separations and failure mechanisms. There was also indirect evidence that contact heating was present in this study involving hot-switching condition that might have caused time-dependent deformation and some softening of the contact material.

References

1. S. Majumder, J. Lampen, R. Morrison and J. Maciel, *IEEE Instrumentation & Measurement Magazine* **6** (1), 12–15 (2003).
2. G. M. Rebeiz and J. B. Muldavin, *IEEE Microwave Magazine* **2** (4), 59–71 (2001).
3. J. McKillop, *Microwave J.* **50** (2), 1–24 (2007).
4. C. Goldsmith, J. Maciel and J. McKillop, *IEEE Microwave Magazine* **8**, 56–60 (2007).
5. D. Hyman and M. Mehregany, *IEEE Trans. Components Packaging Technol.* **22**, 357–364 (1999).

6. J. Schimkat, Contact measurements providing basic design data for microrelay actuators, *Sensors and Actuators A* **73**, 138–143 (1999).

7. G. Gregori, R. E. Mihailovich, J. F. DeNatale and D. R. Clarke, Development of adhesive contact of MEMS-switches upon actuation cycling, in: *Proc. 18th IEEE Conference on Micro Electro Mechanical Systems (MEMS 2005)*, Miami Beach, FL, USA, pp. 439–442 (2005).

8. D. Maugis, *Contact Adhesion and Rupture of Elastic Solids*. Springer Verlag, Berlin (2000).

9. D. Maugis and H. M. Pollock, *Acta Metall.* **32**, 1323–1334 (1984).

10. L. Chen, Physics of Microcontacts for MEMS Relays, *PhD Dissertation*, Northeastern University, Department of Electrical and Computer Engineering, Boston, MA (2007).

11. L. Chen, N. E. McGruer, G. G. Adams and Y. Du, *Appl. Phys. Lett.* **93**, 053503 (2008).

12. D. J. Dickrell and M. T. Dugger, *IEEE Trans. Components Packaging Technol.* **30**, 75–80 (2007).

13. R. E. Mihailovich and J. F. DeNatale, A model for the electrical degradation of metal micro-contacts during many-cycle operation, in: *Proc. 2004 ASME/STLE International Joint Tribology Conference*, Long Beach, CA, USA, pp. 1–4 (2004).

14. S. Majumder, N. E McGruer and G. G. Adams, in: *Proc. IEEE MEMS 2005 Conf.*, Miami Beach, FL, USA, pp. 215–218 (2005).

15. K. W. Gilbert, S. Mall, K. D. Leedy and B. Crawford, in: *Proc. 54th IEEE Holm. Conference on Electrical Contacts*, pp. 137–144 (2008).

16. H. Lee, R. A. Coutu, S. Mall and K. D. Leedy, *J. Micromech. Microeng.* **16**, 557–563 (2006).

17. J. L. Ebel, D. J. Hyman and H. S. Newman, *IEEE Microwave Magazine* **8**, 76–88 (2007).

18. H. Kwon, D. J. Choi, J. H. Park, H. C. Lee, Y. H. Park, Y. D. Kim, H. J. Nam, Y. C. Joo and J. U. Bu, in: *Proc. IEEE 20th International Conference on Micro Electro Mechanical Systems (MEMS 2007)*, pp. 231–134 (2007).

19. G. E. Dieter, *Mechanical Metallurgy*, 3rd edn. McGraw-Hill, Boston (1986).

20. W. R. Chang, Contact, Adhesion, and Static Friction of Metallic Rough Surfaces, *PhD Dissertation*, University of California, Berkeley (1986).

21. R. Holm, *Electric Contacts: Theory and Application*, 4th edn. Springer Verlag, Berlin (1967).

22. E. Rabinowicz, *Friction and Wear of Materials*, 2nd edn. John Wiley & Sons, New York (1995).

23. B. D. Jensen, L. W. Linda, K. H. Chow, K. Sautou, J. L. Volakis and K. Kurabayashi, *J. Microelectromechanical Systems* **14**, 935–946 (2005).

24. J. A. Greenwood and J. B. P. Williamson, *Proc. Royal Soc. (London) Series A* **246**, 13–31 (1958).

25. Z. C. Leseman, A novel method for testing freestanding nanofilms using a custom MEMS load cell, *PhD Dissertation*, University of Illinois Urbana-Champaign, Urbana-Champaign, IL (2006).

26. T. Bannuru, Effects of Alloying on Mechanical Behavior of Noble Metal Thin Films for Microelectronic and MEMS/NEMS Applications, *PhD Dissertation*, Lehigh University, Bethlehem, PA (2008).

27. G. Gregori and D. R. Clarke, *Appl. Phys. Lett.* **87**, 54101 (1–3) (2005).

Characterization and Adhesion of Interacting Surfaces in Capacitive RF MEMS Switches Undergoing Cycling

Seung Min Yeo [a], **Andreas A. Polycarpou** [a,*], **Spyros I. Tseregounis** [a],
Negar Tavassolian [b] **and John Papapolymerou** [b]

[a] Department of Mechanical Science and Engineering, University of Illinois at Urbana–Champaign, Urbana, IL 61801, USA
[b] School of Electrical and Computer Engineering, Georgia Institute of Technology, Atlanta, GA 30332, USA

Abstract

The performance and reliability of capacitive-type Radio Frequency Microelectromechanical Systems (RF MEMS) switches are affected by interfacial interactions such as strong attractive adhesion forces that could cause the switch to get temporarily or permanently stuck, thus rendering the device nonfunctional. Such adhesion forces could be induced and progressively increase during cycling by mechanical surface changes, such as increased real contact area or electrical effects such as increased electrostatic forces by dielectric charging. Both effects can occur simultaneously, which makes the characterization of capacitive switch behavior more complex than metal-to-metal switches. In this work, capacitive RF MEMS switches with Ti-on-silicon nitride contact surfaces were fabricated and tested for different numbers of cycles. After cycling, the Ti beam and Si_3N_4 dielectric contact surfaces were exposed and examined topographically (using Atomic Force Microscopy), chemically (using Time-Of-Flight Secondary Ion Mass Spectrometry) and nanomechanically (using nanoindentation) to measure surface changes induced by cycling. Such surface changes could explain adhesion behavior and failure of capacitive RF MEMS switches. Topographical roughness analysis showed that physical surface changes occurred early (during the first few hundred cycles). Chemical and nanomechanical analyses also showed surface property changes with cycling, in agreement with the roughness analysis. Therefore, RF MEMS switch surface changes occur with cycling and such changes are readily detectable and could adversely affect (increase) interactive adhesion forces and thus render the switch inoperable.

Keywords

Capacitive RF MEMS switch, silicon nitride dielectric, adhesion, AFM, roughness, TOF-SIMS, nanoindentation

* To whom correspondence should be addressed. Tel.: (217) 244-1970; Fax: (217) 244-6534; e-mail: polycarp@illinois.edu

Adhesion Aspects in MEMS/NEMS
© Koninklijke Brill NV, Leiden, 2010

1. Introduction

Radio frequency microelectromechanical systems (RF MEMS) switches are an emerging technological advancement in wireless microwave communication systems such as mobile phones, satellite, automotive, and space-based radar systems. This is primarily due to their relatively small size with virtually no mass, negligible power consumption, low insertion losses, and high linearity. RF MEMS switches can be placed either in series or shunt configurations, and their contact surface can be either metal-to-metal-type or capacitive-type [1]. Metal-to-metal contact switches (typically Au (or Al-alloy)-on-Au (or Al-alloy)) are suitable for use from 0 Hz (DC) to 6 GHz RF signal frequencies, and are used for low current ($\leqslant$10 mA) applications. Capacitive switches with metal-to-dielectric contacts (typically Au (or Ti)-on-silicon nitride) are used at frequencies of 10 GHz and above [2]. The advantages of metal-to-metal contact switches include high fabrication yield and low cost compared to capacitive switches, where a thin dielectric layer is additionally deposited between a movable beam and signal line. However, capacitive switches are preferred in the construction of tunable or reconfigurable switching circuits because they can also function as capacitors in the circuit and not only as simple, on-off switches as metal-to-metal contact switches do [3].

Commercialization of RF MEMS switches and the substitution for conventional gallium arsenide (GaAs) field-effect transistor (FET) and PIN diode switches is hindered by reliability problems and unpredictable switch failures during operation. Such failures are due partly to the fact that as the surface-to-volume ratio increases in MEMS devices, the surface forces become more dominant than the inertial effects, and thus significantly affect the performance and reliability of MEMS devices. Adhesion of the beam to the bottom dielectric surface (sometimes referred to as stiction in the literature) is the primary failure mechanism in MEMS devices in general and capacitive switches [4] in particular. Adhesion forces are mainly attributed to two phenomena; (1) mechanical deterioration of the contact surface with cycling [5], and (2) electrical charge trapping within the dielectric layer which induces electrostatic attraction [6]. One of the main difficulties in capacitive RF MEMS switch research is that these two effects are coupled and occur at the same time, which makes characterization of adhesion behavior of RF MEMS switches more complex, compared to metal switches, for example. Ref. [7] provides an overview of the major interfacial adhesion and tribological issues in MEMS.

The literature on capacitive RF MEMS switches has focused on the alleviation of electrical charging effects. It has been shown that a dielectric layer deposited at low temperature (150°C) exhibits lower electrical charging effects compared to higher deposition temperature (250°C). Thinner dielectric films also showed reduced charging effects [8]. The effects of surface topography and roughness on dielectric charging were studied using capacitors [9, 10] and coupon samples in a switch simulator [4]. Dielectric surface smoothening after repeated contacts increased the electric field within the dielectric, which led to more charge trapping and an eventual pull-down adhesion failure [4].

Research on metal-to-metal contact switches is abundant. A difference from capacitive switches is that in metal switches, electrical contact resistance (ECR) measurements can be readily used to determine performance and reliability of metal-to-metal contact switches [11, 12]. ECR measurements show that surface topography changes during early switch cycling due to plastic deformation of the contact surfaces [12, 13]. The electrical current passing through the contact area also had an effect on the behavior of metal contact switches [14, 15]. For example, in Ref. [15], cold switching (no current flowing through the contact area) showed significantly longer lifetime compared to hot switching (current flowing through the contact area), which suggested that failures were not a result of the physical stress of switching alone, but had to do with the current passing through the contacts as well. Patton and Zabinski [14] showed that contact surface change patterns were different at low and high currents. Necking or ductile-type separation by asperity melting occurred at high currents while brittle-type separation by asperity creep dominated at low currents.

Even though electrical charging plays a significant role in RF MEMS switch operation (which could be recoverable as the charge dissipates with time), adhesion behavior and potential nonrecoverable switch failures are eventually determined by the deterioration of the physical interface contact between the beam and the dielectric surface. Physical parameters that could be changing with switch cycling include topographical, chemical and nanomechanical surface properties. These parameters have been investigated using identical MEMS switches. Coupon-type testing (e.g., using similar materials as real switches and piezoactuators for cycling) may reduce the complexity of coupled electrical and mechanical/surface effects, but it is unable to reproduce the coupled problem as seen in realistic RF MEMS switches since it does not simulate the exact contact conditions. In this research, capacitive RF MEMS switches have been fabricated and then operated for different numbers of cycles. Subsequently, the contact surfaces have been analyzed in order to examine the physical and chemical changes that occurred with cycling. Surface characterization included detailed topographical, chemical and nanomechanical properties measurements.

2. Fabrication and Cycling Experiments

2.1. Fabrication of Capacitive RF MEMS Switches

A typical RF MEMS switch consists of a free-standing plate suspended by beams above a coplanar waveguide (CPW), and under this beam, a thin dielectric layer is present. When a direct current (DC) bias is applied between the CPW center conductor (which also carries the RF signal) and the surrounding ground plane, the beam is attracted to the dielectric electrostatically, and makes complete contact with the dielectric, which results in shorting of the RF signal. Therefore, by actuating the beam electrostatically, the device behaves as an RF shunt switch for signals in the GHz frequency range. The choice of materials used for the dielectric layer could significantly affect the performance and reliability of the capacitive switch in terms of electrical charging effects [16]. The two most common dielectric materials used in capacitive RF MEMS switches are plasma-enhanced chemical vapor deposited (PECVD) silicon dioxide and silicon nitride. Silicon dioxide has a lower trap density than silicon nitride, which implies that devices made with silicon dioxide dielectric layers should be less prone to charge trapping, i.e., longer lifetime. However, PECVD silicon nitride has higher dielectric constant values of 6–9, when compared to PECVD silicon dioxide of 4.1–4.2. This difference ensures sufficient pull-down force and high capacitance when the switch is activated [17]. In this study, silicon nitride was used for the fabrication of the switches.

The switch structures used in this work were fabricated on high-resistivity silicon substrates (525 µm thick) with a 2-µm-thick oxide isolation layer. The CPW signal lines were then fabricated by evaporating Ti/Au (20 nm/200 nm). Using PECVD, Si_3N_4 layer was deposited and patterned to form the dielectric layer between the beam and the signal line. A 2-µm-thick photoresist was spin coated onto the substrate and patterned to create the sacrificial layer for forming the initial gap height, g of the beam. A Ti/Au/Ti (20 nm/200 nm/20 nm) seed layer was evaporated and patterned, and then Au was electroplated to a thickness of 2 µm on the top of the Ti layer of the beam and the CPW line. Lastly, the sacrificial photoresist layer was removed using a liquid resist stripper to release the beam structures, and the resulting devices were dried using CO_2 supercritical drying process.

Scanning Electron Microscopy (SEM) images showing a typical fabricated capacitive RF MEMS switch are depicted in Fig. 1(a) and 1(b). This type of capacitive switch is categorized as fixed-fixed end, air-bridge type shunt switch which has a serpentine-shape spring supporting the center beam structure as depicted in Fig. 1(a). In the magnified image of Fig. 1(b), the rougher electroplated gold material is shown on the top of the beam surface, and the Si_3N_4 dielectric layer (dark part) covers the gold RF signal path (bright color part). The size of the beam is $100 \times 200 \ \mu m^2$ with 2 µm thickness and 2-µm air gap, g. When the switch is actuated, contact occurs between the Ti surface of the upper beam and the silicon nitride surface of the dielectric layer, and the surface analysis was conducted on these two contact areas. Figure 1(c) depicts a simplified schematic of the switch layered structure.

2.2. Cycling Experiments

To examine the evolutionary changes and deterioration of the contacting surfaces with cycling, identical capacitive switches (i.e., of the same structural design and fabricated on the same wafer) were tested up to different predetermined numbers of cycles, namely 10^2, 10^3, 5×10^3, 10^4 and 5×10^4 cycles by presetting the upper limit of the number of cycles in each experiment. The nature of these experiments is such that identical switches needed to be used. Each wafer had 19 identical switches (of the exact same design) and due to fabrication yield issues, less than 10 switches were typically fully functional. Because of this limitation, typically one and in some cases two experiments at each cycle were performed. As shown later in this work

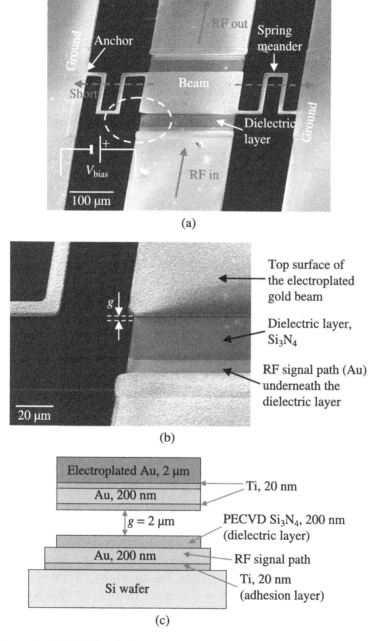

Figure 1. Capacitive RF MEMS switch; (a) overall SEM image showing a complete switch, (b) SEM image zoom-in showing the air gap g and the silicon nitride dielectric layer and (c) schematic layered switch structure.

experiments were repeatable (comparing switches tested under identical number of cycles or comparing switches tested at slightly different numbers of cycles).

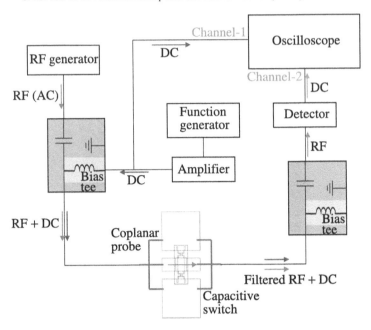

Figure 2. Schematic of capacitive RF MEMS switch cycling experimental setup.

In situ monitoring the switch operation is important to ensure that the beam (or membrane) is fully actuated and making complete contact with the bottom (dielectric) surface. In the case of metal-to-metal contact switches, *in situ* ECR is simultaneously measured with cycling to monitor changes in the real contact area, thus relating it to switch performance. In the case of capacitive switches, one applies an RF signal along with the DC bias to the switch and measures the modulated RF signal resulting from the switch actuation [18].

Figure 2 depicts the experimental setup that was used to cycle the capacitive switches and monitor their actuation and performance. A DC square-wave signal generated using a 20-MHz function/arbitrary waveform generator (Agilent 33220A) was amplified by a power amplifier (Treck 603). The DC bias was then added to the RF alternating current (AC) signal at the bias tee, and applied to the switch through a coplanar probe on a Cascade probe station. As these two signals are applied to the switch, a short or open circuit in the RF transmission line is achieved by mechanical movement of the upper beam. The switch alternates between high (switch down or in close/contact position) and low capacitance (switch up or in open/noncontact position). When the switch is on (down state), the capacitance between the upper beam and the substrate is sufficiently high for the RF signal to pass through, hence the circuit is grounded (Fig. 1(a)). The output RF signal travels to the second bias tee along with the DC signal where the DC signal is removed and only the RF signal remains. The RF signal is fed into a crystal detector (Agilent 8473C) which provides a DC voltage that is proportional to the modulated RF envelope and representative of the switching behavior of the device under test. This

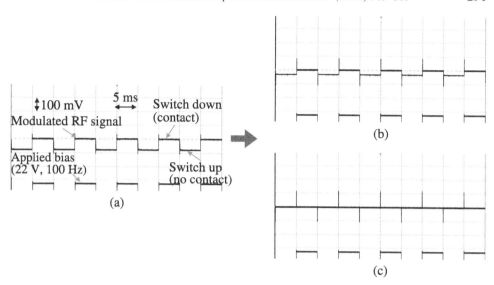

Figure 3. Typical modulated RF signals showing *in situ* performance deterioration with cycling; (a) modulated RF signal (80 mV) and applied bias (22 V) at the beginning of a cyclic experiment, (b) decreased RF signal of a deteriorated switch with cycling and (c) zero RF signal of a stuck switch.

DC voltage is captured in an oscilloscope (Agilent 54622A). Typical oscilloscope output signals are depicted in Fig. 3.

Figure 3(a) shows the modulated RF signal (about 80 mV peak-to-peak amplitude) and the applied bias (22 V) at the beginning of a typical cyclic experiment. Figure 3(b) shows a decreased modulated RF signal (about 30 mV) compared to the initial signal, which implies that the switch performance has deteriorated with cycling. The zero amplitude RF signal in Fig. 3(c) signifies that the beam is stuck to the bottom dielectric surface resulting in shorting of the RF signal. Therefore, by examining these *in situ* RF signals during cycling experiments, it is possible not only to monitor the *in situ* behavior of the capacitive switch, but also to differentiate between working switches and stuck switches. Some of the switches were nonfunctional, e.g., they got stuck as soon as the voltage was applied at the beginning of the test. These switches are subsequently referred to as 'stuck' switches. All other tested switches were not stuck until the end of the test, including the highest number of cycles achieved: 5×10^4 cycles. Attempts to cycle switches beyond 5×10^4 cycles were unsuccessful (switches got stuck).

During measurements, the RF signal was 15 GHz with 0.316 mW RF power, and the frequency of the square-wave input was 100 Hz. The applied voltage was 22 V which was 10% higher than the pull-in voltage of 20 V, determined using a capacitance–voltage characteristic test. Although the contact force depends on the switch geometry and the applied voltage, it is typically in the range of 50–200 μN for typical operating conditions of RF MEMS switches [1]. All tests were conducted in atmospheric ambient laboratory conditions (23°C and 32% relative humidity).

Table 1.

Experimental measurements using different switches under different experimental conditions

Switch no.	RF signal before switching (mV)	Switching time (s)	Number of cycles	RF signal after switching (mV)	Stuck or not
1	230	1	10^2	50	No
2	210	10	10^3	30	No
3	230	50	5×10^3	25	No
4	200	100	10^4	35	No
5	100	500	5×10^4	60	No
6	200	<1	<100	0	Stuck
7	210	<1	<100	0	Stuck
8	190	<1	<100	0	Stuck

Under such relatively low relative humidity conditions, capillary (meniscus) effects are not expected to be significant, as reported in [19]. Table 1 summarizes the switching time and RF signal data before and after cyclic testing for each switch tested. The RF signal power was reduced drastically during the earlier switching cycles, and for some switches, it became zero as soon as the bias voltage was applied (which meant that the beam got stuck to the dielectric layer). The decreased RF signal voltage means that the gap distance between the membrane and dielectric surface has decreased (thus increased capacitance), which could be attributed to the increased interfacial adhesion force. Smoothening or degradation of the contacting surfaces increases both adhesion and dielectric charging. To what extent either of these two phenomena is more dominant is difficult to assess. What is known for certain however is that surface roughness smoothening also affects electrical charging in addition to causing increased adhesion. Specifically smoother surfaces cause more electrical charging, in agreement with [10]. Also, it should be noted that in some switches where the RF signal was significantly reduced to around 15% of its original value, if the test was stopped for some time (5–10 s), the RF signal was restored to about 70% of its original value. Therefore, the large reduction of RF signal and stuck switch occurrence during early cycling could be (at least in part) due to recoverable electrical charging. As the focus of this study was the detailed description of geometrical and physical surface changes with switch cycling, no further studies on probability of failure of RF MEMS switch and rate of failure will be reported in this work.

Once the switching experiments were completed, the switches were opened for the ensuing detailed surface examination. Using a probe station, the top beam of the switch was broken off to expose the two contacting surfaces for measurements. Using four micromanipulators, one end of the upper beam was cut and flipped over to expose contact surfaces of both the beam and the dielectric. A typical switch that was opened for measurements is depicted in Fig. 4.

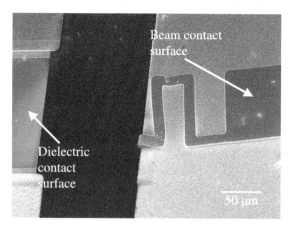

Figure 4. Destructively opened RF MEMS switch exposing the Ti-coated beam surface and the dielectric surface where contact was made.

3. Surface Analysis

3.1. Surface Topographical Analysis

3.1.1. AFM Measurements

Examination of the topographical changes of the contacting surfaces (beam and dielectric) was performed in order to shed light on the contact severity and changes with increasing number of cycles. Switches that were operated for different numbers of cycles (Table 1) as well as new untested and stuck switches were measured using a tapping mode AFM (Veeco Dimension 3100), after opening and exposing the contacting surfaces, as described above. Based on earlier experience using Au metal-to-metal switches [5], where isolated surface damage was seen with cycling, we first needed to establish the level of contact uniformity across the whole beam area. Larger AFM scans of 20×20 µm^2 were performed on both the dielectric and beam surfaces at six to eight different locations to ensure measurements on almost the whole contact surface area of 100×200 µm^2. On comparing these measurements, it was found that they were almost identical indicating that uniform contact across the whole beam surface had occurred during cycling. Specifically for each switch, the standard deviation of the Root-Mean-Square roughness (S_q) was 0.046 nm ($S_q = 1$ nm) and 0.080 nm ($S_q = 8$ nm) for the beam and dielectric surfaces, respectively. Contact over the whole surface area was not surprising since the large beam surface is supported with flexible anchors over the dielectric surface, thus ensuring uniform contact.

To investigate the detailed roughness/topography of the contact surfaces, smaller scan areas of 5×5 µm^2 were analyzed for both the dielectric and beam surfaces. At least 6 measurements for each surface were performed to ensure repeatability. Figure 5 depicts representative AFM measurements. The beam surfaces are significantly smoother with $S_q = 1$ nm compared to the Si$_3$N$_4$ surfaces of the bottom dielectric layer with $S_q = 8$ nm. The beam topographies depicted in Fig. 5(a) exhibit

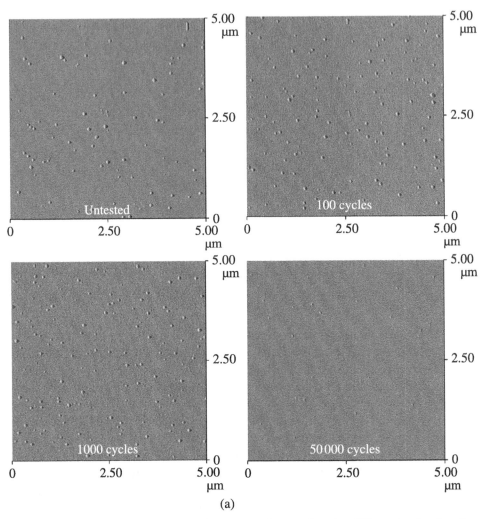

(a)

Figure 5. 5 μm × 5 μm AFM images of contact surfaces undergoing different levels of cycling; (a) beam surface (20 nm, Ti) and (b) dielectric surface (200 nm, Si_3N_4).

isolated peaks which result from the fabrication process. The height of these peaks as well as their density reduces with cycling; this being apparent after 50 000 cycles. Referring to Fig. 5(b) for the dielectric surfaces, no visual roughness changes with cycling were observed.

In summary, based on the large AFM scan surface analysis, contact occurred uniformly throughout the switch surface. Based on small scan roughness measurements, some roughness changes were observed with cycling and these changes were further investigated using detailed roughness parametric analysis.

3.1.2. Parametric Surface Roughness Analysis
Unlike the case of cycling contact of Au metal switches where significant surface damage was observed even after only a few tens of cycles [5], in the capacitive

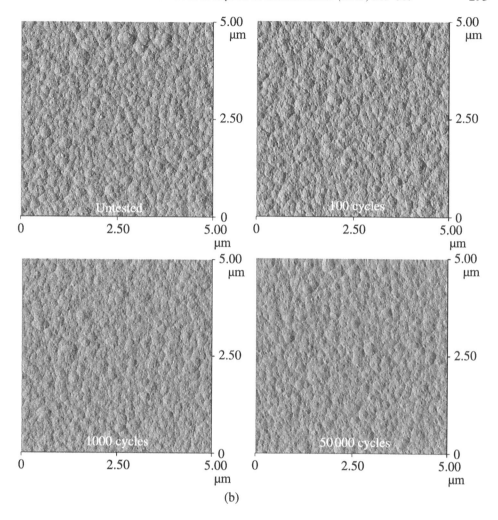

Figure 5. (Continued.)

switches investigated in this work, no drastic changes were seen with the AFM images. In such cases, detailed surface topography parameters have been utilized to
quantify minute surface changes (similar to those occurring in burnished surfaces
[20] and in sub-nanometer surface roughness effects in magnetic storage samples
[21]). Such extracted surface roughness parameters are also needed as input parameters in contact and adhesion models [22, 23]. The roughness properties of the
dielectric surface may have an impact on electrical charging [9, 10] as well, which
could induce adhesion failures in capacitive switches.

As articulated in the literature during the last half century, various roughness parameters have been developed to better define surface topography. A comprehensive
report was issued in 1993 describing the so-called 'Birmingham-14' parameters:
a set of 14 parameters that fully characterize a surface. These parameters are based

on areal surface measurements (referred to as 2-D parameters) and include amplitude, spatial, hybrid and functional parameters. Some of these parameters were extended from 1-D parameters or developed specifically for 2-D surfaces [24]. 2-D areal roughness parameters account for both mapping of geometric features over an area, and also provide insight into the physical and functional behavior of contact surfaces [20, 25, 26]. The digitized AFM data for both the beam and dielectric surfaces were imported into a graphics user interface program and the Birmingham-14 parameters were extracted and compared. For brevity, only four of the Birmingham-14 parameters will be discussed here (the definition and method of calculation for each of these parameters are explained in [20, 24–26] and are also summarized in the Appendix). These parameters are the root-mean-square roughness S_q, the skewness S_{sk}, the developed interfacial area ratio S_{dr} and the surface bearing index S_{bi}.

Figures 6 and 7 show the four extracted parameters as a function of number of cycles for the beam and dielectric surfaces, respectively. The widely used parameter, S_q, which represents the roughness amplitude, decreases for both the beam and

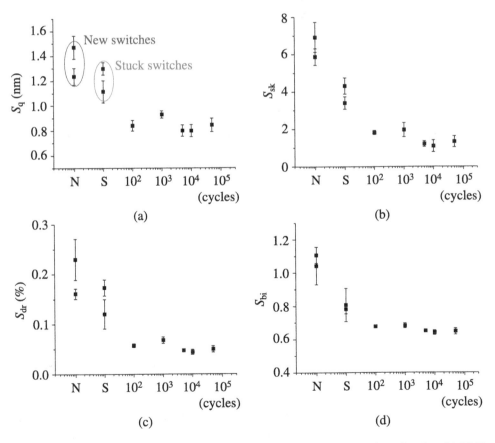

Figure 6. Beam surface extracted roughness parameters as a function of number of cycles; (a) RMS roughness, (b) skewness, (c) developed interfacial area ratio and (d) surface bearing index.

the dielectric surfaces with cycling, which means both surfaces get smoothened (or flattened) *via* contact. In the case of the beam surfaces, two separate new/untested switches and two switches that were nonfunctional (switches got stuck as soon as the voltage was applied) were measured and the mean and $\pm$ one standard deviation are depicted in Fig. 6(a). Clearly there is some variability from switch-to-switch but all four switches that did not experience cycling have the highest S_q values of about 1.4 nm and 1.2 nm for the new and stuck switches, respectively. With cycling, S_q decreases even after only 100 cycles to about 0.8 nm. Interestingly, the beam roughness did not change significantly after the initial reduction, i.e., from 10^2 to 5×10^4 cycles. Roughness reduction or asperity flattening is typically associated with an increase in the real contact area, which consequently may increase the possibility of switch failure by adhesion [23].

Skewness (S_{sk}) is a measure of the asymmetry of the surface height deviations about the mean plane, and gives some indication of the existence of any spiky features or deep valleys on the surface (Appendix). A symmetric, Gaussian surface has zero skewness and for an asymmetrical surface height distribution, positive skew-

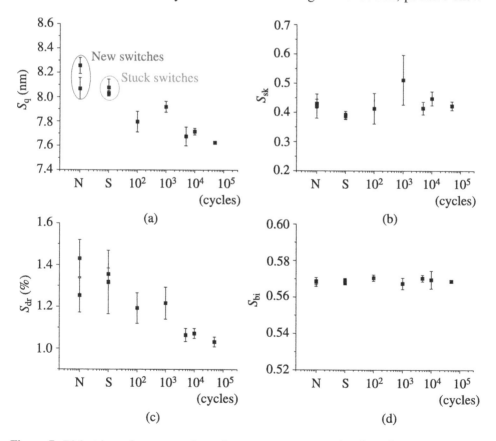

Figure 7. Dielectric surface extracted roughness parameters as a function of number of cycles; (a) RMS roughness, (b) skewness, (c) developed interfacial area ratio and (d) surface bearing index.

ness indicates spiky features on the surface and negative skewness deep trenches or valleys. Referring to Fig. 6(b), the skewness values for the untested switches were highly positive in the range of 6 to 8 due to the spiky features on the surface, as shown in the AFM images of Fig. 5(a). Stuck switches show a reduction in skewness to around 4, which shows that just an initial contact (no cycling) was sufficient to slightly flatten the peaks of the asperities. Similar to the S_q trends, the skewness also decreases to a value of 2 after 10^2–10^3 cycles and eventually to a steady-state value of around 1. This finding confirms that initial switch cycling has a significant effect on the surface topography and it quickly reaches steady-state, which could be attributed to permanent plastic deformation of the topmost contacting asperities, in agreement with [12].

An important hybrid parameter that contains both amplitude and spatial information is the developed interfacial area ratio (S_{dr}) which was specifically developed to characterize the topography of 2-D surfaces. S_{dr} is the ratio of the increment of the interfacial area of a surface over the sampling area [25] (see Appendix). In other words, if the total area of the surface composed of all of its summits and valleys is the Total Surface Area, and the projected total area is called the Nominal Reference Plane, then Areal Difference = Total Surface Area − Nominal Reference Plane. The developed interfacial area ratio is the ratio of this Areal Difference over the Nominal Reference Plane. Therefore, the more features the surface has, the larger the S_{dr} would be. The S_{dr} for the beam surface (Fig. 6(c)) is about 0.2 and 0.15 for both new and stuck switches and decreases to 0.05 with cycling, exhibiting similar behavior as S_q and skewness.

A functional parameter, the Surface Bearing Index (S_{bi}) also captures minute changes on the beam surface with cycling. S_{bi} is defined as the ratio of S_q to the surface height that corresponds to a 5% bearing area [25] (Appendix), and indicates the bearing property of a surface (i.e., is derived from the bearing area curve). For a variety of engineering surfaces, S_{bi} is larger than zero and ranges from 0.3 to 2, and a Gaussian surface has a value of 0.608. S_{bi} values larger than 1 for the new switch beam surfaces (Fig. 6(d)) show that the surface height at 5% bearing area is larger than S_q, which means there are some high spiky peaks on the beam surface. As the switch cycles, these peaks are flattened and S_q decreases, whereas the 5% bearing area height does not decrease as much. Thus, S_{bi} reduces to a stable value of 0.65.

The extracted values for the rougher dielectric surfaces are depicted in Fig. 7. As the dielectric roughness is near-Gaussian with a skewness value of around 0.4 and is unchanged after cycling, the functional parameter S_{bi} also remains constant at 0.57. S_q of the new switches is in the range of 8.0–8.3 nm with a small standard variation within each switch. Even though S_q of the stuck switches does not decrease significantly, the variation in these measurements is significantly larger, indicating that when the switch got stuck some small contact area was significantly altered whereas other areas were unaffected. With cycling, S_q decreased to 7.8 nm and settled at 7.6 nm after 5×10^3 cycles.

The selected Birmingham-14 parameters that were discussed above are representative parameters from the amplitude, hybrid and functional groups and best show the changes that occurred with cycling. The remaining Birmingham-14 parameters also affirm the observations described above. For example, the mean radius of asperity curvature (R) (a hybrid parameter) on both the beam and dielectric surfaces increases with cycling, which means that the spiky peaks on the contacting surfaces get flattened with contact (smoothened). From all the parameters investigated, only two parameters, the mean radius of asperity (R) and the core fluid retention index (S_{ci}) showed an increasing trend with cycling whereas all other parameters showed a decreasing trend or no apparent change with cycling. The isotropy index (γ), valley fluid retention index (S_{vi}), and valley void volume (S_v) did not exhibit any changes with cycling, neither on the dielectric nor on the beam surfaces. Lastly, note that parameter changes with cycling were more clearly observed on the beam surfaces compared to the dielectric surfaces, which could be attributed to the fact that the dielectric surface was significantly rougher, and hence more insensitive to the surface topography changes compared to the smoother beam surfaces.

Based on the detailed roughness parametric analysis both the beam and the dielectric surfaces exhibited some surface smoothening especially during the initial cycles, reaching steady-state values quickly. On the one hand, reaching steady-state values is good for the switch's performance (indicating that no further geometrical surface deterioration occurs), but on the other hand, smoothening of the surfaces will result in higher interfacial adhesion forces. Also, from the electrical point-of-view, smoothened surfaces increase the electric field within the dielectric layer, which leads to charge trapping and electrostatic attraction between the two contact surfaces [4].

3.2. TOF-SIMS Chemical Analysis

Interfacial adhesion interactions are affected by both chemical and physical phenomena between interacting surfaces. In regards to chemical interactions, ionic and covalent chemical bonds play a significant role in the adhesion behavior of MEMS switches [27]. Also, the presence of hydrocarbons on the contacting surfaces showed lower adhesion failures in capacitive RF MEMS switches [4]. The significance of surface 'contaminants' on such surfaces has been known, see for example [28]. To investigate the chemical composition of the contacting surfaces and their changes with switch cycling, a Time-Of-Flight-Secondary-Ion-Mass-Spectrometer (TOF-SIMS) (PHI TRIFT III, Physical Electronics) was utilized.

TOF-SIMS utilizes a pulsed primary ion beam to desorb and ionize species from a sample surface. The resulting secondary ions are accelerated into a mass spectrometer, where they are mass analyzed by measuring their time-of-flight from the sample surface to the detector. There are two different modes of analysis; (1) mass spectra are acquired to determine the elemental and molecular species on a thin (few Å to 1–2 nm) surface layer, called static analysis, and (2) depth profiles are used to determine the distribution of different chemical species as a function of depth from the surface, called dynamic analysis [29]. In this work, the static mode TOF-SIMS was used with 22 keV-pulsed Au^+ as a primary ion beam bombarding the sample surfaces. The analysis size for each sample was 50×50 μm^2 which was sufficiently large to represent the chemical species on the switch's surface, and the ion capturing time was 5 min for each measurement. The advantage of TOF-SIMS is that it can identify and distinguish chemical species which have the same atomic mass. For instance, it is widely known that Si, N_2 and C_2H_4 have the same mass, 28, but the exact mass of each species is in fact slightly different: Si: 27.9769, N_2: 28.0061 and C_2H_4: 28.0313. Thus, TOF-SIMS can identify and determine each of these chemical species corresponding to the exact mass number with 0.0001 resolution.

The untested and cycled switch surfaces (Table 1) were analyzed using TOF-SIMS and representative mass spectra from the 5×10^3 cycled switch can be seen in Fig. 8. The major chemical species detected on the beam surface, Fig. 8(a) and the dielectric surface, Fig. 8(b) are listed in each graph. An interesting observation from both spectra is that Si was abundant on both surfaces, including the beam surface (top surface layer is Ti), including Si on the untested switches, while no Ti was found on the dielectric surface (made of silicon nitride). The carbon–nitrogen compound C_3H_7NO had the most number of counts (largest amount) on both surfaces. The chemical compositions of both beam and dielectric surfaces were similar even though the nominal material for each surface is different. It is postulated that the reason for this is the fact that the static mode TOF-SIMS investigates only the topmost nm thick surface, and this topmost surface layer may be dominated by foreign compounds derived from the fabrication processes. The sacrificial photoresist layer used in fabrication contacts both the beam and the dielectric surfaces and only during the final fabrication steps the sacrificial layer is removed using a liquid resist stripper, and a supercritical drying process. Such operation cannot remove all chemicals present in the photoresist, and may leave some residue on the topmost surface. This is an important finding, complemented also by the analysis of the cycled switches where the beam surface (expected to be Ti) contains only a small amount of Ti. This phenomenon can alter the switch behavior from that expected for a pure Ti/Si_3N_4 interface. We used static TOF-SIMS analysis in this work since the contacting surfaces involve only the topmost nm-scale surface layer and no significant plastic surface deformation changes occurred, as supported by the roughness analysis discussed earlier.

Figure 8 depicts the number of counts of the most significant chemical species measured on the surface. For direct comparison and to minimize variation of the absolute number of counts from experiment-to-experiment, a TOF-SIMS's software feature (Wincadence) was used to calculate the normalized percent intensity of each chemical species on the examined surface. The calculated intensity *versus* cycling for each chemical species was obtained and plotted in Fig. 9 for both the beam and dielectric surfaces. Si shows changes with cycling for both the beam and dielectric surfaces: It decreases for the first thousand cycles, and then increases again. This change is more significant in the case of the beam surface where from 12% (for

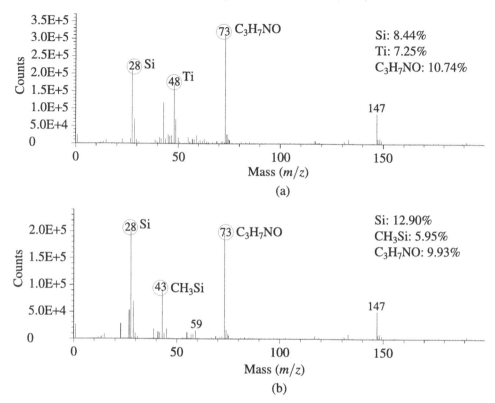

Figure 8. TOF-SIMS spectra of the contact surfaces for 5×10^3 cycled switch; (a) beam surface spectrum and (b) dielectric surface spectrum.

untested switches) it decreases to 8% (after 10^2–10^4 cycles) and then increases to 10% ($>10^4$ cycles). The slight increase after 10^4 cycles could be due to Si transfer from the dielectric surface. Even though the quantity of the SiH fragment is smaller than Si, it also changes in the same manner as Si. The chemical fragment C_3H_7NO shows the opposite trends than silicon. We also see that CH_3Si gradually increases with cycling on the dielectric surface as shown in Fig. 9(b). As pointed out earlier, the amount of Ti detected on the beam surface is only few percent, which could have significant implications in the electrical performance of the switch.

The two main observations from the TOF-SIMS results can be summarized as follows: (a) The amount of Ti detected on the beam surfaces (topmost nanometer layer) is very low (despite the fact that the surface is expected to be nominally pure Ti), and (b) no major chemical changes occur on both the dielectric and the beam surfaces with cycling.

3.3. Nanomechanical Property Measurements

In addition to the roughness and chemical analyses presented above, the nanomechanical material properties of the switches, especially measurements near the topmost surface layers, including changes due to cycling are important. To measure the

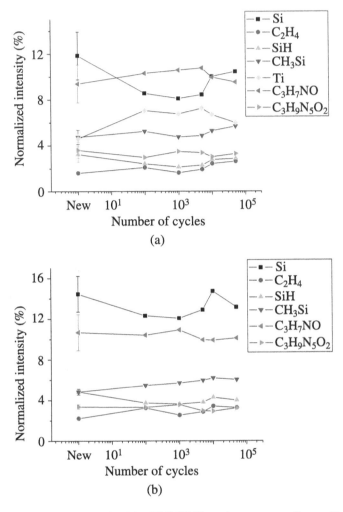

Figure 9. Chemical changes determined by TOF-SIMS on the contact surfaces with respect to the number of cycles; (a) beam surface and (b) dielectric surface.

nanomechanical properties of the contact surfaces, the nanoindentation technique was used. These measurements were limited only to the dielectric surface because once the beam was opened it could not be securely attached to a substrate for measurements. Using a nanoindenter instrument (TriboScope, Hysitron) interfaced with a MultiMode AFM (Veeco), shallow nanoindentation measurements were performed. Using the Oliver–Pharr method [30], hardness (H) and reduced Young's modulus (E_r) values of the dielectric layer were obtained. The indenter tip used for the measurements was a cube-corner tip which is a three-sided pyramid with mutually perpendicular faces and a tip radius of 100 nm. This system allows one to obtain load *versus* displacement curves for a loading-unloading cycle as depicted in Fig. 10(a) and typical scanned images of the residual indentation impressions

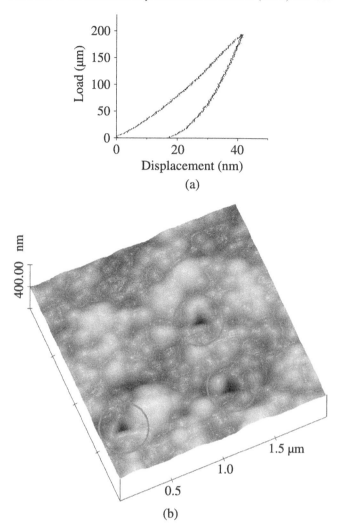

Figure 10. (a) Load–displacement curve during one full cycle of loading and unloading on the dielectric surface of a new switch and (b) residual indentation impressions on the dielectric surface of 5×10^4 cycled switch.

shown in Fig. 10(b) (these images were obtained using the same tip as used for the nanoindentation).

For each dielectric switch surface, 10 different areas were selected for measurements, and for each measurement, H and E_r were calculated at 6–8 different contact depths (from 19 nm to 55 nm). Since the focus of this study was the nanomechanical properties of the topmost surface layers, only near-surface measurements are reported. Figure 11 shows the measurement results with respect to the number of cycles at the shallow contact depth of 19 nm. At deeper contact depths, the properties were slightly lower than the topmost surface properties due to substrate effects. The

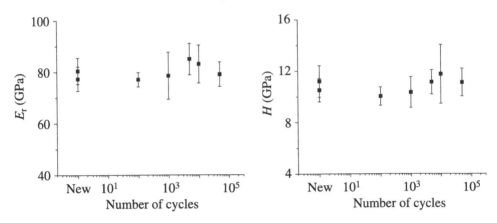

Figure 11. Dielectric surface reduced modulus E_r and hardness H with respect to the switch cycling at a contact depth of 19 nm.

average value of the hardness is about 11 GPa and that of the reduced Young's modulus about 80 GPa. These values are different from values reported in the literature (e.g., H: 14.5 $\pm$ 0.5 GPa, E_r: >145 GPa) [31], because the material properties of PECVD silicon nitride are known to be affected by the fabrication process (e.g., deposition temperature and film thickness [32]). In examining the hardness variation with cycling, we can observe a slight increase. For instance, hardness starts to increase after 100 cycles from 10.03 GPa, and continues to increase up to 11.75 GPa (17% higher) after 10^4 cycles. Hence, surface hardening (*via* modification of the topmost surface layer and not necessarily material/dislocation based hardening) could be one change the dielectric surface undergoes as the switch cycles. Note however that the variability in the measured data, shown in Fig. 11, is such that the observed surface hardening with switch cycling may not be statistically significant. Another observation is that the trend of hardness change is the same as the trend of silicon intensity change on the dielectric surface from the TOF-SIMS results (Fig. 9(b)). The amount of silicon on the dielectric surface may be related to the nanomechanical properties of the contact surface.

In summary and in accordance with the roughness and TOF-SIMS results, the changes in the nanomechanical properties of the dielectric surface with cycling are detectable but small, indicating that only minor geometrical, chemical and mechanical changes occur on the RF MEMS switch surfaces.

4. Conclusions

Capacitive RF MEMS switches with a Ti beam on silicon nitride dielectric layer were fabricated and tested under different numbers of cycles. Both beam and the dielectric surfaces were then analyzed topographically and chemically, and the dielectric surface was also analyzed nanomechanically to measure the evolutionary changes on the contact surfaces with cycling. Analysis of detailed roughness parameters (amplitude descriptor S_q, skewness S_{sk}, hybrid parameter S_{dr} and functional parameter S_{bi}) indicated that mild surface smoothening by plastic deformation of the asperities occurs during the initial cycling of the switch and a steady-state is reached after about 10^3 cycles. TOF-SIMS chemical analysis showed that both the Ti beam and Si_3N_4 dielectric surfaces exhibited similar chemical compositions, which was unexpected. Even after 5×10^4 cycles these chemical species are not significantly altered, which is also confirmed by the roughness analysis. Nanoindentation measurements on the dielectric switch surfaces showed consistent hardness (H) and reduced Young's modulus (E_r) values. Hardness values indicated a slight increase (but not statistically significant) with switch cycling. In general, the topographical, chemical and nanomechanical properties changes of the contacting surfaces of capacitive RF MEMS switches with cycling that we observed were small, indicating no major surface degradation. The contribution of this work is the detailed measurement of geometrical, chemical and mechanical properties of capacitive RF MEMS switches. The main conclusion is that, although small, mechanical and surface properties changes of both the beam and bottom surfaces occur with switch cycling and contribute to switch degradation, *via* smoothening of the surfaces and thus increased adhesion interactions. This switch degradation is in addition to electrical charging degradation that could also occur. Improved switch design and use of improved materials would most likely eliminate electrical charging effects, in which case the long-term switch performance and reliability will be dominated by the surface changes that have been presented in this work.

Acknowledgements

Financial support of the project was provided by the Defense Advanced Research Projects Agency (DARPA) under Grant HR0011-06-1-0046. AFM, SEM, TOF-SIMS and nanoindentation measurements were carried out in the Materials Research Laboratory, University of Illinois, which is supported by the U.S. Department of Energy under Grants DE-FG02-07ER46453 and DE-FG02-07ER46471. The authors also thank Dr. Timothy Spila for his help in TOF-SIMS measurements and interpretations.

References

1. G. M. Rebeiz, *RF MEMS Theory, Design, and Technology*. Wiley, Hoboken, New Jersey (2003).
2. S. T. Patton and J. S. Zabinski, *Proc. SPIE* **6111**, 61110E (2006).
3. A. Q. Liu, A. B. Yu, M. F. Karim and M. Tang, *J. Semiconductor Technol. Sci.* **7**, 166–176 (2007).
4. S. T. Patton and J. S. Zabinski, *Tribology Lett.* **19**, 265–272 (2005).
5. S. Yeo, S. I. Tseregounis, A. A. Polycarpou, A. Fruehling and D. Peroulis, in: *Proceedings of ASME/STLE International Joint Tribology Conference*, Memphis, TN, IJTC2009-15120 (2009).
6. R. W. Herfst, H. G. A. Huizing, P. G. Steeneken and J. Schmitz, in: *Proceedings of the IEEE International Conference on Microelectronic Test Structures*, Austin, TX, pp. 133–136 (2006).
7. S. H. Kim, D. B. Asay and M. T. Dugger, *Nanotoday* **2**, 22–29 (2007).
8. R. Daigler, E. Papandreou, M. Koutsoureli, G. Papaioannou and J. Papapolymerou, *Microelectronic Eng.* **86**, 404–407 (2009).

9. L. Kogut, *J. Micromech. Microeng.* **15**, 1068–1075 (2005).

10. A. B. Yu, A. Q. Liu, Q. X. Zhang and H. M. Hosseini, *J. Micromech. Microeng.* **16**, 2157–2166 (2006).

11. R. A. Coutu Jr, J. R. Reid, R. Cortez, R. E. Strawser and P. E. Kladitis, *IEEE Trans. Components Packaging Technol.* **29**, 341–349 (2006).

12. S. Majumder, N. E. McGruer, G. G. Adams, P. M. Zavracky, R. H. Morrison and J. Krim, *Sensors Actuators A* **93**, 19–26 (2001).

13. S. Majumder, N. E. McGruer and G. G. Adams, in: *Proceedings of the 18th IEEE International Conference on Micro Electro Mechanical Systems*, Miami, FL, pp. 215–218 (2005).

14. S. T. Patton and J. S. Zabinski, *Tribology Lett.* **18**, 215–230 (2005).

15. R. Chan, R. Lesnick, D. Becher and M. Feng, *J. MEMS* **12**, 713–719 (2003).

16. T. Lisec, C. Huth and B. Wagner, in: *Proceedings of the 34th European Microwave Conference*, Amsterdam, Netherlands, pp. 73–76 (2004).

17. C. F. Herrmann, F. W. DelRio, D. C. Miller, S. M. George, V. M. Bright, J. L. Ebel, R. E. Strawser, R. Cortez and K. D. Leedy, *Sensors Actuators A* **135**, 262–272 (2007).

18. C. L. Goldsmith, J. Ehmke, A. Malczewski, B. Pillans, S. Eshelman, Z. Yao, J. Brank and M. Eberly, *IEEE MTT-S International Microwave Symposium Digest*, Vol. 3, pp. 227–230 (2001).

19. X. Xue, A. A. Polycarpou and L. M. Phinney, *J. Adhesion Sci. Technol.* **22**, 429–455 (2008).

20. A. Y. Suh, A. A. Polycarpou and T. F. Conry, *Wear* **255**, 556–568 (2003).

21. A. Y. Suh and A. A. Polycarpou, *Tribology Lett.* **15**, 365–376 (2003).

22. J. A. Greenwood and B. P. Williamson, *Proc. R. Soc. London A* **295**, 300–319 (1966).

23. A. Y. Suh and A. A. Polycarpou, *J. Tribology* **125**, 193–199 (2003).

24. K. J. Stout, P. J. Sullivan, W. P. Dong, E. Mainsah, N. Luo, T. Mathia and H. Zahouani, *The Development of Methods for the Characterization of Roughness in Three Dimensions*. Commission of the European Communities, Brussels–Luxembourg (1993).

25. W. P. Dong, P. J. Sullivan and K. J. Stout, *Wear* **178**, 29–43 (1994).

26. W. P. Dong, P. J. Sullivan and K. J. Stout, *Wear* **178**, 45–60 (1994).

27. S. T. Patton, W. D. Cowan, K. C. Eapen and J. S. Zabinski, *Tribology Lett.* **9**, 199–209 (2004).

28. D. B. Asay and S. H. Kim, *J. Chem. Phys.* **124**, 174712 (2006).

29. http://www.phi.com/techniques/tof-sims.html, accessed in May 24 (2009).

30. W. C. Oliver and G. M. Pharr, *J. Mater. Res.* **7**, 1564–1583 (1992).

31. D. Klaffke and U. Beck, *Tribotest* **8**, 57–72 (2001).

32. Y. Ren and D. C. C. Lam, *Mater. Sci. Eng. A* **467**, 93–96 (2007).

Appendix — Formulae for Calculating the Surface Roughness Parameters

Root Mean Square Roughness (RMS or S_q):

$$S_q = \sqrt{\frac{1}{MN} \sum_{j=1}^{M} \sum_{i=1}^{N} z^2(x_i, y_j)}, \qquad (A.1)$$

M: number of sampling points in the y-direction, N: number of sampling points in the x-direction, $z(x, y)$: 2-D surface roughness as a function of x and y orthogonal coordinates.

Skewness (S_{sk}):

$$S_{sk} = \frac{1}{MNS_q^3} \sum_{j=1}^{M} \sum_{i=1}^{N} z^3(x_i, y_j). \tag{A.2}$$

Developed Interfacial Area Ratio (S_{dr}):

$$S_{dr}\ (\%) = \frac{\overbrace{\text{Total Surface Area} - \text{Sampling Area}}^{\text{Areal Difference}}}{\text{Sampling Area}} \times 100 \tag{A.3}$$

or

$$S_{dr}\ (\%) = \frac{\overbrace{\text{Total Surface Area} - (N-1) \cdot (M-1) \cdot \Delta x \cdot \Delta y}^{\text{Areal Difference}}}{(N-1) \cdot (M-1) \cdot \Delta x \cdot \Delta y} \times 100, \tag{A.4}$$

Δx: sampling interval in the x-direction, Δy: sampling interval in the y-direction.
See [24, 26] for calculating the 'Total Surface Area'.

Surface Bearing Index (S_{bi}):

$$S_{bi} = \frac{S_q}{z_{0.05}} = \frac{1}{h_{0.05}}, \tag{A.5}$$

$z_{0.05}$: surface height at 5% bearing area, $h_{0.05}$: surface height at 5% bearing area normalized with respect to S_q.

Molecular Mobility and Interfacial Dynamics in Organic Nano-electromechanical Systems (NEMS)

Scott E. Sills [a,*] **and René M. Overney** [b]

[a] Advanced Technology Group, Micron Technology, Inc., 8000 S. Federal Way, Boise, ID 83703-0006, USA

[b] Department of Chemical Engineering, University of Washington, Seattle, WA 98195, USA

Abstract

The underpinnings of material properties in nanoscopic polymer systems are reviewed in light of the relaxation modes available for molecular motion. When motion is altered due to the presence of constraints, a rich variety of material and transport behaviors become apparent. On the one hand, external constraints imposed by interfaces and system boundaries generate structural and dynamical anisotropies that can propagate over device-relevant length scales. On the other hand, the ability to cater relaxation behavior through molecularly-engineered internal constraints offers a path to optimize material properties in nanoscopic systems. These two aspects are highlighted throughout the review, and their technological implications are discussed for MEMS/NEMS operations, including frictional and mechanical loading of polymer thin films. In this regard, material performance attributes and feedback for molecular designs are drawn from perturbation techniques that provide access to the energetic signatures and characteristic scales of molecular relaxation. In particular, the importance of the operating time scale in nanoscopic devices is emphasized with examples of both quasi-static and dynamic operations. Material responses have been found to change significantly when the drive velocity or loading rate was either comparable to or in excess of the characteristic molecular frequencies. Correspondingly, intrinsic molecular modes were found to either couple with the external mechanical disturbance, thus establishing channels for energy transport, or to remain passive, leading to apparent material *stiffening*. Findings such as these suggest that comprehensive investigations of the spatial distribution of molecular relaxation spectra in confined systems are necessary in order to match (or mismatch) molecular response times with system operating times. Along this line, this review provides examples for cognitive engineering of functional materials, with molecular structures that are tailored to system constraints and employed in nanoscale device technologies.

Keywords

Nanoscopic systems, functional materials, polymer thin films, molecular mobility, molecular relaxation, confinement, glass transition, MEMS, NEMS, Millipede, electro-optics, photonics, AFM, SM-FM, intrinsic friction

---

[*] To whom correspondence should be addressed. Advanced Technology Group, Micron Technology, Inc., Mail Stop 1-715, 8000 S. Federal Way, Box #6, Boise, ID 83707-0006, USA. Tel.: (208)363-4629; Fax: (208) 368-3002; e-mail: ssills@micron.com

Adhesion Aspects in MEMS/NEMS

1. Introduction

Organic macromolecules, such as oligomers and polymers, offer a synthetic route to tailor interfaces and condensed phases for application-specific material and transport properties. For instance, molecularly engineered micro- and nano-electromechanical systems (MEMS/NEMS) are gaining widespread use as actuators and biological and chemical sensors [1–6]. The *finite-size constraints* that arise in MEMS/NEMS may lead to appreciable surface effects, which must be well managed in the design and fabrication of viable devices. Due to the large surface area to volume ratio of nanoscopic systems, problems involving interfacial phenomena, such as adhesion and friction, are central to NEMS/MEMS development.

This review addresses the delicate balance between material performance and the molecular mobility in nanoscopic polymer systems. Molecular mobility will be addressed as a spectrum of relaxation modes that stems from the underlying degrees of freedom available for conformational transitions. The energetics associated with relaxation mode dynamics can be either solely enthalpic, as in small side-chain rotations, or contain a substantial entropic component, as in the cooperative *crankshaft-type* motions about the molecular backbone [7, 8]. The location of the modes, e.g., surface *vs* bulk modes [7], and the effect of interfacial interactions, such as interfacial tension [9–14], are important contributors to the resulting *apparent* relaxation barriers. Thus in this review, particular emphasis is given to the role of constraints, which may restrict or enable specific relaxation processes. Constraints are classified as either *internal* or *external*; where internal constraints are associated with molecular structure and interactions within the material system, and external constraints are attributed to the interactions that arise in the vicinity of system boundaries, i.e., at interfaces between the material system and its external environment.

The judicious incorporation of internal molecular constraints within a variety of polymer thin film systems is reviewed here as a material design opportunity. For example, molecular constraints are imposed by intermolecular crosslinks to tune the friction and wear characteristics for NEMS applications [15]. Further, attention is given to the influence of external interfacial constraints, which may lead to the creation of anisotropic phase boundaries in spin-cast polymer films [16, 17]. Material anisotropy within phase boundaries is discussed in the context of the glass transition profiles that arise within several tens of nanometers of the polymer–substrate interface [18]. This initially surprising far-field effect is associated with mobility constraints at the polymer–substrate interface, combined with shear-structuring during the spin casting process. The first practical importance of the anisotropic phase boundaries is illustrated with applications in ultra-fast, nano-contact recording [19].

In addition to discussing relaxational constraints at interfaces, the role of molecular relaxation in dissipation processes is also reviewed in the light of sliding and contact operations for thin film NEMS. Nanoscopic contacts can produce stresses of hundreds of mega-Pascals, under loading forces of only a few nano-Newtons.

A contacted polymer surface accommodates these stresses through both viscoelastic modes, and modes that can be attributed to wear or fracture. When a contact is sheared, for example during the scanning or sliding between two surfaces, frictional dissipation within the contact zone is *tribo-rheologically* complex, involving (i) material-specific relaxation modes, and (ii) dynamic adhesive coupling parameters. Simply expressed, any sliding situation is comprised of adhesive and cohesive contributions to energy dissipation. We will focus on recent progress in exploring thermokinetic friction processes, which have provided characteristic signatures of the dissipation mechanism(s), such as activation energies, relaxation times, or critical length scales. Thereby, particular emphasis will be given to critical spatial and temporal interactions, and to material-specific molecular relaxation modes. Cooperative mode behavior will be discussed in the light of glass forming polymer elastomers. The experimental measurement of the so-called correlation length, a length scale for cooperative relaxation, leads to a discussion of cooperatively rearranging regions (CRRs), which characterize the extent to which neighboring chain segments and adjacent molecules participate collectively in the dissipation process. The size of CRRs turns out to be technologically relevant to device performance, and is illustrated with the applications of terabit thermomechanical data storage and optoelectronic devices.

To correlate the electromechanical performance with the molecular design, one must consider the relative time scales between internal relaxation modes and device operations. In the case of a mechanical disturbance, such as frictional sliding or indentation, relaxation processes are known to couple with the mechanical *operation* if the timescale of the external disturbance is on the order of the molecular relaxation time. If the process is *fast*, however, and exceeds the molecular response time(s), the relaxation mode(s) may appear frozen and are unavailable unless stress-activated. This is illustrated with high-rate dynamic loading of ultra-thin polymer films in NEMS for thermomechanical storage. In this discussion, we will provide a molecular picture for plastic yielding, as inferred from dynamic compression loading experiments. To further elucidate this aspect, the role of interfacial constraints and material anisotropy in nanomechanical loading of thin films is reviewed.

To tackle the multifaceted aspects of the molecular mobility and dynamics in confined polymeric systems, we will first (Section 2) review the fundamental underpinnings of molecular mobility, and provide a classification of the types of constraints that affect it. We then explore the manifestation of constraints and *effective* material properties in confined systems, with focus on the formation of structurally modified boundary layers in spin cast polymer thin films. Next (Section 3), we will discuss the molecular competition between interfacial adhesion and internal cohesion in nanoscale sliding operations. The origin of friction is discussed in terms of molecular relaxation and the length-scale over which energy is dissipated. We continue our discussion (Section 4) with an assessment of critical time scales and device performances, with respect to the availability of particular molecular relaxation modes. Emphasis is given to high-rate nano-impact operations in confined polymer films, which are detailed, by example, with thermomechanical data storage NEMS. Finally (Section 5), the review closes with a summary and outlook.

2. Molecular Mobility and the Influence of Constraints

Polymeric systems exhibit properties that are rooted in the enthalpic and entropic nature of the underlying molecular processes. These macromolecules reside within a complex potential energy landscape, comprised of intramolecular interactions between backbone segments and sidechains, and intermolecular interactions with neighboring molecules. The surrounding molecular environment can be *self-similar*, as it is for homogeneous bulk phases, or *dissimilar* when the system is composed of multiple phases (e.g., blends), or contains interfaces (e.g., the substrate surface in ultra-thin films). The resulting *apparent* system properties depend on how the molecules *move* within their energy landscapes. They move through various relaxation modes, which are determined by the underlying degrees of freedom associated with their collective interactions. Constraints that restrict these relaxations, either internal (chemical composition, or conformational structure) or external (dimensional, interfacial), augment the energy landscape and hence the resulting material properties [20].

Internal constraints are inherent to molecular architecture, and generally result from direct bonding or van der Waals interactions. They can be incorporated *a priori* into molecular designs, as a prescription for desired material properties. For example, by reducing the crosslink spacing in polymers, the naturally occurring molecular relaxation may be confined [21]. This approach has been employed towards increasing the yield strength of polymer thin films for enhanced wear characteristics [15], which is discussed here for a terabit storage NEMS application [22].

In the NEMS storage device, the polymer storage medium consists of a thin film of crosslinked polystyrene, which in Fig. 1 exhibits a critical crosslinking spacing δ_c below which the wear mechanism is strongly dominated by the crosslink density [15]. This constraint-induced wear transition occurs when the spacing between crosslinks δ_c matches the length scale required for segmental backbone relaxation ξ_α, or the cooperation length [23]. Cooperative relaxation is generally visualized in terms of the correlated motion of multiple polymer segments, which must relax collectively, in series or unison, in order for a single backbone relaxation event to occur. The spatial domain in which this process occurs has been termed 'cooperative rearranging region' (CRR), which has a volume proportional to ξ_α^3. When the crosslink-spacing falls below the cooperation length, i.e., $\delta_c < \xi_\alpha$, constraints are imposed on the backbone mobility, resulting in a *stiffer* material. Under these conditions, increased hardness and reduced wear are strongly correlated with δ_c. If the concentration of crosslinks is insufficient to interfere with the natural backbone

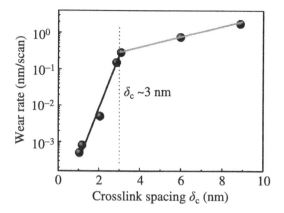

Figure 1. The influence of prescribed internal molecular constraints on resulting material behavior. A crosslinked polystyrene film exhibits a strong reduction in wear when the crosslinking spacing is reduced below a critical space, δ_c, that matches the length scale required for cooperative segmental relaxation, ξ_α. Reprinted with permission from [15]. Copyright (2006) American Chemical Society.

relaxation, i.e., $\delta_c \gg \xi_\alpha$, the crosslinks have little impact on the polymer response and the material behaves similar to the un-crosslinked native material.

In addition to internal constraints, nanoscopic systems are subject to external constraints. External constraints stem from the interactions a molecular system has with its boundaries, which propagate over a finite distance. In polymers, interfacial constraints can have a preponderant effect over a length scale of tens to about a hundred nanometers, far beyond the interfacial pinning regime [16, 17]. External constraints impose field gradients and lead to non-uniformities like material anisotropies, entropic cooling, and even local structuring [8, 24, 25].

Deviations from bulk thermodynamic properties are not uncommon when polymer films are sufficiently thin that modifications of the conformational chain entropy at the substrate and free surface dominate bulk contributions. Considerable experimental evidence suggests that segmental mobility can be enhanced in ultrathin polymer films. For example, the dependence of the glass transition temperature, T_g, on film thickness has been extensively studied [18, 26–33] and reviewed [34]. In most cases, there is an apparent depression in the polymer's T_g with decreasing film thickness [26–29, 34, 35]. Such behavior is generally attributed to an increased free volume due to an easement in chain mobility constraints at the polymer–air interface, coupled with relatively weak polymer–substrate interactions.

When the supporting substrate is removed and free-standing films with two polymer–air interfaces are considered, the T_g reduction in a thin film may be enhanced relative to substrate-supported films [36]. This is apparent in the work of Torkelson's group [36, 37], where the substrate-supported PS films in Fig. 2 reveal an overall T_g reduction of $\sim$30°C with respect to the bulk, and the free-standing PS films in their later work [36] indicate a T_g reduction of up to $\sim$60°C compared to the bulk. While the T_g reduction is roughly twice in free-standing films, it is inter-

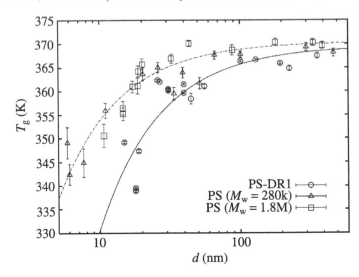

Figure 2. The effect of relaxed constraints at the polymer–air interface on the T_g reduction in polymer thin films. Film thickness, d, dependence of T_g determined for substrate-supported PS films with only one free polymer–air interface (dye-labeled PS ($M_w \sim 10k$) — circles; PS ($M_w = 280k$) — triangles, PS ($M_w = 1.8M$) — squares). Reprinted with permission from [37]. Copyright (2007) by the American Physical Society.

esting that the interfacial distance or length-scale for confinement remains similar to substrate-supported films. In both cases, interfacial confinement effects become apparent when the film thickness is reduced below ~100 nm. Since the reduction in T_g can be considered a measure of increased mobility, the larger T_g reduction in free-standing films seems to indicate that the free surface dominates any substrate interface effects. Thus, if enhanced mobility at the free-surface is the primary contributor to the T_g suppression, then substrate effects must be sufficiently small in both (i) length-scale, i.e., less than 100 nm, and (ii) magnitude. In other words, constraints at the substrate interface appear insufficient to compensate for the T_g reduction from the free surface.

Based on this interpretation, analytical techniques that sample the entire film thickness are expected to observe a continuous decrease in the average film T_g as the film thickness is reduced. This is because the contribution of the mobile surface layer to the overall average dynamics of the film becomes weighted heavier as the amount of bulk material in the sample is reduced. However, it is important to note that observations of a lower T_g in thinner films do not confirm *a priori* a spatially-dependent T_g profile that is lower near the polymer–air interface, and reaches a bulk value at some depth. Alternatively, the finding that a second free-surface does not affect the overall length scale for interfacial confinement suggests that the presence of the free-surface is not the sole contributor to the bulk-deviating T_g behavior. Other factors, such as the film preparation method, may contribute significantly to the interfacial T_g profiles as well.

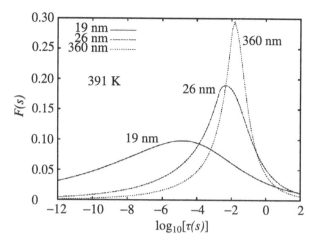

Figure 3. Dispersion of the α-relaxation times in interfacially confined polymer films at 391 K. The distribution function $F(s)$ of the α-relaxation times, τ, determined by dielectric spectroscopy of dye labeled PS thin films is found to broaden with decreasing film thickness. Reprinted with permission from [37]. Copyright (2007) by the American Physical Society.

One approach to gain further insight into the molecular origins of the interfacial T_g profiles is to study how the distribution of α-relaxation times is influenced by the film thickness. With dielectric spectroscopy measurements on dye-labeled polystyrene thin films sandwiched between aluminum electrodes, Priestley *et al.* [37] observed that the relaxation time of the α-process decreased with decreasing film thickness, and that the α-relaxation time distribution broadened. This is illustrated in Fig. 3, where the distribution of α-relaxation times increases from two decades for a 360 nm thick film to eight decades for a film that is only 19 nm thick. The reduction of the α-relaxation time is consistent with a reduction in T_g, and its decrease with film thickness was interpreted based on the same free-surface model discussed above. It was argued that the evaporation of an aluminum electrode onto the free polymer surface (which is necessary for the dielectric measurements) does not mask or alter the free surface effects that are associated with the polymer–air interface [37]. Considering the extremely complex interfacial phenomena associated with the evaporation of metal atoms onto polymer films [38–42], this assumption remains to be verified. Nevertheless, in context of a free-surface model, where the effects of the mobile surface layer do not propagate over appreciable length scales and only provide a local contribution to a reduction in the average film T_g, such a binary layer system would be expected to lead to a bifurcation of the α-relaxation spectrum into a bimodal distribution of relaxation times, with each mode representing contributions from each the surface layer and the unperturbed sub-surface layer. Since this is not the case, the broadening of the α-relaxation spectrum suggests a spatial distribution of the molecular relaxation phenomena within the thin film, which is ultimately manifested in a continuous gradient of material properties throughout the thickness of the film.

Ellison and Torkelson [43] prepared multilayered films of PS and showed a surface-normal spatial distribution of the local T_g within the film, based on a fluorescence study. Their findings suggest that the principal length scale defining interfacial confinement is the distance over which a local perturbation to the molecular relaxation behavior, e.g., at the free surface and or substrate interface, affects the cooperative dynamics within the polymer. This apparent dispersion length for cooperative relaxation was estimated to be a few tens of nanometers [43]. One assumption implicit in the multilayer film analysis is that the stacked thin films do not maintain any memory of their original casting configuration. In other words, post-thermal annealing is sufficient to alleviate any residual stresses and shear structuring that may be present within each layer of the film, in proximity of the original casting substrates. In this scenario, the molecules within each layer of the film stack relax during annealing and reconfigure within the continuum of the interfacial fields that originate at the free surface and underlying substrate.

One approach that provides insight into the role of the film preparation process on the T_g distribution in thin films is based on local thermomechanical sampling involving scanning force microscopy (SFM). This method is known as shear modulation force microscopy SM-FM [20, 30, 31], which is a local, nanoscopic analog to dynamic mechanical analysis (DMA). SFM-based techniques are highly sensitive to molecular relaxations within condensed organic matter [7, 18, 31, 44], and thus, SM-FM is well suited for local relaxation studies with a penetration depth sensitivity on the order of 1 nm. It avoids the complications associated with techniques that average over the entire film thickness. In various studies [20, 30, 31], it has been shown that SM-FM is not sensitive to relaxations attributed to de Gennes' mobile surface layer model [45, 46], which occur within an angstrom distance from the free surface [47]. This is illustrated in Fig. 4(a), where the T_g values measured with SM-FM for thick (>100 nm) polystyrene films are in excellent agreement with conventional macroscopic techniques, such as differential scanning calorimetry (DSC) and electron spin resonance (ESR).

For thin, substrate-confined films (<100 nm), however, SM-FM results revealed non-monotonic T_g profiles, as shown in Fig. 4(b) [18]. The profiles were discussed in light of a two-phase rheological boundary layer model, Fig. 4(b). Based on this model, the interfacial interactions lead to the formation of a less dense ultrathin *sublayer* (~10 nm) adjacent to the interface, towards which molecules from the outer layer diffuse. The thickness of the sublayer is characterized, in part, by the molecular dimensions and the interaction potential at the interface. The coupled effects of shear-induced structuring during spin casting and anisotropic diffusion during annealing create an *intermediate regime* between the sublayer and bulk phase. The overall boundary may extend two orders of magnitude beyond the polymer's persistence length; and molecular restructuring within the boundary may be thermally stable well above T_g [18]. The non-monotonic T_g profiles obtained by SM-FM have been recently confirmed by spectroscopic ellipsometry studies in a polymer

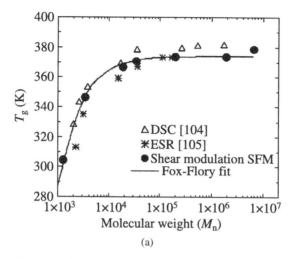

(a)

Figure 4. (a) Direct T_g comparison by SM-FM (shear modulation SFM), differential scanning calorimetry (DSC) [104], and electron spin resonance (ESR) [105] on polystyrene. Reprinted with permission from [53]. Copyright (2000), American Chemical Society. (b) (top) Impact of interfacial constraints on relaxation properties of polystyrene thin films ($M_w = 12k$). The glass transition temperature, T_g, exhibits both an increase and decrease relative to the unconfined bulk, depending on the film thickness, δ, regime (intermediate *vs* sublayer, SL). (bottom) Anisotropic structural model associated with the observed $T_g(\delta)$ profile. Reprinted with permission from [23]. Copyright (2005), American Institute of Physics.

photoresist for extreme ultraviolet photolithography (EUV) [48], as well as in conjugated polymers used in optoelectronic devices [49].

Although still lacking unambiguous verification, the most plausible explanation for the non-monotonic T_g profile within the interfacial boundary region is the compounded effect of spin casting and diffusion within the structurally heterogeneous polymer matrix. It has been argued that models which account for the relative roles of both the polymer–substrate interaction and the air–polymer interface [50, 51] provide only moderate agreement with experimental data [49]. In fact, models that rely solely on the influence of the air–polymer interface [52] contradict SM-FM data even for thick ($\gg$100 nm) films, which, as pointed out above, are in excellent agreement with results from complementary techniques. If the free surface mobility is altered to such a degree that it impacts T_g over tens of nanometers beneath the interface, SM-FM data should divert significantly from DSC and other bulk averaging methods such electron spin resonance (ESR) [53], which is contrary to the data presented in Fig. 4(a).

A possible explanation for the qualitatively similar decreasing T_g with film thickness between both free-standing films and their substrate-supported counterparts could be found in the film preparation history. In the free-standing film study [36], the films were first spun cast on cleaved mica substrates, and then transferred (floated) to perforated free-suspension supports. This transfer process cannot be expected to anneal the films to such a degree that they lose any structural hetero-

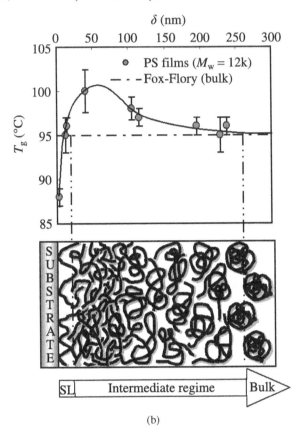

(b)

Figure 4. (Continued.)

geneity imparted by the spin casting process. In fact, it is possible that the transfer process itself altered the thin films, e.g., by swelling in the interfacial region, producing a film with higher interfacial mobility, which, in turn, gives rise to a more prominent film-averaged T_g depression compared to films left on their original casting substrates. While a dominant drop in T_g in the ultra-thin film region has been reported by all the studies reviewed here, the less significant T_g increase in the intermediate region was only resolved by three groups [18, 48, 49], possibly due to signal resolution limitations.

The non-monotonic T_g profile, in particular the location of the local maximum, has already found practical significance as a design parameter in ultra-thin film applications, such as ultrahigh density thermomechanical recording [19]. By incorporating internal constraints within the polymer thin films, the T_g profile associated with external system constraints can be modified. The profiles can be shifted toward or away from the substrate interface, or amplified or suppressed, by changing the molecular weight [54], the side-chain or backbone repeat unit structure [55], or the crosslink density [18]. Thus, with judicious placements or omissions of internal constraints, it is possible to regulate system responses to external constraints. This is the essence of molecular engineering, where material designs for nanoscopic applications must consider the coupled nature of external and internal constraints.

3. Friction and Molecular Relaxation at Polymer Interfaces

Material responses to external perturbations, such as shear forces, are ultimately manifested through the relaxation processes that are associated with molecular mobility. As discussed above, molecular mobility may be influenced by both internal and external constraints. The shear resistance to sliding of a polymeric elastomer has been shown to be highly susceptible to the rheological properties of the material [23, 56–58], and, thus, to stem from a rich spectrum of molecular relaxation modes. The available surface (or near-surface) modes couple with the external sliding process, and thereby withdraw energy from the macroscopic motion, which is experienced as frictional resistance to sliding. The degree of coupling depends on the interfacial qualities between the contacting materials, such as the degree of conformity, the combined modulus, the interfacial energy, and the cyclic work of adhesion. While these interfacial qualities have been studied extensively for many systems, the intrinsic molecular modes through which energy is transferred and dissipated have rarely been addressed. Frictional dissipation has been typically described in terms of heat loss, without specification of the spatiotemporal evolution of the molecular processes that occur within the material during the sliding process.

Recent studies involving polymer elastomers [23, 59, 60] have demonstrated that frictional dissipation within organic materials is intimately related to specific molecular relaxation behavior. A molecular understanding of the dissipative process is faced with the triborheological complexity [58] in order to distinguish between shearing at the elastomer–rigid body interface (adhesive yield), and shearing at a newly formed interface within the bulk elastomer (cohesive yield). Sliding friction considers sequential bonding and debonding events between the polymer surface and the countersurface. When a shear force is imposed, the contact stores elastic energy until it overcomes the adhesion energy, causing the propagation of a shear crack which releases the contact [61]. Cohesive yield occurs when the adhesion bond between the contacting surfaces is strong relative to the internal strength of the polymer. In this case, the polymer yields internally through viscoelastic mechanisms associated with internal friction and chain pull-out, or through viscoplastic modes like chain scission or crazing.

Both adhesion debonding and cohesive yielding are activated processes with a velocity- and temperature-dependent behavior that is well described by Eyring's theory of reaction rates [62]. If the adhesion force, F_{adh}, exceeds the cohesion force, F_{coh}, the contact undergoes cohesive yield, and *vice versa*. However, the adhesion–cohesion relation is not straightforward. Cohesion will depend on the internal pressure within the elastomer, which under equilibrium conditions must equal

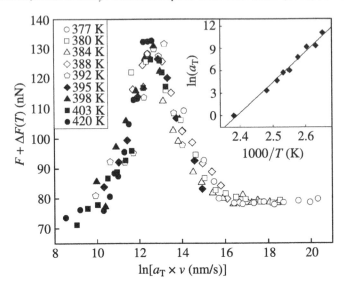

Figure 5. Molecular relaxation dependent friction behavior. Friction force-velocity, $F(v)$ master curve for polystyrene ($M_w = 96.5$ kg/mol, $M_e \sim 20$ kg/mol) above $T_g = 373$ K. Inset: Slope of the Arrhenius shift factor $a_T(T)$ reveals an apparent activation energy of 81 kcal/mol, which identifies the α-relaxation responsible for frictional dissipation. Reprinted with permission from [23]. Copyright (2005), American Institute of Physics.

the contact pressure. Furthermore, the adhesion–cohesion relation depends on the ratios of the sliding velocity to the characteristic molecular timescales for adhesion and cohesion interactions. Thus, the external drive velocity becomes a critical factor, and distinguishing adhesion *versus* cohesion contributions to friction requires some characteristic signature of the dissipation mechanism, e.g., the activation energy, relaxation time, or length scale of the process.

Sills *et al.* demonstrated that the characteristic signatures of the dissipation process can be determined by the superposition of friction-velocity isotherms and from the critical velocity corresponding to the maximum in the friction force [23]. This is illustrated in Fig. 5 with SFM measurements on an entangled polystyrene (PS) melt. An activation barrier of 81 kcal/mol (3.5 eV) is deduced from the *apparent* Arrhenius behavior of the thermal a_T shift factor in the inset Fig. 5. The value coincides with the 80–90 kcal/mol activation energy for the α-relaxation process [63], i.e., segmental relaxation of the PS backbone. Similarly, an activation barrier of 7 kcal/mol was determined for glassy polystyrene films, which corresponds to the hindered rotation of the phenyl ring side chains about their bond with the *frozen* backbone [59]. In these cases, the activation barrier overcome during the course of frictional sliding corresponds directly with molecular relaxation in the bulk elastomer.

The friction peak in Fig. 5 is analogous to a spectroscopic peak in frequency space, which is characterized by the competition between material and experimental time-scales, the so-called Deborah number, *De*. At slower sliding velocities, the

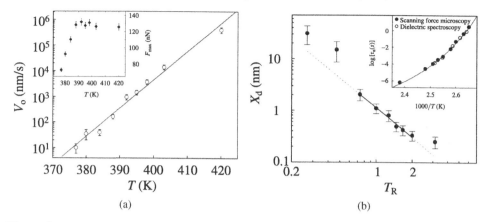

Figure 6. Friction spectroscopic analysis provides input about (a) the temperature range over which cooperative phenomena occur (i.e., the glass forming process), and (b) the a direct quantitative measure of the cooperative length scale when combined with dielectric spectroscopy [23]. (a) illustrates with the velocity at the friction peak V_o and the friction peak intensities F_{max} (inset) from Fig. 5. (b) The length scale of cooperation X_d for the α-relaxation of polystyrene is obtained from friction and dielectric data (inset) in terms of the reduced temperature, $T_R = (T - T_g)/(T_c - T_g)$ where T_g is the calorimetric glass transition temperature (373 K) and T_c is the crossover temperature of the dynamic glass transition (388 K). The power law fit $X_d \sim T_R^{-\phi}$ over the range $0.7 < T_R < 2.0$ (solid line) reveals an exponent of $\phi = 1.89 \pm 0.08$, which is in good correspondence with mode coupling theories [106]. Reprinted with permission from [23]. Copyright (2005), American Institute of Physics.

contact stress stored in the soft material is capable of relaxing (through internal friction modes) before an asperity can slip to the next contact site. In this region, the friction force increases logarithmically with velocity, consistent with activated molecular relaxation in the soft material. Above the critical velocity, the probing tip is driven to the next contact site before the material can respond internally through viscoelastic relaxation. Hence, with increasing velocity, the tip experiences fewer and fewer dissipative relaxation events per jump and the friction force decreases.

The extent to which neighboring chain segments and adjacent molecules participate collectively in the dissipation, or relaxation, process is deduced from the velocity at the friction peak in Fig. 6(a). Grosch [56] and Ludema and Tabor [57] were the first to combine the velocity at the friction peak with the frequency for the maximum viscoelastic loss, to access the characteristic dissipation length. However, with their macroscopic techniques, they could only suspect dissipation through segmental relaxation. With the single asperity SFM approach [23], the length scale $X_d(T)$ of the α-relaxation is directly determined by combining the velocity at the friction peak $V_o(T)$ with α-relaxation times $\tau_\alpha(T)$ from dielectric spectroscopy, via: $X_d(T) = V_o(T) \cdot \tau_\alpha(T)$. For an entangled polystyrene melt, the dissipation length in Fig. 6(b) grows from the segmental scale to $\sim$2 nm, following a power law behavior that is consistent with predictions for cooperative motion during α-relaxation events [23].

Since Adam [64], structural relaxation near the glass transition has been visualized in terms of a correlated motion of polymer segments or domains, giving rise to dynamic heterogeneities [65–69]. While the existence of structural and dynamic heterogeneity around T_g could be directly inferred from isothermal multidimensional nuclear magnetic resonance (NMR), dielectric spectroscopy, photobleaching, dynamic light scattering, and quasi-elastic neutron scattering studies [68], estimating the size of the cooperatively rearranging regions (CRRs) (typically 1–3 nm) required model assumptions [66, 68, 70, 71]. In contrast, the above thermokinetic SFM approach of Sills *et al.* [23] provides a model-independent experimental evaluation of the CRR size by combining frictional dissipation and relaxation time data from independent experiments. The CRR sizes measured with thermokinetic SFM have been verified by a *self-concentration* extension to the original Adam–Gibbs theory [72].

The nanometer size of the CRRs deserves particular attention in light of emerging nanotechnological MEMS/NEMS applications. The size of the CRRs in PS near the glass transition grows from single molecular segments to domains up to ~10 nm. Compared to structures in modern device technologies, e.g., ultra-thin films and nanocomposites involving sub-100 nm dimensions, one can expect a competition between material and device length scales. As discussed above, material properties in dimensionally confined systems are likely to be modified from their original bulk values. The impact of dimensional constraints on the CRR thermodynamic growth rate, ϕ, will dictate the thermal range over which the glass transition occurs. Constraints leading to enhanced coordination (high ϕ — more restricted mobility) would increase the local T_g; while finite size effects that prevent coordination (low ϕ — less restricted mobility) would reduce the local T_g. Given the technological importance of T_g, the continued evolution of thin film applications stands to benefit from accurate characterization of $\xi_\alpha(T)$ and ϕ in confined geometries. Not only frictional dissipation but all transport processes depend on the intra- and inter-molecular degrees of freedom for molecular motion. For example, a recent report on thermomechanical data storage in polymer films recognized a minimum strain requirement that prevents data density scaling below a 10 nm half-pitch. The limitation arises from the fact that, in the scaling limit, the smallest indented datum bit requires a deformed volume that cannot be smaller than the size of a CRR [73].

Another example where cooperative effects play an important role is found in organic nonlinear optical (NLO) materials that are actively pursued for applications in photonic devices. Practical device applications require NLO materials to maintain both high macroscopic electro-optical (EO) activity and thermal stability at operating temperatures. High macroscopic EO activity can be achieved by acentrically ordering a system containing a high density of strong dipole chromophores *via* electric field poling at elevated temperatures [74]. Thermal stability requires the system to have internal constraints that prevent collapse of the acentric order at operating temperatures. In a recent study involving self-assembling organic molecular NLO glasses, molecular cooperations were found essential to the ordering process [8].

One can expect similar implications in other thin film applications, including, for example, charge carrier transport in organic electronic devices.

4. Transient Processes: Dynamic Loading of Interfacially Constrained Polymer Films

So far, *slow* processes have been discussed in terms of the molecular relaxations; inertial effects have been negligible and rate effects have been limited to the availability of thermally active relaxation mechanisms. If the process is *fast* however, as in pulsed dynamic loading, relaxation modes which are thermally active (compliant) under quasi-static conditions may become restricted and the material loses some "molecular" channels for energy dissipation. In other words, at high strain rates, a molecular system behaves as if it is thermally cooled. This is due to a transition in the Deborah number, *De*, from $De < 1$ to $De > 1$, i.e., the external drive time becomes shorter than the characteristic relaxation time. Under these conditions, phenomenological properties, such as yield stress and failure mode, may shift to reflect an apparent stiffening of the material.

Dynamic loading in nanoscopic systems becomes important in MEMS/NEMS where high-speed operations are critical to meeting processing timescales. A primary example is ultrahigh-density thermomechanical data storage (TDS) in Millipede NEMS [22, 75]. TDS is a nanoscopic analog to the punch cards used in early digital computers; it relies on writing, reading, and erasing nanometer sized data bits in thin polymer films. In essence, TDS recording is a high speed (kHz/probe), elastic–viscoplastic nanoindentation process, Fig. 7. Each indented bit represents a metastable state of the deformed volume, and will either initiate spontaneous dewetting (film instability) or strive for recovery of the initial unstressed state (bit

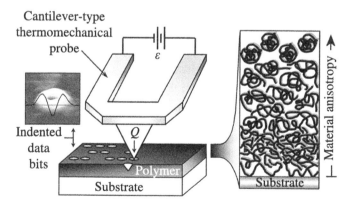

Figure 7. Illustration of a fast nanoindentation process conducted on an interfacially constrained polymer film. The thermomechanical SFM probe is resistively heated above the glass transtion temperature of the polymer, and electrostatically actuated to indent the polymer film. This pulsed indentation process is the basis for the ultra-fast terabit data storage scheme in Millipede MEMS. Reprinted from [20] with kind permission from Springer Science + Business Media, copyright (2006).

instability) [76]. The delicate balance between these instability nodes constitutes one optimization scenario in the design of polymeric storage media. The polymeric storage media must be designed to achieve the narrow range of physicochemical properties necessary for high data density, fast data rates, high durability, long shelf life, and low power consumption. The ideal polymer should be easily deformable for bit writing; however, the written bits must be stable against thermal degradation and wear.

Initial models for ultra-fast nanoindentation of polymers considered only adiabatic heating as a *melting* mechanism which enabled formation of an indented datum bit [77]. The importance of strain rate and inertial effects during thermomechanical recording in Millipede MEMS was reported by Sills *et al.* [19] in thin films of unentangled polystyrene (PS), and later confirmed by Gotsmann *et al.* in crosslinked SU8 epoxy films [78]. Typical strain rates of 10^3–10^6 s^{-1} exceed those of quasi-static indentation, and fall within the regime of impact dynamics [79]. The main difficulty with impact studies is that the inertial and strain-rate effects are usually coupled [80]. Inertial effects are propagated through the material as stress waves; while strain-rate effects are attributed to a transition from thermally activated mechanisms to linear viscous mechanisms [81]. The characteristics of wave propagation inevitably depend on the strain-rate dependence of the material properties.

Under impact loading, the stress–strain behavior of most polymers, including crosslinked epoxy resins [82, 83], exhibits an initial elastic response followed by yielding, strain softening, and then viscoplastic flow at a nearly constant *flow* stress [84–87]. At large strains, many polymers exhibit significant strain hardening following softening. An example of a typical stress–strain diagram for high rate compression loading of a polymer is illustrated in Fig. 8(a). For a given polymer, the extent of strain softening, and the range over which viscoplastic flow occurs depends on the nature of the entropic chain stretching and alignment, and on the amount of adiabatic heat generated from internal molecular friction. In macroscopic systems, the effects of adiabatic heating cannot be neglected, due to the low thermal diffusivity of polymers [86, 87]. However, for nanoscopic systems, it has been argued that the role of adiabatic heating is minimal, due to the small deformation volumes involved [78].

The occurrences of strain softening and viscoplastic flow have important implications on coupled inertial effects. The propagation speed of a plastic stress wave is $c_p(\sigma) = (1/\rho_0 \, \partial\sigma/\partial\varepsilon)^{1/2}$ [79], which approaches zero above the flow stress. ρ_0 is the density of the unloaded material, and $\partial\sigma/\partial\varepsilon$ is the slope of the stress–strain curve at a given strain and strain rate. Thus, plastic stress waves generated during nanoimpact rapidly attenuate [79] (exponentially [88]) as they propagate from the impact site, while the energy carried by the pressure pulse is dissipated through viscoplastic deformation processes [88]. Indeed, intrinsic strain softening has been recognized as the primary cause for plastic localization phenomena in deformation of glassy polymers [89].

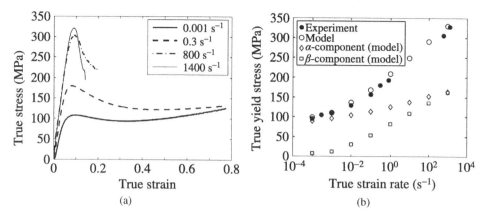

Figure 8. The contribution of multiple relaxation processes to the impact response of polymers. (a) Representative stress–strain behavior of PMMA in uniaxial compression at low, moderate, and high strain rates. (b) PMMA yield behavior in uniaxial compression as a function of strain rate, model prediction and experiment. In addition to the total yield stress, the numerically predicted α and β components of the yield stress show that the β-process becomes more significant with increasing strain rates. Reprinted from [87] with permission from Elsevier, copyright (2006).

Many molecular theories have been proposed to describe the thermal and strain rate effects on the stress–strain behavior of polymers [86, 87, 90–96]. These theories consider the yield behavior as an activated molecular relaxation process. Fotheringham and Cherry [95, 96] introduced the concept that yielding involves the cooperative motion of many polymer chain segments, and hence the physical existence of an activation volume for the yield process, e.g., the volume of a CRR. In these formalisms, the onset of inelastic deformation occurs, once a collective ensemble of chain segments overcomes local resistances to particular relaxation modes. The initial yield depends on pressure (load), temperature, and strain rate.

Based on the well-known time–temperature superposition principle, Bauwens-Crowet *et al.* [92] established the strain rate–temperature superposition principle for stress–strain behavior in polymers. According to both principles, an increase in rate is equivalent to a decrease in temperature. The Bauwens-Crowet approach is directly analogous to the velocity–temperature superposition employed in the study of frictional phenomena, which, as discussed above, enables the evaluation of characteristic activation energies. The difference between the two applications lies in the external drive time of the system; in mechanical loading, the drive time is the strain rate, whereas in kinetic friction, the drive time is the sliding velocity.

In the spirit of Williams–Landel–Ferry (WLF) [97], Richeton *et al.* [86] extended the Eyring cooperative yield model to accurately depict experimental yielding phenomena in bulk material samples for a wide range of temperatures and strain rates, including the glass transition, as well as dynamic loading. Their work on polycarbonate (PC), poly(methyl methacrylate) (PMMA), and poly(amide-imide) (PAI)

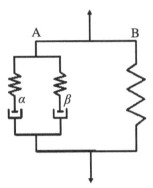

Figure 9. Rheological model for rate-dependent thermoplastic behavior. (A) Two Maxwell elements in parallel characterize the α- and β-molecular relaxation processes, and (B) a nonlinear Langevin spring describes the entropic chain stretching in the post-yield region.

revealed yield activation energies that are consistent with the β-relaxation process and activation volumes on the order of 0.01 nm^3. However, in a recent report based on nanoimpact studies in highly crosslinked PS, Altebaeumer *et al.* [73] suggest that the yield behavior in nanoscopic systems relies solely on the α-relaxation process, citing an activation volume that is two orders of magnitude larger, 1.5–3.5 nm^3. While these studies suggest that either primary or secondary relaxation mechanisms lead to the onset of plastic yield, its likely that the two processes are not mutually exclusive.

The relative contributions of multiple relaxation modes to the rate-dependent elastic–viscoplastic deformation of glassy polymers were considered by Mullikan and Boyce [87]. Based on the Ree–Eyring extension [98] to the general Eyring theory, they describe the material yield and flow characteristics with two rate-activated (Maxwell) processes acting in tandem with a nonlinear entropic (Langevin) spring, Fig. 9. The two Maxwell elements represent the α- and β-relaxation processes in the amorphous polymer, and the Langevin spring is essential to capture the post-yield strain softening and hardening associated with the entropic stretching of the polymer chains. This approach leads to a parallel arrangement of two activated spectra of relaxation times, each with a different stress-dependence, which is no longer related to the total stress, but only to part of the total stress. Such a model accurately predicts the strain rate–temperature dependence of PC and PMMA yield behavior; however, the effects of adiabatic heating remain to be incorporated for improved accuracy in the post-yield regime [87]. The contributions of the each relaxation mode to the yield process were inferred with viscoelastic characterization combined with compression testing over a wide range of strain rates. At low strain rates and high temperatures, the β-process is relatively compliant and most, if not all, of the intermolecular resistance to deformation is associated with α-relaxation. However, at high strain rates and low temperatures, the β-relaxation becomes constrained and the intermolecular resistance to yield and flow is composed of both α- and β-components. This is shown in Fig. 8(b), where at high strain rates of 10^3 s^{-1}, the contribution of the β-process to the overall yield stress is 50% for PMMA (a value of 23% was reported for PC). Further, in addition to the α- and β-processes in amorphous polymers, Govaert *et al.* recognized the contribution of chain transport due to defect migration in crystalline polymers [99]. The most significant outcome of these studies is the understanding that secondary molecular motion, distinct from the molecular processes of the α-transition, has an independent contribution to the observed yield behavior of polymers, particularly at the high strain-rates experienced during nanoimpact operations.

Dynamic loading and nanoimpact operations in polymer thin film systems are further complicated by the presence of supporting substrates and by material anisotropy within the rheological boundary layer. In confined systems, a rigid boundary interacting with the stress field during indentation, e.g., bit writing, alters the stress and strain distributions, leading to bulk-deviating mechanical responses [100–103]. For indentations in compliant films, increased rim heights are observed when elastic strain and plastic flow are constrained, or shielded, by a rigid substrate [100–103]. In the case of rigid films on compliant substrates, the plastic yield of the underlying substrate accommodates an enhanced *sink-in* of the surface around indentation sites [101].

The nanoimpact studies of Sills *et al.* [19] in un-entangled PS films revealed substantial strain shielding at the substrate interface, which led to undesired rim formation during indentation. Rim formation was exacerbated due to the reflection of the stress and strain fields at the rigid interface. This results in a local amplification of the stress, whereas the strain response is quenched at the rigid boundary, leading to stress-activated relaxation and an additional upward material displacement at the free surface. The magnitude of the response depends on both the film thickness and the modulus mismatch between the polymer film and its supporting substrate. This is apparent in Fig. 10, where for film thicknesses (δ) exceeding $\sim$100 nm, the ratio of the rim height to the indentation depth ($\overline{\zeta/d}$), displays a constant value of approximately 0.2, which reflects the bulk material response. For film thicknesses below 100 nm, the rim height increases with decreasing film thickness, due to elastic strain shielding and stress reflection at the rigid silicon substrate. When the rigid substrate was replaced by a crosslinked polymer with a modulus similar to the indented PS film, the strain shielding effects were alleviated because the elastic deformation front was no longer quenched at the substrate interface.

In a similar nanoimpact study, Altebaeumer and coworkers [73] confirmed that the same behavior was observed in highly crosslinked PS thin films. In the context of geometric self-similarity, they noted that the scaling parameters which relate the lateral and verticle dimensions of the residual indentation become a function of film thickness. Altebaeumer *et al.* [73] also observed a breakdown of self-similarity for indentation depths below 1 nm. In the limit of vanishingly shallow indents, regardless of film thickness, the minimum radius of plastic deformation, C_0, approached

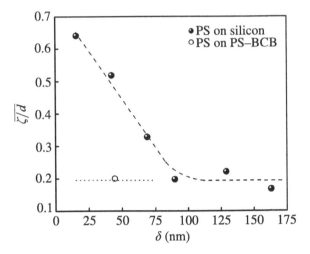

Figure 10. Substrate contribution to strain shielding in thin polystyrene films (ζ is the rim height, d is the indentation depth, and δ is the film thickness; dotted lines are guides). Reprinted from [19] with kind permission from Springer Science + Business Media, copyright (2005).

a constant value of 38 nm, which is consistent with the value of 42 nm from the initital work of Sills *et al.* [19]. Altebaeumer *et al.* [73] interpreted the existence of C_0 in terms of a minimum strain requirement necessary to induce plastic yield with the domain of a single CRR, along with a geometric contribution from the indenting probe.

Material anisotropy within the rheological boundary layer was also shown to have a significant influence on the resulting elastic-viscoplastic behavior under nanoimpact conditions. The rheological gradient discussed above was attributed to shear induced structuring during the spin casting process and to anisotropic diffusion during annealing. Entropic cooling of the polymer chains in the vicinity of the substrate results from a previous stress history. Consequently, the pre-stretched chains may shift the local stress–strain behavior further toward the strain hardening regime. For the PS films with the non-monotonic T_g profile in Fig. 4, the indentation studies in [19] suggested that material anisotropy within the boundary layer leads to a distribution of contact pressures between two asymptotic scenarios: (i) a compliant near-surface with a rigid sub-surface and (ii) a rigid near-surface with a compliant sub-surface [19]. This is apparent in Fig. 11, where the elastic modulus profiles from the nanoimpact studies match the T_g-profiles in Fig. 4. Viewing the glass transition as a mobility barrier, an increase in T_g offers resistance to molecular mobility, and the associated *stiffness* is accompanied by an increase in the modulus. Hence, the individual thermal and mechanical responses coincide.

Regarding device performance, increased rim heights were observed for film thicknesses between 60–120 nm because elastic strain and plastic flow are constrained by the more rigid sub-surface. When the film thickness falls below ~60 nm, decreased rim heights are expected because the indentation load is accommodated

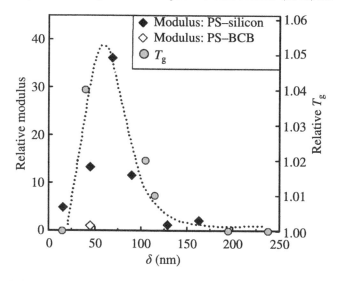

Figure 11. Interfacial thermal and mechanical profiles for thin polystyrene films are consistent with the anisotropic boundary model in Fig. 4. For films thicker than 150 nm, the material responds like the bulk. For film thicknesses between 60–120 nm the relative modulus of the near-surface increases with decreasing film thickness, until a maximum is reached for ∼60 nm films. The relative modulus then decreases with a continued reduction in film thickness. Reprinted from [19] with kind permission from Springer Science + Business Media, copyright (2005).

by the strain or flow of a more compliant sub-surface. Thus, the film thickness can be optimized to minimize rim formation. Alternatively, internal constraints could be incorporated into the molecular design [15, 18] to shift the transition between the opposing structural gradients. In either scenario, we strive to reach a delicate balance between a molecular system and its interactions with its boundaries.

5. Summary and Outlook

This review on molecular mobility, interfacial dynamics, and motion in constrained polymeric systems addressed apparent material and transport properties in context of the underlying relaxation mechanisms that enable molecular motion. Such considerations are critical to on-going engineering efforts regarding miniaturization in mechanics, electronics, optoelectronics, and in the fields of bioengineering and biomimetics. Polymer-based nanotechnological applications, such as the NEMS process for terabit thermomechanical storage, rely on very specific relaxation and transition properties in sub-100 nm systems. Nanoscale material design challenges were discussed in the light of critical length scales, namely, the device dimension and the length scale associated with molecular-level processes.

The basis for macromolecular motion was discussed in terms of various relaxation modes, which may be restricted (or unrestricted) by the inclusion (or omission) of nanoscopic constraints. In this light, two types of constraints were considered: internal constraints that are inherent to molecular structure and chemical functionality, and external constraints that are generated in the vicinity of interfaces or system boundaries. The implementation of internal constraints for engineering design opportunities was illustrated with a crosslinked system, in which the spacing between crosslinks was adjusted to regulate the length scale required for cooperative segmental relaxation. This molecular engineering approach enabled control of the yield and wear characteristics of polymer thin films.

The influence of external constraints on the relaxational behavior in condensed ultra-thin materials was illustrated with local variations in the glass transition temperature. Various groups reported far-reaching interfacial effects on the glass transition temperature. Most groups reported a monotonic decrease in the glass transition temperature for film thicknesses decreasing below ~100 nm. In addition, a less pronounced increase in the local T_g was found within an intermediate film thickness regime located a few tens of nanometers from the substrate interface, which was attributed to diffusion within an interfacially constrained heterogeneous polymer matrix. Near-surface sensitive T_g studies involving SM-FM revealed a non-monotonic interfacial T_g profile and led to the conclusion that the film preparation history, i.e., spin casting, was likely the origin for this extraordinary long-range field-effect in thermally annealed polymer films. This unique non-monotonic T_g profile has already been recognized for engineering applications in polymer thin films for photolithography, optoelectronic devices, and thermomechanical recording. Regarding thermomechanical recording NEMS, we reviewed how material anisotropy within the interfacial region dictates contact pressures during nanoimpact operations. This was contrasted with substrate constraints in terms of stress and strain shielding at the rigid interface.

The interplay between external and internal constraints was highlighted as a design opportunity. Concerning local mobility, the T_g profiles can be modified by selectively tuning internal material constraints, *via*, for instance, the molecular weight, the side-chain or backbone repeat unit structure, or the crosslink density. On the other hand, external constraints may be tailored for a desired material response. This was illustrated with modulus-matching techniques, where generating a quasi-continuous modulus gradient between a polymer thin film and its supporting substrate offers enhanced interfacial stress transmission and improved stability and durability of the interface. Thus, efforts to develop a balance between external and internal constraints offer a route to optimizing nanoscopic systems.

Optimization scenarios were further extended to embrace the external operating environment, such as electromechanical perturbations. Regarding mechanical disturbances in the form of sliding and mechanical loading, the time scale of the external process relative to the intrinsic material response time was found critical to material performance. When the drive velocity or the loading rate is comparable to characteristic molecular frequencies, i.e., the Deborah Number $De \sim 1$, molecular relaxation processes can effectively couple with external mechanical operations. In this case, a material is capable of responding through internal molecular motion. From a device performance perspective, operating times that match molecular timescales may lead to energy consuming resonance effects, and system designs (internal and external constraints) may have to be adjusted with molecularly engineered damping modes. For example, friction is greatest when the sliding velocity matches the relaxation time(s) of available relaxation mode(s). This was illustrated for polymer melts near the glass transition, where nanoscopic friction revealed a highly cooperative relaxation process. In one of the first direct studies of this relaxation process, the length scale over which collective molecular motion occurs, i.e., the size of cooperatively rearranging regions (CRRs), could be determined. In the bulk, this dimension ranges from a collection of monomeric segments on the sub-nanometer scale up to tens of nanometers, involving several molecules. The size of the CRRs was found to be critical for many modern device technologies, for example, cooperative relaxation was tied to chromophore alignment in organic nonlinear optical (NLO) materials, and to a minimum strain requirement associated with nanoscopic yield phenomena during thermomechanical recording. One can expect similar implications in other thin film applications, including, for example, charge transport in organic electronic devices.

Furthermore, a particular class of operations was discussed when operational drive times exceed the intrinsic material relaxation time, i.e., $De \gg 1$, which is the case for dynamic or impact loading. In this case, relaxation modes that are thermally available under quasi-static conditions became constrained, and the molecular systems behaved as if they were thermally cooled. Consequently, phenomenological properties, such as friction, yield stress, and failure mode, may shift to reflect apparent material stiffening. This was illustrated for impact loading of polymer materials, where secondary molecular motion, distinct from the molecular processes of the α relaxation, had an independent contribution to the observed yield behavior.

In summary, modern device technologies such as MEMS/NEMS involving polymers or other macromolecules require a careful analysis of the relevant spatio-temporal scales. When systems are reduced to the sub-100 nm scale, the external field-effects associated with interfacial and dimensional constraints often modify the intrinsic molecular relaxation behaviors. Consequently, structural, material, and transport properties may deviate significantly from those of the bulk material, leading to internal gradients and material anisotropy over relevant length scales. A basic understanding of the spatial distribution of molecular relaxation spectra within the confined systems and means for its control are necessary to match (or mismatch) molecular response times to system operating times. Ultimately, achieving the desired performance goals requires materials with molecular structures that are specifically tailored to function within the system constraints. To this end, the ability to cater molecular relaxation behavior through molecularly-engineered internal constraints offers a path to optimize materials for nanoscopic systems.

References

1. B. Peng, Y. Zhu, I. Petrov and H. D. Espinosa, *Sensor Letters* **6**, 76 (2008).

2. C. Liu, *Adv. Mater.* **19**, 3783 (2007).
3. S. A. Wilson, R. P. J. Jourdain, Q. Zhang, R. A. Dorey, C. R. Bowen, M. Willander, Q. U. Wahab, M. A. H. Safaa, O. Nur, E. Quandt, C. Johansson, E. Pagounis, M. Kohl, J. Matovic, B. Samel, W. van der Wijngaart, E. W. H. Jager, D. Carlsson, Z. Djinovic, M. Wegener, C. Moldovan, R. Iosub, E. Abad, M. Wendlandt, C. Rusu and K. Persson, *Mater. Sci. Eng. Reports* **56**, 1 (2007).
4. J. Kusterer, P. Schmid and E. Kohn, *New Diamond and Frontier Carbon Technology* **16**, 295 (2006).
5. Z. G. Zhou and Z. W. Liu, *J. Bionic Eng.* **5**, 358 (2008).
6. D. Collard, S. Takeuchi and H. Fujita, *Drug Discovery Today* **13**, 989 (2008).
7. D. B. J. Knorr, T. Gray and R. M. Overney, *J. Chem. Phys.* **129**, 074504 (2008).
8. T. Gray, T.-D. Kim, D. B. Knorr Jr, J. Luo, A. K. Y. Jen and R. M. Overney, *Nano Lett.* **8**, 754 (2008).
9. Y. F. Ding, S. Pawlus, A. P. Sokolov, J. F. Douglas, A. Karim and C. L. Soles, *Macromolecules* **42**, 3201 (2009).
10. G. Martin, C. Barres, P. Sonntag, N. Garois and P. Cassagnau, *Mater. Chem. Phys.* **113**, 889 (2009).
11. J. P. Killgore and R. M. Overney, *Langmuir* **24**, 3446 (2008).
12. T. Gray, J. Killgore, J. D. Luo, A. K. Y. Jen and R. M. Overney, *Nanotechnology* **18**, 044009 (2007).
13. X. S. Hu, Z. Jiang, S. Narayanan, X. S. Jiao, A. R. Sandy, S. K. Sinha, L. B. Lurio and J. Lal, *Phys. Rev. E* **74**, 010602 (2006).
14. M. G. Munoz, M. Encinar, L. J. Bonales, F. Ortega, F. Monroy and R. G. Rubio, *J. Phys. Chem. B* **109**, 4694 (2005).
15. B. Gotsmann, U. T. Duerig, S. Sills, J. Frommer and C. J. Hawker, *Nano Lett.* **6**, 296 (2006).
16. R. M. Overney, D. P. Leta, L. J. Fetters, Y. Liu, M. H. Rafailovich and J. Sokolov, *J. Vac. Sci. Technol. B* **14**, 1276 (1996).
17. C. Buenviaje, S. Ge, M. Rafailovich, J. Sokolov, J. M. Drake and R. M. Overney, *Langmuir* **15**, 6446 (1999).
18. S. E. Sills, R. M. Overney, W. Chau, V. Y. Lee, R. D. Miller and J. Frommer, *J. Chem. Phys.* **120**, 5334 (2004).
19. S. E. Sills, R. M. Overney, B. Gotsmann and J. Frommer, *Tribology Lett.* **19**, 9 (2005).
20. S. Sills and R. M. Overney, in: *Applied Scanning Probe Methods III*, B. Bushan and H. Fuchs (Eds). Springer-Verlag, Heidelberg, Germany (2006).
21. S. Takashi, U. Tadashi and S. Kensuke, *J. Polym. Sci. B* **44**, 1958 (2006).
22. G. K. Binnig, G. Cherubini, M. Despont, U. T. Duerig, E. Eleftheriou and P. Vettiger, in: *Springer Handbook of Nanotechnology*, B. Bhushan (Ed.). Springer-Verlag, Heidelberg, Germany (2004).
23. S. Sills, T. Gray and R. M. Overney, *J. Chem. Phys.* **123**, 134902 (2005).
24. A. Walther, K. Matussek and A. H. E. Muller, *ACS Nano* **2**, 1167 (2008).
25. S. M. Fielding, *Soft Matter* **3**, 1262 (2007).
26. J. L. Keddie, R. A. L. Jones and R. A. Cory, *Europhys. Lett.* **27**, 59 (1994).
27. J. A. Forrest, K. Dalonki-Veress and J. R. Dutcher, *Phys. Rev. E* **56**, 5705–5716 (1997).
28. J. H. Kim, J. Jang and W. C. Zin, *Langmuir* **17**, 2703 (2001).
29. L. Singh, P. J. Ludovice and C. L. Henderson, *Thin Solid Films* **449**, 231 (2004).
30. S. Ge, Y. Pu, W. Zhang, M. Rafailovich, J. Sokolov, C. Buenviaje, R. Buckmaster and R. M. Overney, *Phys. Rev. Lett.* **85**, 2340 (2000).
31. R. M. Overney, C. Buenviaje, R. Luginbuhl and F. Dinelli, *J. Therm. Anal. Calorimetry* **59**, 205 (2000).

32. J. A. Forrest and K. Dalnoki-Veress, *Adv. Colloid Interface Sci.* **94**, 167 (2001).

33. M. Campoy-Quiles, M. Sims, P. G. Etchegoin and D. D. C. Bradley, *Macromolecules* **39**, 7673 (2006).

34. C. B. Roth and J. R. Dutcher, *J. Electroanalytical Chem.* **584**, 13 (2005).

35. J. A. Forrest, *Eur. Phys. J. E* **8**, 261 (2002).

36. S. Kim, C. B. Roth and J. M. Torkelson, *J. Polym. Sci. B* **46**, 2754 (2008).

37. R. D. Priestley, L. J. Broadbelt, J. M. Torkelson and K. Fukao, *Phys. Rev. E* **75**, 061806 (2007).

38. A. Hooper, G. L. Fisher, K. Konstadinidis, D. Jung, H. Nguyen, R. Opila, R. W. Collins, N. Winograd and D. L. Allara, *J. Am. Chem. Soc.* **121**, 8052 (1999).

39. G. L. Fischer, A. E. Hooper, R. L. Opila, D. L. Allara and N. Winograd, *J. Phys. Chem. B* **104**, 3267 (2000).

40. G. L. Fischer, A. V. Walker, A. E. Cooper, T. B. Tighe, K. B. Bahnck, H. T. Skriba, M. D. Reinard, B. C. Haynie, R. L. Opila, N. Winograd and D. L. Allara, *J. Am. Chem. Soc.* **124**, 5528 (2002).

41. A. V. Walker, T. B. Tighe, O. M. Cabarcos, M. D. Reinard, B. C. Haynie, S. Uppili, N. Winograd and D. Allara, *J. Am. Chem. Soc.* **126**, 3954 (2004).

42. S. Sills, K. Unal, L. D. Bozano, J. Frommer and J. C. Scott, *J. Vac. Sci. Technol. B* **25**, 421 (2007).

43. C. J. Ellison and J. M. Torkelson, *Nature Mater.* **2**, 95 (2003).

44. T. Gray, J. Killgore, J. Luo, A. K.-Y. Jen and R. M. Overney, *Nanotechnology* **17**, S1 (2006).

45. P.-G. de Gennes, *Eur. Phys. J.* **E2**, 201 (2000).

46. K. Dalnoki-Veress, J. A. Forrest, P.-G. de Gennes and J. R. Dutcher, *J. Phys. IV* **10**, 221 (2000).

47. P.-G. de Gennes, Personal communication, R. M. Overney (2000).

48. S. Marceau, J. H. Tortaia, J. Tilliera, N. Vourdasb, E. Gogolidesb, I. Raptisb, K. Beltsiosc and K. v. Werdend, *Microelectronic Eng.* **83**, 1073 (2006).

49. M. Campoy-Quiles, M. Sims, P. G. Etchegoin and D. D. C. Bradley, *Macromolecules* **36**, 7673 (2006).

50. J. H. Kim, J. Jang and W. Zin, *Langmuir* **16**, 4064 (2000).

51. H. K. Zhou, H. K. Shi, F. G. Zhao and B. Yota, *J. Microelectronics* **33**, 221 (2002).

52. J. L. Keddie, R. A. L. Jones and R. A. Cory, *Europhys. Lett.* **27**, 59 (1994).

53. C. Buenviaje, F. Dinelli and R. M. Overney, in: *Interfacital Properties on the Submicron Scale*, J. Frommer and R. M. Overney (Eds), ACS Symp. Ser., Vol. 781, p. 76. Oxford University Press, Oxford (2000).

54. C. B. Rotha, A. Pound, S. W. Kamp, C. A. Murray and J. R. Dutche, *Eur. Phys. J. E* **20**, 441 (2006).

55. C. J. Ellison, M. K. Mundra and J. M. Torkelson, *Macromolecules* **38**, 1767 (2005).

56. K. A. Grosch, *Proc. R. Soc. London Ser. A* **274**, 21 (1963).

57. K. C. Ludema and D. Tabor, *Wear* **9**, 329 (1966).

58. S. Sills, R. M. Overney, K. Vorvolakos and M. Chaudery, in: *Nanotribology: Friction and Wear on the Atomic Scale*, E. Gnecco and E. Meyer (Eds), p. 659. Springer-Verlag, Heidelberg, Germany (2007).

59. S. Sills and R. M. Overney, *Phys. Rev. Lett.* **91**, 095501 (2003).

60. J. A. Hammerschmidt, W. L. Gladfelter and G. Haugstad, *Macromolecules* **32**, 3360 (1999).

61. A. R. Savkoor, *Wear* **8**, 222 (1965).

62. H. Eyring, *J. Chem. Phys.* **3**, 107 (1935).

63. G. D. Patterson, C. P. Lindsey and J. R. Stevens, *J. Chem. Phys.* **70**, 643 (1979).

64. G. Adam, *J. Chem. Phys.* **43**, 139 (1965).

65. P. G. Debenedetti and F. H. Stillinger, *Nature* **410**, 259 (2001).

66. M. D. Ediger, *Annu. Rev. Phys. Chem.* **51**, 99 (2000).

67. R. Richert, *J. Phys.: Condens. Matter* **14**, R703 (2002).
68. H. Sillescu, *J. Non-Crystalline Solids* **243**, 81 (1999).
69. M. H. Cohen and G. S. Grest, *Phys. Rev. B* **20**, 1077–1098 (1979).
70. U. Tracht, M. Wilhelm, A. Heuer, H. Feng, K. Schmidt-Rohr and H. W. Spiess, *Phys. Rev. Lett.* **81**, 2727 (1998).
71. H. Sillescu, R. Bohmer, G. Diezemann and G. Hinze, *J. Non-Crystalline Solids* **307**, 16 (2002).
72. D. Cangialosi, A. Alegría and J. Colmenero, *Phys. Rev. E* **76**, 011514 (2007).
73. T. Altebaeumer, B. Gotsmann, A. Knoll, G. Cherubini and U. Duerig, *Nanotechnology* **19**, 475301 (2008).
74. F. Kajzar, K.-S. Lee and A. K.-Y. Jen, *Adv. Polym. Sci.* **161**, 1 (2003).
75. P. Vettiger, G. Cross, M. Despont, U. Drechsler, U. Duerig, W. Heberle, M. I. Lantz, H. E. Rothuizen, R. Stutz and G. K. Binnig, *IEEE Trans. Nanotechnol.* **1**, 39 (2002).
76. I. Karapanagiotis, D. F. Evans and W. W. Gerbrich, *Polymer* **43**, 1343 (2002).
77. B. Gotsmann and U. Duerig, in: *Applied Scanning Probe Methods: Industrial Applications*, B. Bhushan and H. Fuchs (Eds), Vol. IV, p. 215. Springer, Heidelberg (2004).
78. B. Gotsmann, H. Rothuizen and U. Duerig, *Appl. Phys. Lett.* **93**, 093116 (2008).
79. J. A. Zukas, T. Nicholas, H. F. Swift, L. B. Greszczuk and D. R. Curran, *Impact Dynamics*. Wiley, New York, NY (1982).
80. L. L. Wang, *Chinese J. Mech. A* **19**, 177 (2003).
81. C. Y. Chiem, in: *Shock-Wave and High-Strain-Rate-Phenomena in Materials*, M. A. Meyers, L. E. Murr and K. P. Staudhammer (Eds), p. 69. Marcel Dekker, New York, NY (1992).
82. K. W. Thomson and L. J. Broutman, *J. Mater. Sci.* **17**, 2700 (1982).
83. A. Trojanowski, C. Ruiz and J. Harding, *J. Phys. IV (France)* **7**, C3 (1997).
84. S. M. Walley, J. E. Field, P. H. Pope and N. A. Safford, *Philos. Trans. Roy. Soc. London Ser. A* **328**, 1 (1989).
85. A. V. Lyulin, B. Vorselaars, M. A. Mazo, N. K. Balabaev and M. A. J. Michels, *Europhys. Lett.* **71**, 618 (2005).
86. J. Richeton, S. Ahzi, K. S. Vecchio, F. C. Jiang and R. R. Adharapurapu, *Intl J. Solids Structures* **43**, 2318 (2006).
87. A. D. Mulliken and M. C. Boyce, *Intl J. Solids Structures* **43**, 1331 (2006).
88. N. Cristescu, *Dynamic Plasticity*. North-Holland Publishing Co., Amsterdam (1967).
89. L. E. Govaert, P. H. M. Timmermans and W. A. M. Brekelmans, *J. Eng. Mater. Technol.* **122**, 177 (2000).
90. H. Eyring, *J. Chem. Phys.* **4**, 283 (1936).
91. R. E. Robertson, *J. Chem. Phys.* **44**, 3950 (1966).
92. C. Bauwens-Crowet, J. C. Bauwens and G. Homes, *J. Polym. Sci. A* **7**, 176 (1969).
93. A. S. Argon, *Philosophical Magazine* **28**, 839 (1973).
94. P. B. Bowden and S. Raha, *Philosophical Magazine* **29**, 149 (1974).
95. D. Fotheringham and B. W. Cherry, *J. Mater. Sci.* **11**, 1368 (1976).
96. D. Fotheringham and B. W. Cherry, *J. Mater. Sci.* **13**, 951 (1978).
97. M. L. Williams, R. F. Landel and J. D. Ferry, *J. Am. Chem. Soc.* **77**, 3701 (1955).
98. T. Ree and H. Eyring, *J. Appl. Phys.* **26**, 793 (1955).
99. L. E. Govaert, P. J. de Vries, P. J. Fennis, W. F. Nijenhuis and J. P. Keustermans, *Polymer* **41**, 1959 (2000).
100. T. Y. Tsui and G. M. Pharr, *J. Mater. Res.* **14**, 292 (1999).
101. T. Y. Tsui, J. Vlassak and W. D. Nix, *J. Mater. Res.* **14**, 2204 (1999).
102. N. X. Randall, C. Julia-Schmutz and J. M. Soro, *Surface Coatings Technol.* **108–109**, 489 (1998).

103. D. E. Kramer, A. A. Volinsky, N. R. Moody and W. W. Gerberich, *J. Mater. Res.* **16**, 3150 (2001).
104. P. Claudy, J. M. Letoffe, Y. Chamberlain and J. P. Pascault, *Polymer Bulletin* **9**, 208 (1983).
105. P. L. Kumler, S. E. Keinath and R. F. Boyer, *J. Macromol. Sci. Phys.* **B13**, 613 (1977).
106. C. Bennemann, C. Donati, J. Baschnagel and S. Glotzer, *Nature* **399**, 246 (1999).

Part 5

Adhesion Mitigation Strategies

Microscale Friction Reduction by Normal Force Modulation in MEMS

W. M. van Spengen [a,b,*], G. H. C. J. Wijts [a], V. Turq [a,c] and J. W. M. Frenken [a]

[a] TU Delft, 3mE-PME, Mekelweg 2, 2628CD Delft, The Netherlands
[b] Leiden University, LION, Niels Bohrweg 2, 2333CA Leiden, The Netherlands
[c] Now with Université de Toulouse, UPS, INP, Institut Carnot Cirimat, 118 route de Narbonne,
F-31062 Toulouse cedex 9, France

Abstract

Friction in MEMS-scale devices is troublesome because it can result in lateral stiction of two sliding surfaces. We have investigated the effect of modulation of the normal force on the friction between two sliding MEMS surfaces, using a fully MEMS-based tribometer. We have found that the friction is reduced significantly when the modulation is large enough. A simple model is presented that describes the friction reduction as a function of modulation frequency as well. Using this technique, lateral stiction-related seizure of microscopic sliding components can be mitigated.

Keywords

MEMS, friction, modulation, vibration

1. Introduction

Although high wear can be disastrous in MEMS (micro-electromechanical systems) technology, friction is an important issue as well. The typical forces that can be generated with MEMS-based actuation are in the nano- to microNewton range. With adhesion forces of the same order of magnitude and friction coefficients between 0.1 and up to more than 1, MEMS devices that have not been carefully designed for their tribological properties can easily become stuck.

Macroscopically, friction is usually mitigated by the use of oils, but on the scale of MEMS, the viscous drag of oil is so large that the devices either get stuck, or the energy dissipation becomes unacceptably high [1].

Other methods have been explored for friction reduction instead, notably the use of self-assembled monolayers (SAMs) [2–5]. These layers can reduce friction significantly, from a friction coefficient of up to 2.3 for bare silicon to around 0.08 for an optimised SAM layer [6]. However, the durability of these layers is question-

---

* To whom correspondence should be addressed. E-mail: W.M.vanSpengen@tudelft.nl

Adhesion Aspects in MEMS/NEMS
© Koninklijke Brill NV, Leiden, 2010

able: they tend to be rubbed off during use [7, 8], after which the friction is again high and poses a reliability threat. Hard coatings such as tungsten, diamond-like carbon (DLC) and silicon carbide have been proposed as well [9–11], but, although the wear rates are reported to be significantly lower, the friction coefficient remains relatively high.

In this paper, we discuss a different solution, i.e., the application of high-frequency vibrations in the direction normal to the surface.

2. High-Frequency Vibrations for Friction Reduction

The application of high-frequency vibrations to ease the friction of a sliding contact was proposed in 1959 in a paper by Fridman and Levesque [12], of course with macroscopic systems in mind. They showed that vibrations normal to the sliding surfaces can temporarily lessen the effective normal load, hence reducing friction according to Amontons' law $F_F = \mu F_N$, where F_F is the friction force, F_N is the normal force, and μ is the friction coefficient. This concept was further studied in the 1960's by Tolstoi [13] and Lenkiewicz [14], and it was established that vibrations could lower the effective friction coefficient by more than 85%. The technique has not been used extensively because wear rates are, of course, not influenced as much as in the case of the use of oil. However, the technique is regarded as a standard way of controlling macroscale friction, and is often also called 'dither' [15].

On the atomic scale, the energy barriers for sliding over single atoms are small, and atomic-scale stick–slip was first observed by Mate *et al.* [16] with a tungsten tip on a graphite lattice. Atomic-scale stick–slip frictional behaviour was theoretically described by Prandtl [17], Tomlinson [18], Zhong and Tomanek [19], and Tomanek *et al.* [20]. That thermal vibrations on the atomic scale can reduce friction significantly was envisioned by Feynman [21], and described and measured by Gnecco *et al.* [22], Sang *et al.* [23], Jinesh *et al.* [24], Krylov *et al.* [25] and others, by an effect that has been called 'thermolubricity'.

Apart from naturally occurring thermal fluctuations, also intentionally applied vibrations can lower friction on the atomic scale in two ways. The effect of vibrations on the friction in a microscopically small contact was studied experimentally by Dinelli *et al.* [26] using a friction force microscope (FFM) [16]. They found that not only can friction be reduced on this scale by fast out-of-plane vibrations, but also the behaviour is affected by the capillary condensed liquid layer between the microscopic tip and the sample. In this case, the tip really loses contact with sample, which is also the case in the work of Jeon *et al.* [27].

A second situation, of more fundamental importance, was found in which external vibrations can lower the energy barriers for sliding so much that one enters the thermolubricity regime without losing contact completely: frictionless sliding while still in contact! This work has been pioneered by Socoliuc *et al.* in ultra-high vacuum [28].

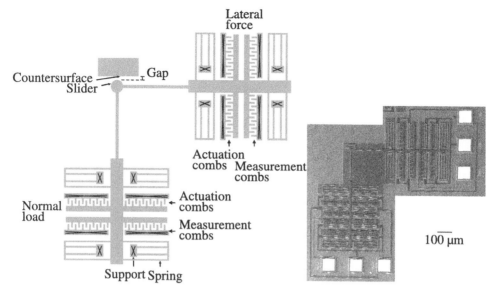

Figure 1. The Leiden MEMS tribometer: schematic layout (left) and SEM (Scanning Electron Micro-scope) micrograph (right).

3. Experiment: Normal Force Modulation in MEMS

For the MEMS friction reduction experiment, we used the Leiden MEMS tribometer (Fig. 1), produced in the MEMSCAP PolyMUMPS process [29]. This device is mechanically rather similar to the tribometers of Senft and Dugger [30] and Tas *et al.* [31], but the motion is read out in a different way, and provides much more detailed information on the interaction between the surfaces than reported in previous work. It has been used to study adhesion [32] and friction [33] between typical polycrystalline silicon MEMS sidewall surfaces.

Two comb drives provide the force to move two perpendicular, suspended beams, which together can move a slider in two orthogonal directions. The slider can be pressed with its sidewall to a countersurface, after which it can be used to make a sliding motion against this countersurface with adjustable normal force. The actual normal and lateral positions of the slider with respect to the countersurface are measured by monitoring the capacitance change of a second set of comb drive fingers with attofarad accuracy (indicated as 'measurement combs' in Fig. 1) [34].

A typical measurement of the lateral forces between the round slider and the countersurface is given in Fig. 2, presented as a 'friction loop'. We have observed repeatable, irregular stick–slip motion, and wear at high normal loads [33]. The energy dissipated by the friction is given by the area enclosed by the loop. The static friction coefficient μ_s is the ratio of lateral force to normal force at the moment the device starts slipping, while the dynamic friction coefficient μ_d is undetermined because the slider moves in a stick–slip fashion. Arguably, there is no good measure

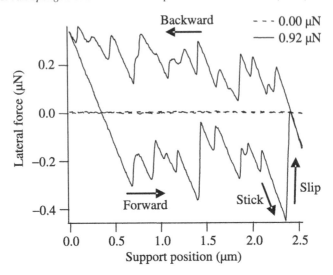

Figure 2. Typical friction loop measured with the tribometer.

for μ_s in this particular measurement either, because in the stick–slip motion, the first slip is not at the highest lateral force encountered in the loop. We, therefore, define μ_s as the highest lateral force encountered, a measure that is prone to statistical errors. In this particular device under the conditions of temperature $= 24.4°C$ and relative humidity $= 24\%$, $\mu_s = 0.45$, and the average friction $\mu_{av} = 0.21$. The tribometer makes one forward and backward sliding cycle per 0.5 seconds in this experiment. The slips are the moments at which energy is being dissipated: the potential energy stored in the stuck device under tension is converted to the kinetic energy of the slip, which is converted to phonons that are launched into the structure at the moment of impact.

To investigate the effect of vibrations on the friction encountered, we have systematically varied the normal force by modulating it at different frequencies and amplitudes. Two basic sets of data have been acquired: one in which the *average* normal load is kept constant (50 nN), and one in which the *minimum* normal load is kept constant (also 50 nN) (Fig. 3). Note that the actual normal force on the surfaces is marginally higher than the applied normal load due to adhesion between the surfaces. The sliding distance was 1.2 µm back and forth, and one loop was acquired in 0.5 s. All the different modulation conditions have been summarized in Table 1.

Typical results for both types of modulation are given in Fig. 4. At low modulation voltage amplitudes, no effect is seen. Only when a significant modulation is applied, the effect becomes noticeable, as a reduction of the area of the friction loop (less energy dissipation means less friction). Figure 5 shows the effect of the modulation amplitude on the friction, with an actuation frequency of 500 Hz. At 5 V_{pp} modulation, with the average normal load kept constant, the friction reduction is so large that the friction basically drops to zero! We associate this regime with the

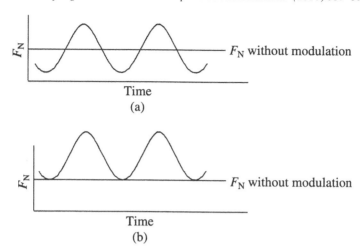

F_N without modulation

Time

(a)

F_N without modulation

Time

(b)

Figure 3. Two modulation types have been employed: (a) average normal load constant, and (b) minimum normal load constant.

Table 1.
Measurement conditions

Amplitude (V_{pp})	Frequency (Hz)	Type of modulation
0.01 (0.56 nN$_{pp}$)	20	Constant average normal load
0.1 (5.6 nN$_{pp}$)	50	Constant minimal normal load
0.2 (11.2 nN$_{pp}$)	100	
0.5 (28 nN$_{pp}$)	500	
1.0 (56 nN$_{pp}$)		
2.0 (112 nN$_{pp}$)		
5.0 (280 nN$_{pp}$)		

There were 56 different settings in total, subscript pp means peak-to-peak. Both the driving amplitude for the vibrations in Volt, and the calculated resultant variation in normal load (in parentheses) are given.

situation in which the contact between the surfaces is temporarily broken and reformed at every modulation cycle. We also see that the friction remains constant in the case where the minimum normal load is kept constant, even though the average normal load is increasing significantly with increasing modulation. This means that the surfaces can slip during the lowest force part of the modulation cycle.

Important also is the behaviour as a function of modulation frequency. To obtain low friction, the modulation frequency should be much higher than the frequency of the original stick–slip process. Only then the energy-dissipating slip events are successfully suppressed. We see this in Fig. 6. The figure shows the friction loops as a function of frequency. Only at high modulation frequencies, the surfaces lose contact often enough to prevent stick–slip behaviour. Figure 7 shows the energy

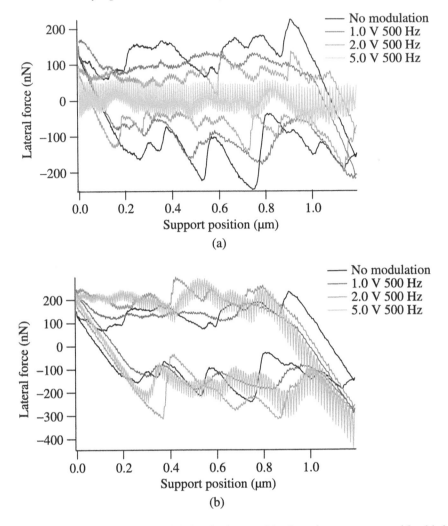

Figure 4. Results of normal force modulation in the case (a) where the average normal load is kept constant, and (b) where the minimum normal load is kept constant. All curves are 300 sliding cycle averages to average out the noise in the measurements.

dissipation as a function of frequency. The energy dissipation drops steadily with increasing modulation frequency and is not quite zero even at 500 Hz.

4. Discussion

As in the macroscopic and the atomic scale cases, we have found that on the scale of MEMS, modulation of the normal force can significantly reduce the friction experienced by the sliding contact. Important to notice is that the physical processes governing the friction reduction on the macroscopic and the atomic scales are completely different: on the macroscopic scale, it is the (intermittent) full loss of contact,

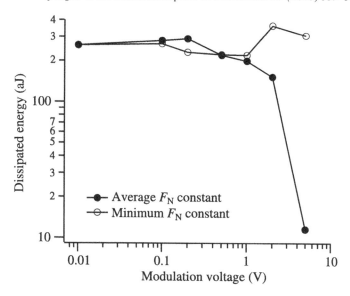

Figure 5. At high modulation amplitude, the energy dissipation, and hence the friction, drops sharply. The energy dissipation has been calculated by determining the full area of the loop.

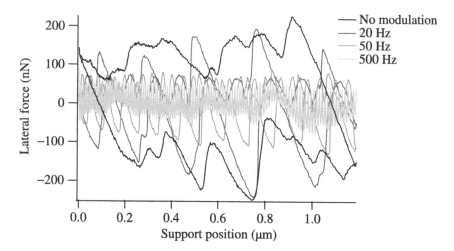

Figure 6. The effect of modulation frequency on the friction loop when the average normal load is constant at 5.0 V_{pp} modulation amplitude. Only at high modulation frequencies, the system vibrates fast enough to go from stick–slip behaviour to smooth sliding.

while on the atomic scale, super-slipperiness has been observed even if the surfaces are fully in contact, albeit with strong periodic reduction in normal force.

That we are not dealing with the atomic scale case in our experiment is already obvious from the amplitude of the modulation that we have to apply to observe the friction reduction. At 500 Hz and 5 V_{pp} modulation (the typical case for full loss of friction), there is a significant pulling force on the contact. After the adhesion

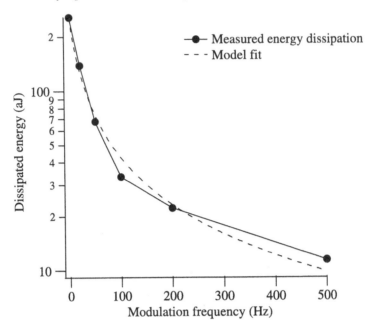

Figure 7. The energy dissipation as a function of modulation frequency for the case that the average normal force is kept constant at 5 V_{pp} modulation.

is subtracted (measured separately as discussed in [32], being with these samples under these conditions only a few nN), we end up with a relatively high net force trying to separate the two surfaces. The pulling force at 5 V_{pp} modulation is 90 nN at the minimum of the modulation period. At 2 V_{pp} the pulling force is 16 nN on the surfaces at the lowest point of the modulation cycle, but in that case significant friction is still observed, even though the pulling force is already larger than the static adhesion. Based on the applied forces alone, at 2 V_{pp}, loss of contact would be expected during 15% of the cycle, while for 5 V_{pp}, this is 38%.

We certainly do not have a contact with a net pressure of the surfaces towards each other when the friction vanishes, like in the atomic-scale case of Socoliuc *et al.* [28]. Instead, it may be that the dynamic breaking of the capillary neck of the condensed water is observed, which is always present at typical relative humidities between hydrophilic surfaces [35–37] and that this makes up the largest part of the dynamic adhesion force between them. This would mean that the surfaces are effectively in contact for a much longer time than anticipated based on the modulation amplitude and the measured adhesion force alone. The physical principle will be investigated later and it is expected that it is related to the work reported by Gao *et al.* [38], Gourdon and Israelachvili [39], and Jinesh *et al.* [40] on the temporal behaviour of confined, ordered liquid layers between contacting surfaces on the nanometer scale. Other MEMS related work in this area is given in [41–43].

The friction reduction effect only occurs if the modulation lowers the normal force during part of the cycle. From Fig. 5 it is seen that there is a critical amplitude of the normal force modulation below which there is only partial cancellation of the friction force, and from Fig. 7 it is learned that the frequency is also an important parameter.

The dependence of the dissipated energy on modulation frequency can be ratio-nalized with a simple model, which nicely reproduces the general features of the observations. If one assumes that slip will always occur as soon as the lateral force exceeds the static friction coefficient μ_s times the normal force F_N, and that the slip will make the lateral force go to zero, the average friction force $F_{F,av}$ is half the maximum static friction force:

$$F_{F,av} = \frac{1}{2}\mu_s F_N. \tag{1}$$

If l is the sliding distance, the corresponding dissipated energy E is

$$E = lF_{F,av} = \frac{1}{2}l\mu_s F_N. \tag{2}$$

When one starts pulling, the 'reference position' is moved. This is the position where the slider would have been if there would have been no contact, and to which it is connected by the springs with spring coefficient k_{spring} to the rest of the MEMS tribometer. If the reference position moves by x, before the first slip event occurs, the lateral force F_L on the stuck contact increases as

$$F_L = xk_{spring}. \tag{3}$$

F_L reaches its maximum when it reaches $\mu_s F_N$ at position x_0, after which it starts its first slip, so its first slip occurs at

$$x_0 = \frac{\mu_s F_N}{k_{spring}}. \tag{4}$$

If the expected number of stick–slip events n during the sliding over the distance l is much larger than unity, it is given by

$$n = \frac{l}{x_0} = \frac{k_{spring}}{\mu_s F_N}l, \tag{5}$$

being the average number of times that the friction force reaches its maximum value, after which the contact slips and the friction falls back to zero. When F_N is modulated, sliding will occur every cycle at the point where the momentary value of $\mu_s F_N$ falls below F_F. If F_N goes to zero, as in the case of loss of contact, sliding will *certainly* occur at every modulation cycle. The number of slip events n hence will be increased, resulting in a lower $F_{F,av}$. It is estimated that this increase in the number of slips will simply be equal to the number of times that the modulated nor-mal force is minimal (over the distance l), which is lf/v, where f is the modulation frequency and v the sliding velocity, so that l/v is the time required to traverse the distance l. The factor R by which the average friction force is reduced can now be

estimated as the ratio between the original number of stick–slip cycles in the un-
modulated case, described in equation (5), and the increased number of stick–slip
events in the modulated case.

$$R = \frac{(k_{\text{spring}}/(\mu_{\text{s}} F_{\text{N}}))l}{fl/v + (k_{\text{spring}}/(\mu_{\text{s}} F_{\text{N}}))l} = \frac{1}{1 + f\mu_{\text{s}} F_{\text{N}}/(vk_{\text{spring}})} \tag{6}$$

and

$$E = \frac{1}{2} Rl\mu_{\text{s}} F_{\text{N}}. \tag{7}$$

We find that the ratio, and hence the dissipated energy according to equations (2)
and (7), has an approximately $1/(f + \text{constant})$ behaviour, and the fit based on this
relationship has been plotted in Fig. 7 along with the data. The stochastic nature
of the stick–slip behaviour in the MEMS device has been neglected, as well as
the relative phase of the original stick–slip events and the modulation. Also the
possibility of sliding before the modulation reaches its minimum and the fact that
less 'natural' slip events will occur when the force build-up is frustrated as the
modulation frequency increases have not been taken into account, but the agreement
is already excellent.

The implementation of the friction control method described in this experiment
can follow different routes. Electrostatically induced vibrations are convenient,
but will certainly not be applicable to all MEMS-based devices. Other actuation
schemes can be envisioned as well, e.g., the actuation of the whole die by a small
piezo-transducer.

The effect of normal force modulation on the wear rate of a MEMS device
with sliding surfaces remains to be investigated, and will certainly be material-
dependent. The highest normal force encountered between the surfaces will in most
cases be higher than without actuation, which may crush the highest contact asperities even further than they would normally be. On the other hand, the surfaces
are only stressed under impact, not anymore under shear, while shear often creates
more damage. It may be easier to optimize coatings for MEMS devices for impact
resistance than shear resistance. The effect of normal force modulation on wear
should certainly be studied in future work.

5. Conclusions

We have shown that friction between MEMS surfaces can be reduced significantly
by applying a modulation of the normal force, in which the average normal force
remains constant, but the actual value is fluctuating. The minimum modulation frequency to obtain super-slipperiness is related to the frequency of the stick–slip
events in the system and for full friction reduction the modulation frequency should
be chosen significantly higher than the highest natural frequency of slip events.
Similarly, the amplitude of the modulation should be so high as to enforce periodic
loss of contact between the surfaces, although only a short contact time is required

every cycle. The effect of the modulation frequency on the friction reduction is described well by a simple model.

By optimizing the frequency and amplitude of the modulation, the friction can be reduced to a very low value that will in most MEMS devices not cause any friction-related seizure of sliding components (lateral stiction).

Acknowledgements

This work was financially supported by the Dutch NWO-STW foundation in the 'Veni' program under ref no. LMF.7302. The third author would also like to thank the French Foreign Affairs Ministry for a Lavoisier fellowship and the Dutch FOM foundation for financial support.

References

1. K. Deng, G. P. Ramanathan and M. Mehregany, *J. Micromech. Microeng.* **4**, 226 (1994).
2. U. Srinivasan, M. R. Houston, R. T. Howe and R. Maboudian, *J. Microelectromech. Syst.* **7**, 252 (1998).
3. X. Y. Zhu and J. E. Houston, *Tribology Lett.* **7**, 87 (1999).
4. W. R. Ashurst, C. Yau, C. Carraro, R. Maboudian and M. T. Dugger, *Sensors Actuators A* **91**, 239 (2001).
5. W. R. Ashurst, C. Yau, C. Carraro, R. Maboudian and M. T. Dugger, *J. Microelectromech. Syst.* **10**, 41 (2001).
6. R. Maboudian, W. R. Ashurst and C. Carraro, *Sensors Actuators A* **82**, 219 (2000).
7. S. T. Patton, W. D. Cowan, K. C. Eapen and J. S. Zabinski, *Tribology Lett.* **9**, 199 (2000).
8. D. A. Hook, S. J. Timpe, M. T. Dugger and J. Krim, *J. Appl. Phys.* **104**, 034303 (2008).
9. R. Bandorf, H. Lüthje and T. Staedler, *Diamond Relat. Mater.* **13**, 1491 (2004).
10. S. Sundararajan and B. Bhushan, *Wear* **217**, 251 (1998).
11. G. Fleming, S. S. Mani, J. J. Sniegowski and R. S. Blewer, US patent 6,290,859 (2001).
12. H. D. Fridman and P. Levesque, *J. Appl. Phys.* **30**, 1572 (1959).
13. D. M. Tolstoi, *Wear* **10**, 199 (1967).
14. W. Lenkiewicz, *Wear* **13**, 99 (1969).
15. B. Armstrong-Hélouvry, P. Dupomt and C. C. De Wit, *Automatica* **30**, 1083 (1994).
16. C. M. Mate, G. M. McClelland, R. Erlandsson and S. Chiang, *Phys. Rev. Lett.* **59**, 1942 (1987).
17. L. Prandtl, *Z. Angew. Math. Mech.* **8**, 6 (1928).
18. G. A. Tomlinson, *Phil. Mag.* **7**, 905 (1929).
19. W. Zhong and D. Tománek, *Phys. Rev. Lett.* **64**, 3054 (1990).
20. D. Tománek, W. Zhong and H. Thomas, *Europhys. Lett.* **15**, 887 (1991).
21. R. Feynman, *J. Microelectromech. Syst.* **2**, 4 (1993).
22. E. Gnecco, R. Bennewitz, T. Gyalog, Ch. Loppacher, M. Bammerlin, E. Meyer and H.-J. Güntherodt, *Phys. Rev. Lett.* **84**, 1172 (2000).
23. Y. Sang, M. Dubé and M. Grant, *Phys. Rev. Lett.* **87**, 174301 (2001).
24. K. B. Jinseh, S. Y. Krylov, H. Valk, M. Dienwiebel and J. W. M. Frenken, *Phys. Rev. B* **78**, 155440 (2008).
25. S. Y. Krylov, K. B. Jinesh, H. Valk, M. Dienwiebel and J. W. M. Frenken, *Phys. Rev. E* **71**, 065101 (2005).

26. F. Dinelli, S. K. Biswas, G. A. D. Briggs and O. V. Kosolov, *Appl. Phys. Lett.* **71**, 1177 (1997).

27. S. Jeon, T. Thundat and Y. Braiman, *Appl. Phys. Lett.* **88**, 214102 (2006).

28. A. Socoliuc, E. Gnecco, S. Maier, O. Pfeiffer, A. Baratoff, R. Bennewitz and E. Meyer, *Science* **313**, 207 (2006).

29. See www.memscap.com/en_mumps.html

30. D. C. Senft and M. T. Dugger, *Proc. SPIE* **3224**, 31 (1997).

31. N. R. Tas, C. Gui and M. Elwenspoek, *J. Adhesion Sci. Technol.* **17**, 547 (2003).

32. W. M. van Spengen and J. W. M. Frenken, *Tribology Lett.* **28**, 149 (2007).

33. W. M. van Spengen, E. Bakker and J. W. M. Frenken, *J. Micromech. Microeng.* **17**, S91 (2007).

34. W. M. van Spengen and T. H. Oosterkamp, *J. Micromech. Microeng.* **17**, 828 (2007).

35. R. Maboudian and R. T. Howe, *J. Vac. Sci. Technol. B* **15**, 1 (1997).

36. W. M. van Spengen, R. Puers and I. De Wolf, *J. Micromech. Microeng.* **12**, 702 (2002).

37. M. P. De Boer, *Expl. Mech.* **47**, 171 (2007).

38. J. Gao, W. D. Luedtke and U. Landman, *J. Phys. Chem. B* **102**, 5033 (1998).

39. D. Gourdon and J. Israelachvili, *Phys. Rev. E* **68**, 021602 (2003).

40. K. B. Jinesh and J. W. M. Frenken, *Phys. Rev. Lett.* **96**, 166103 (2006).

41. Z. Wei and Y.-P. Zhao, *Chin. Phys.* **13**, 1320 (2004).

42. K. Komvopoulos, *J. Adhesion Sci. Technol.* **17**, 477 (2003).

43. Y.-P. Zhao, L. S. Wang and T. X. Yu, *J. Adhesion Sci. Technol.* **17**, 519 (2003).

Microchannel Induced Surface Bulging of a Soft Elastomeric Layer

Abhijit Majumder, Anurag Kumar Tiwari, Krishnarao Korada and Animangsu Ghatak [*]

Department of Chemical Engineering, Indian Institute of Technology, Kanpur, UP 208016, India

Abstract

When a wetting liquid fills in microchannels embedded inside a thin elastomeric layer, the surface of the layer does not remain smooth but bulges out in the vicinity of the channels. The height of the bulge depends on the deformability of the layer and the surface tension of liquid; in addition, it depends also on the vertical location of the channel from the surface of the layer and the channel diameter. While, for liquids of low viscosity $\sim$500 cP bulging occurs instantaneously, for liquids of high viscosity $\sim$4000 cP, it occurs over a period of time. Concomitant to bulging of the layer, the cross section of the channel alters too in its shape and size suggesting that the elastic energy penalty associated with the bulging of the layer is supplied by the interfacial energy of the liquid–air and liquid–solid interfaces. Local bulging of the layer alters also its local deformability which is demonstrated by contact mechanics experiments: indentation with a spherical indenter yields a non-circular contact area, the shape and size of which vary with the depth of indentation. Thus, sub-surface microchannels can be suitably used for generating surface patterns on the elastomer and also for modulating its modulus.

Keywords

Microchannel, bulging, indentation, modulation of modulus, non-circular contact area

1. Introduction

Nano to micro-scale patterning of the surface of a soft elastomeric layer is important for variety of practical and technological applications, e.g., for creating super-hydrophobic surfaces [1, 2]; generating structural colour [3]; biological [4] and chemical sensors [5]; scaffolds for tissue engineering [6]; micro-electromechanical systems [7]; patterned adhesives [8]; and so on. Many of these applications demand not just patterning of the material surface but hierarchical patterning and in many cases it is not just physical texturing of the material but spatial modulation of one or more physical properties [9] also becomes important. These applications demand novel fabrication methods which can generate patterns with controlled geometric

---

[*] To whom correspondence should be addressed. E-mail: aghatak@iitk.ac.in

Adhesion Aspects in MEMS/NEMS

lengthscales yet economical to implement over a large area. Beyond traditional lithographic routes, several bottom-up approaches have been devised which have several advantages [10–16]: they do not require any mask or any sophisticated and expensive equipment, are not limited by the optical wavelength of light used for developing the photoresist, involve fewer steps and do not require expensive and often toxic chemicals to be handled. Here we present a novel technique in which surface undulations can be created by embedding microchannels in the bulk of an elastic film and filling in these channels with a liquid that wets its surface. Such a liquid forms concave meniscus inside the channel so that pressure in the liquid remains sub-atmospheric. We show that this pressure alters the stress field in the elastic wall around the channel which finally results in buckling of the thin skin above the channel. The liquid pressure alters also the shape of the channel and its cross-sectional area. In fact, the wetted curved surface of the channel increases, releasing interfacial energy which can increase the elastic energy of the layer. The buckling phenomenon leads also to modulation of the shear modulus of the layer which does not remain isotropic but varies along different directions. As a result, when a hemi-spherical indenter is brought in contact with this film the contact area does not remain circular but depending on the orientation of the channel, a near-elliptic contact area appears whose major axis lies along the length of the channel. For films with very thin skin, contact area with sharp corners appears, the height to width ratio (h_c/w_c) of which decreases with increase in the contact load. In fact, h_c/w_c ratio varies differently during loading and unloading signifying different stress profiles in the film in the vicinity of the channel. This simple method can be useful for generating complex patterns in a controlled manner.

2. Experimental

We made thin elastomeric films of crosslinked poly(dimethylsiloxane) (Sylgard 184 elastomer, a Dow Corning product) (PDMS) of shear modulus $\mu = 1.0$ MPa and thickness ranging from $h = 110$–2000 μm. The film remains strongly bonded to the rigid substrate. Microchannels of circular cross sections of diameter $d = 50$–900 μm are embedded at different vertical distances or skin thicknesses $t = 10$–100 μm from the surface of the film [17], as shown in Fig. 1. Channels of circular cross section are prepared by using nylon filaments and steel rods of circular cross sections as templates which are placed in a pool of the pre-polymer liquid mixed with the crosslinker (10:1 by weight). The vertical location of the channels is fixed by using spacers of known heights. The pre-polymer liquid is then crosslinked between two uniformly separated parallel plates: one of the plates is plasma oxidized in order create surface active groups while the other one is coated with a self-assembled monolayer (SAM) of octadecyltrichlorosilane molecules. As a result, after curing the pre-polymer liquid at 80°C for about an hour, the top plate is easily withdrawn while the elastomeric film remains strongly bonded to the bottom plate [18, 19]. The material of construction of the templates is such that it does not adhere to the

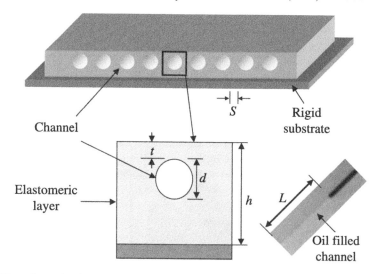

Figure 1. The schematic of a typical elastomeric film of thickness h with multiple, uniformly spaced microchannels of diameter d buried at a vertical distance t from the surface. The film remains bonded to the rigid substrate.

PDMS so that it can be easily withdrawn from the crosslinked film on application of a gentle pull [17]. Straight microchannels, which span through the whole width L of the films are thus generated as monolithic structures inside PDMS elastomeric films.

3. Surface Deformations

These channels are filled with liquids which wet the surface of the channel wall completely, e.g., silicone oils of different viscosities $\eta = 100$–1000 cP, surface tension $\gamma = 20$ mN/m. These liquids fill in the channels by capillary action, therefore no pump is required for this purpose. Figure 2(a) shows the top view of the partially filled channel. The molecular weights of these oils are such that they do not diffuse into the crosslinked network within the time duration of most of our experiments. This was confirmed by placing a block of PDMS in a pool of the oil which did not swell the PDMS even after a very long time. Interestingly, as the liquid fills in the channel, the thin skin covering it spontaneously bulges out forming surface undulations, with maximum height of the undulation δ occurring at the location of least thickness t of the skin. While for lower viscosity of oil, $\eta \sim 100$–400 cP the equilibrium deformation is reached within $\sim$5–10 min, for the ones with higher viscosity, $\eta \sim 400$–4000 cP the film deforms over a longer period of time $\sim$4–5 hours. While this deformation can be felt by sliding a finger over the surface of the film and can be characterized qualitatively by optical profilometry, its cross-sectional image is obtained by cutting a slice of the film using a sharp razor blade and viewing it under an optical microscope. Analysis of the channel cross section by this process requires that after filling it with oil, the channel be permanently set at its

Top view

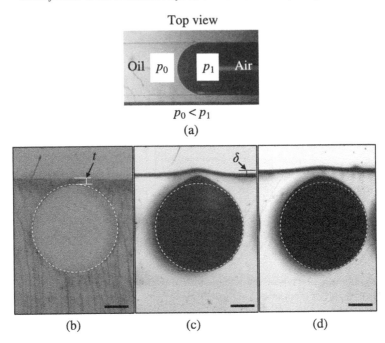

$p_0 < p_1$

(a)

(b)   (c)   (d)

Figure 2. (a) Top view of a microchannel filled with oil shows concave meniscus of oil. (b)–(d) Optical micrographs of the cross section of the film show the bulging effect of the thin skin above the channel. (b) Empty channel embedded inside an elastomeric film at a vertical distance $t = 30$ μm from the surface. (c)–(d) Deformation of the channel cross sections in two different elastomeric films with skin thicknesses $t = 25$ and 40 μm respectively, due to filling with oil. Scale bar represents 200 μm.

deformed form; however, the silicone oil without any functionalized groups in its molecular structure is not amenable to crosslinking. Therefore, in order to examine the deformation of the channel, it is filled with Sylgard 170 elastomer (a Dow Corning product) mixed with the curing agent followed by curing it at ~100°C. Since, viscosity of Sylgard 170 pre-polymer liquid is ~4000 cP, sufficient time is given for the equilibrium deformation to be reached before crosslinking the liquid. In Fig. 2(b)–(d) we present typical cross-sectional views of the films which show an empty channel and its deformed form after it is filled with the oil. Notice that here the elastic film remains bonded to the rigid plate, so that it can deform only on its free surface. However, if we use channel embedded free film, which remains supported only at its edges, bulging occurs on its both surfaces, albeit to different degrees depending on the thickness of the skin.

Optical profilometry of the surface of these films shows that δ increases with diameter d of the channel but decreases with skin thickness t of the film. By carrying out experiments with a large number of films with different channel diameters $d = 50$–900 μm and skin thickness $t = 10$–80 μm, it has earlier been shown that δ scales [18] as $\delta \sim \frac{d}{\sqrt{t}}$. In the absence of any external mechanical field, the bulging possibly occurs because of surface effects, to be specific, because of the wetting of the PDMS surface with oil. Since, the PDMS surface is oleophilic, it leads to negative Laplace pressure at the meniscus, i.e., at the interface of the oil in the channel and atmosphere; as a result, the pressure in oil p_0 decreases below the atmospheric pressure p_1 by $\Delta p = p_1 - p_0 = 4\gamma_1/d$. The deformability of the matrix, i.e., the channel wall, however, brings in complications. For example, the liquid inside the channel is subjected to two opposing forces: the force at the open ends given as $(p_1 - p_0)\pi d^2/4$ which should squeeze in more liquid inside the channel and the force along the length of the channel: $(p_1 - p_0)\pi d L$ which tends to squeeze out the liquid; here L is the length of the channel. Since in experiments, $L \gg d$, it is the latter force that should dominate leading to squeezing out of the liquid, but such an effect would be accompanied also by the bending of the channel wall requiring large elastic energy penalty. In fact, the channel indeed gets compressed, but only very slightly in the portion where wall thickness is infinitely large; however, it bulges out close to the thin skin above the channel.

Figure 3 shows the optical micrograph of the cross section of a PDMS film in which a channel of diameter $d = 710$ μm is embedded to the maximum depth so that the skin thickness is maintained at $t = 50$ μm. The channel is filled with Sylgard 170 pre-polymer liquid mixed with crosslinker and is allowed to cure. While

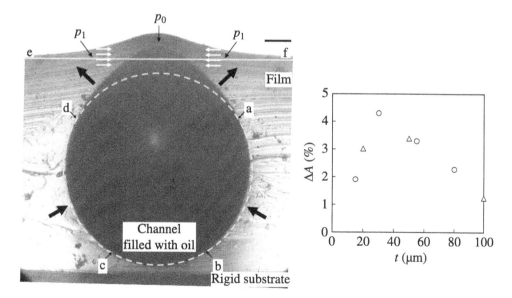

Figure 3. The cross section of an elastomeric film embedded with the microchannel filled with Sylgard 170. The diameter of the channel and the skin thickness are $d = 570$ μm and $t = 50$ μm, respectively. The dotted circle represents the trace of the empty channel and the solid line e–f represents the undeformed free surface of the film. The black arrows indicate that the channel gets compressed along a–b and c–d and it expands along d–a. Along b–c, the channel wall remains unaltered. The scale bar represents 100 μm. Percentage increase in cross-sectional area of the channel is plotted against the thickness t of the thin skin above the microchannel. The symbols (○, △) represent channel diameters $d = 550$ μm and 690 μm, respectively.

the dotted circle represents the trace of the undeformed channel, the black portion represents the deformed cross section of the channel. The figure clearly shows that the channel gets compressed along a–b and c–d; it bulges out along d–a; no significant deformation is observed along b–c. This observation suggests that the differential pressure $p_1 - p_0$ acts differently in different portions of the channel wall. For example, in the thin skin covering the channel, the pressure varies gradually from p_0 at the wall of the channel to p_1 at the PDMS–air interface. Hence, there is a distribution of pressure which may be assumed to be linear, so that, the average pressure at the skin covering the channel $(p_0 + p_1)/2$ is less than the atmospheric pressure. Thus, the skin acts like a thin plate being subjected to a compressive axial stress of $(p_1 - p_0)/2$ which results in its buckling.

Buckling of thin plates and rods follows a set of functional forms known as elliptic curves generated by solving the following generic equation and the corresponding boundary conditions [20]:

$$\frac{d^2\theta}{ds^2} + a\sin\theta = 0; \quad \text{at } s = 0, \theta = \theta_0 \text{ and } \frac{d\theta}{ds} = \frac{d\theta_0}{ds}, \tag{1}$$

where s defines the contour length of the centreline of the plate and $\theta(s)$ defines angle of the tangent to the centreline; parameter a is the ratio of the axial load per unit width of the plate and its bending rigidity, θ_0 and $\frac{d\theta_0}{ds}$ are, respectively, the tangent to the centreline and rate of change of tangent with the contour length. While equation (1) accounts only for small bending of a thin plate, more involved analysis for large bending deformation of a finitely thick elastic block has been derived. However, for the sake of simplicity, we examine if the surface profile of the deformed film matches with that of an elastica in the light of equation (1). In Fig. 4, the traces of the surface profiles for films of thickness $h = 60, 120, 300$ and 500 μm all with embedded channels of diameter $d = 50$ μm are used to fit to equation (1). For all these cases the channels remain maximally buried within the film so that the skin thickness of the film covering the channel increases from $t = 10$ to 450 μm for the above set of films. Curve 1 shows that when the skin thickness is small, e.g., $t = 10$ μm, it bulges out to form a narrow peak which is not captured well by equation (1). However, for films of larger skin thickness, e.g., $t = 70, 250$ and 450 μm, the deformation profiles are represented reasonably well by equation (1) as shown by curves 2–4 implying that the surface deformation of films may indeed occur *via* buckling of the skin. While equation (1) approximates the buckling phenomenon rather well, it does not incorporate the effect of thickness of the thin skin which in the present problem does not remain uniform but varies spatially. This variation becomes more prominent for thinner skins, and as a result equation (1) fails to capture the corresponding optical profiles.

4. Deformation of Microchannel

Equation (1) is not adequate for another reason: close examination of the channel cross section suggests that it not only deforms from being circular but its perimeter

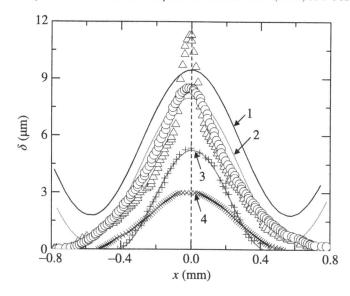

Figure 4. Optical profilometry traces the surface of the film showing undulations in the vicinity of the channel. Symbols △, ○, + and × represent experiments in which channels of diameter $d = 50$ μm are embedded in films of thickness $h = 60, 120, 300$ and 500 μm, respectively. Curves 1–4 show the trace of deformation of these different films fitted to elliptic equations.

also changes, contrary to what is expected of pure buckling for which the contour length of the plate or, in this case, the thin skin should remain unaltered. The perimeter can increase or decrease depending on the skin thickness and the diameter of the channel. Concomitant to alteration of the perimeter, the cross-sectional area of the channel changes too which implies that some quantity of liquid is either squeezed in or out because of wetting of the channel. The plot in Fig. 3 shows the percentage change in area $\%\Delta A = (\Delta A / A_0) \times 100$ for films with microchannels of diameter $d = 550$ μm and skin thickness $t = 15$–80 μm, and $d = 700$ μm and $t = 20$–50 μm. $\%\Delta A$ increases with increase in skin thickness until an intermediate thickness is reached, beyond which it decreases. The plot further shows that $\%\Delta A$ values for different channel diameters follow similar trend implying that extent of area expansion is possibly determined solely by the thickness of the skin above the channel. One consequence of increase in channel area is that it increases also the perimeter of the channel. For example, percentage increase of the perimeter of the channel can be deduced as, $\%\Delta l \sim \%\Delta A / 3$ which when wetted by the liquid results in excess interfacial energy $(\gamma_s + \gamma_l - \gamma_{sl})\Delta l$ per unit length of the channel. Here $\gamma_s = \gamma_l = 22$ mJ/m^2 is the surface energy of PDMS surface and silicone oil respectively and $\gamma_{sl} = 0$ mJ/m^2 is the interfacial energy of the wetted interface; Δl represents extension in perimeter of the microchannel. Since, the perimeter multiplied by the length of the channel yields its internal surface area, Δl represents also the increase in curved surface area of the channel per unit length. This excess energy can contribute to the elastic energy associated with the bulging deformation of the channel wall. Thus finite positive values of $\% \Delta A$ suggest that bulging deformation of the thin skin occurs not only due to the Laplace pressure at the meniscus of the liquid inside the channel but also due to conversion of surface energy to the elastic energy of the channel wall.

5. Contact Mechanics Experiment

Besides causing surface deformation, the liquid filled sub-surface microchannels alter also the effective local modulus of the elastic film in the vicinity of the channels. While it is expected that the presence of the buried microchannel should decrease the local modulus of the film enhancing its deformability, the bulging induced by the liquid inside the channel can further amplify it. The alteration in local deformability can be probed by contact mechanics experiments as presented in Fig. 5. Here the elastic film remains strongly bonded to the rigid substrate and a soft hemispherical indenter made of PDMS of radius of curvature $R = 1.442$ mm is pressed against it in displacement controlled mode. When the indenter is loaded against the smooth portion of the film, devoid of any subsurface microstructure, a circular contact area appears the diameter of which increases with the depth of indentation. However, when indented at the location of the sub-surface microchannels filled with liquid, a non-circular contact area appears, the geometric shape and size of which vary with the skin thickness above the channel and the depth of indentation. Figure 6 shows the optical micrographs of the contact area in experiments in which

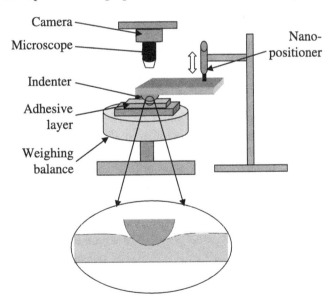

Figure 5. Schematic of the experiment in which an indenter is brought in contact with the film in a displacement controlled mode using a nano-positioner. The corresponding load is measured using a weighing balance interfaced with a computer and the contact area is visualized with a microscope fitted with a digital camera.

Microchannel

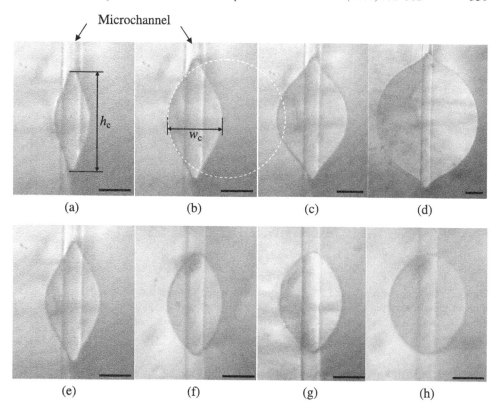

Figure 6. Optical micrographs show the contact area of the indenter and the elastic film with a single microchannel of diameter $d = 50$ μm embedded inside. The channel is filled with silicone oil of viscosity 380 cP and surface tension 20 mN/m. Micrographs (a)–(d) represent the experiments in which we subjected a film with skin thickness $t = 5$ μm to indentation depths $\delta = 6, 11, 21, 101$ μm, respectively. Micrographs (e)–(h) represent films with skin thickness $t = 5, 35, 45, 80$ μm, respectively; the indentation depth is maintained at $\delta = 11$ μm. The scale bar is 100 μm.

the hemispherical lens is symmetrically aligned with sub-surface channels of diameter $d = 50$ μm buried to different skin thicknesses $t = 50$–100 μm while the depth of indentation δ is varied from 6 to 100 μm. The series of optical micrographs show that the presence of the microchannel alters the local mechanics as a result of which non-circular contact area with narrow corners appears. For smaller indentation depths, the effect of the channel becomes more pronounced than that of the smooth portion of the film: as a result, the height of the contact is significantly larger than its width. With increase in the indentation depth, the fraction of contact with the smoother portion of the film increases and as a result the h_c/w_c ratio decreases. In Fig. 7, we plot h_c/w_c as a function of δ_i/d for films in which the skin thickness varies: $t = 5$–25 μm. A large value of h_c/w_c ratio shows that the film is more deformable along the axial direction of the channels than perpendicular to it. We can rationalize this observation by considering the JKR (Johnson, Kendall and Roberts) theory [21] of adhesion which suggests that for contact of a hemi-

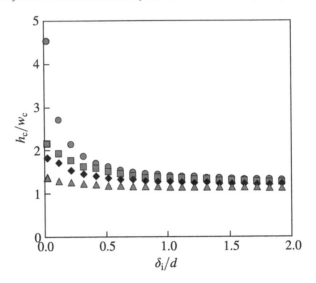

Figure 7. The ratio of height and width, h_c/w_c as obtained from optical micrographs in Fig. 6 for films of various skin thicknesses: $t = 5, 25, 45$ and 80 μm (represented by symbols ●, ■, ◆ and ▲, respectively) is plotted with respect to depth of indentation normalized with the diameter of the microchannels δ_i/d.

spherical indenter with a semi-infinite elastic half-space, at zero contact load, the radius of the contact area varies with effective elastic modulus E of the adhering bodies as $a = E^{-1/3}$. Hence for two different effective elastic moduli, the ratio of contact area is $a_1/a_2 = (E_2/E_1)^{1/3}$. Drawing analogy to this problem, the effective shear modulus of the film in the vicinity of the channel can be obtained in terms of the ratio $h_c/w_c : E_h = E_w(w_c/h_c)^3$, where E_h is the effective shear modulus of the layer in the vicinity of the channel for contact along the direction of the axis of the channel and E_w is the same in a direction perpendicular to it. Using representative numbers, e.g., $(h_c/w_c) \sim 4.5$ for skin thickness $t = 5$ μm, E_h is estimated to be $E_w/1000$. The ratio h_c/w_c decreases asymptotically to 1 with increase in thickness of the skin, and thus, the ratio E_h/E_w increases with increase in the skin thickness. A more exact theoretical analysis will be required to predict correctly the effective elastic modulus of the film in the vicinity of the channels.

6. Summary

We have presented here an example of wetting induced deformation of a soft elastomeric substrate in which surface energy gets converted to elastic energy. In particular, we have shown that the soft, thin wall of a microchannel, embedded inside a soft, flat and elastic layer, e.g., PDMS bulges out when the channel is filled with a liquid, e.g., silicone oil which wets its surface. But the same does not happen when a non-wetting liquid like water is used, as it does not fill in the channel spontaneously. In the context of wetting liquid, the bulging occurs by the interplay of Laplace pressure at the meniscus of the liquid and the excess interfacial energy due to wetting; coupling of these two effects leads to buckling of the thin skin in the vicinity of the channel with concomitant change in the shape of the channel cross section and the channel area. These effects become more pronounced for films with thinner skin; however, beyond an optimum thickness these effects diminish, implying that the deformation is maximally pronounced for film with intermediate skin thickness. The local deformation of the film results also in spatial modulation of the shear modulus and consequent deformability of the layer. It appears that the local deformability increases at the location of the liquid filled channels as observed in contact experiments in which indentation using a spherical contactor does not produce a circular area of contact at the interface but an elongated one with narrow corners oriented along the length of the channel. The contact area becomes more and more circular with increase in depth of indentation implying more pronounced effect of the flat portion of the layer. This phenomenon may be important for patterning of soft surfaces, chemo-mechanical signal transduction and so on.

Acknowledgements

AG acknowledges the research grants of The Department of Science and Technology, India for this work.

References

1. W. Barthlott and C. Neinhuis, *Planta* **202**, 1 (1997).
2. J.-Y. Shiu, C.-W. Kuo and P. Chen, *Proc. SPIE* **5648**, 325 (2005).
3. Z.-Z. Gu, H. Uetsuka, K. Takahashi, R. Nakajima, H. Onishi, A. Fujishima and O. Sato, *Angew. Chem. Int. Ed.* **42**, 894 (2003).
4. N. A. Peppas, J. Z. Hilt, A. Khademhosseini and R. Langer, *Adv. Mater.* **18**, 1345 (2006).
5. H.-J. Galla, *Angew. Chem. Int. Ed.* **31**, 45 (1992).
6. R. Langer and J. P. Vacanti, *Science* **260**, 920 (1993).
7. P. L. Burn, A. Kraft, D. R. Baigent, D. D. C. Bradley, A. R. Brown, R. H. Friend, R. W. Gymer, A. B. Holmes and R. W. Jackson, *J. Am. Chem. Soc.* **115**, 10117 (1993).
8. A. Ghatak, L. Mahadevan, J. Y. Chung, M. K. Chaudhury and V. Shenoy, *Proc. R. Soc. Lond. A* **460**, 2725 (2004).
9. S. Okazaki, *J. Vac. Sci. Technol. B* **9**, 2829 (1991).
10. Y. Xia, E. Kim, X.-M. Zhao, J. A. Rogers, M. Prentiss and G. M. Whitesides, *Science* **273**, 347 (1996).
11. K. Y. Suh, Y. S. Kim and H. H. Lee, *Adv. Mater.* **13**, 1386 (2001).
12. E. Kim, Y. Xia and G. M. Whitesides, *Nature* **376**, 581 (1995).
13. D. J. Elliot, *Integrated Circuit Mask Technology.* McGraw-Hill, New York (1985).
14. G. Reiter, *Phys. Rev. Lett.* **68**, 75 (1992).
15. A. Sharma and G. Reiter, *J. Colloid Interface Sci.* **178**, 383 (1996).
16. G. Reiter, R. Khanna and A. Sharma, *Phys. Rev. Lett.* **85**, 1432 (2000).
17. M. K. S. Verma, A. Majumder and A. Ghatak, *Langmuir* **22**, 10291 (2006).
18. A. Majumder, A. Ghatak and A. Sharma, *Science* **318**, 258 (2007).

19. A. Ghatak, L. Mahadevan and M. K. Chaudhury, *Langmuir* **21**, 1277 (2005).
20. A. E. H. Love, *Mathematical Theory of Elasticity*. Dover Publishing Ltd., New York (1944).
21. K. L. Johnson, K. Kendall and A. D. Roberts, *Proc. R. Soc. Lond. A* **324**, 301 (1971).

Fabrication of Novel Superhydrophobic Surfaces and Water Droplet Bouncing Behavior — Part 1: Stable ZnO–PDMS Superhydrophobic Surface with Low Hysteresis Constructed Using ZnO Nanoparticles

Bin-Bin Wang [a], **Jiang-Tao Feng** [a], **Ya-Pu Zhao** [a,*] **and T. X. Yu** [b]

[a] State Key Laboratory of Nonlinear Mechanics, Institute of Mechanics,
Chinese Academy of Sciences, Beijing 100190, China
[b] Department of Mechanical Engineering, Hong Kong University of Science and Technology,
Clear Water Bay, Kowloon, Hong Kong SAR, China

Abstract

A superhydrophobic surface has many advantages in micro/nanomechanical applications, such as low adhesion, low friction and high restitution coefficient, etc. In this paper, we introduce a novel and simple route to fabricate superhydrophobic surfaces using ZnO nanocrystals. First, tetrapod-like ZnO nanocrystals were prepared *via* a one-step, direct chemical vapor deposition (CVD) approach. The nanostructured ZnO material was characterized by scanning electron microscope (SEM) and X-ray diffraction (XRD) and the surface functionalized by aminopropyltriethoxysilane (APS) was found to be hydrophobic. Then the superhydrophobic surface was constructed by depositing uniformly ZnO hydrophobic nanoparticles (HNPs) on the Poly(dimethylsiloxane) (PDMS) film substrate. Water wettability study revealed a contact angle of $155.4 \pm 2°$ for the superhydrophobic surface while about $110°$ for pure smooth PDMS films. The hysteresis was quite low, only $3.1 \pm 0.3°$. Microscopic observations showed that the surface was covered by micro- and nano-scale ZnO particles. Compared to other approaches, this method is rather convenient and can be used to obtain a large area superhydrophobic surface. The high contact angle and low hysteresis could be attributed to the micro/nano structures of ZnO material; besides, the superhydrophobic property of the as-constructed ZnO–PDMS surface could be maintained for at least 6 months.

Keywords

Superhydrophobic surface, contact angle hysteresis, ZnO, PDMS, surface modifications

Notations

R radius of contact area

δ average distance between liquid and solid molecules

---

* To whom correspondence should be addressed. Fax: +86-10-6256-1284; e-mail: yzhao@imech.ac.cn

ε restitution coefficient

θ contact angle

γ surface/interface tension

σ line tension

Subscripts

L liquid

S solid

s stable

V vapor

List of Abbreviations

APS aminopropyltriethoxysilane

CA contact angle

CAH contact angle hysteresis

CD constant diameter

CVD chemical vapor deposition

HNPs hydrophobic nanoparticles

MEMS microelectromechanical systems

NEMS nanoelectromechanical systems

PDMS poly(dimethylsiloxane)

SCCM standard cubic centimeter per minute

SEM scanning electron microscope

XRD X-ray diffraction

1. Introduction

Numerous micro/nanotribological and micro/nanomechanical applications, such as in micro/nanoelectromechanical systems (MEMS/NEMS) require surfaces with low adhesion and friction [1]. As the size of these devices decreases, the surface forces tend to dominate over the volume forces, and adhesion and 'stiction' constitute a challenging problem for proper operation of these devices. This makes the development of nonadhesive surfaces crucial for many of these applications. With this background, the so called 'superhydrophobic' or 'ultrahydrophobic' surfaces and their corresponding properties have attracted the attention of many researchers all over the world. Superhydrophobic surfaces have been applied in MEMS field especially in electrowetting research. Kakade *et al.* [2] fabricated superhydrophobic multiwalled carbon nanotube bucky paper showing fascinating electrowetting behavior. The droplet behavior can be reversibly switched between superhydrophobic Cassie–Baxter state to hydrophilic Wenzel state by the application of an electric field, especially below a threshold value. Bahadur and Garimella analyzed the influence of applied voltage in determining and altering the state of a static droplet resting on a superhydrophobic surface [3]. In general, surfaces with a static contact angle (CA) higher than 150° are defined as superhydrophobic surfaces [4–8]. Large CA or limited contact area reduces the adhesion or friction between liquid droplets and solid surfaces. Thus the CA is a measure of adhesion between water and solid surface. However, in some cases, contact angle hysteresis (CAH) is more important than maximum CA since CAH is directly related to driving force for a liquid drop [9]. For example, liquid flow requires low solid–liquid friction. That is, in addition to high contact angle, a superhydrophobic surface should also have very low water CAH. The condition of CAH < 10° may be a good suggestion for claiming superhydrophobicity [6, 10].

A third wetting characteristic for a superhydrophobic surface, in addition to a high CA and low CAH, has been proposed by some researchers [11, 12], i.e., the ability of the surface to bounce off droplets. It has been shown that this ability is related to energy barriers associated with the transition between the Cassie [13] and Wenzel [14] wetting states. From energy consideration, the Cassie–Baxter state is often metastable. In practice, this means that the drop will remain in a Cassie–Baxter state only if it is subjected to small external perturbations. However, as for the definition of a superhydrophobic surface, there is no consensus whether restitution coefficient ε should be considered. Furthermore, ε first increases with increasing velocity, and then stays at a constant value (stable restitution coefficient ε_s) almost independent of the velocity. Thus, $\varepsilon_s > 0.8$ may be a good suggestion from the viewpoint of rebound of droplets from a superhydrophobic surface.

However, the static CA of a superhydrophobic surface is higher than 150° generally, but the impact of hysteresis is still controversial. And the ability of a surface to bounce off droplets has received relatively less attention. In the current decade, there has been an explosion of publications describing how surface topography can be used to control wettability. A large number of ways to produce superhydrophobic surfaces have been investigated using many and different materials [15]. Thus, zinc oxide (ZnO), an excellent candidate for the fabrication of electronic and optoelectronic nanodevices, was also studied. Many efforts have been focused on the fabrication of ZnO based superhydrophobic films. Badre *et al.* successfully prepared ZnO films with well-controlled morphologies by electrochemical deposition [16]. In their work, a seed layer of nanocrystallites of ZnO was prepared from which ZnO nanowires were grown. Then, a treatment with alkylsilane yielded superhydrophobic surfaces. Badre *et al.* [17] also prepared highly water-repellent surfaces from arrays of ZnO nanowires by treatment with stearic acid. They attributed the superhydrophobic properties to a micro-nano binary structure combined with a low surface free energy. Berger *et al.* [18] fabricated ZnO crystals arrays by employing hydrothermal approach and the hydrophobicity of the films could be regulated by regulating zinc ion precursor concentration. The CA on these surfaces increased from 110° to 156° with decreasing concentration of zinc ion precursor, demonstrating controllable wetting behavior. Liu *et al.* [19] fabricated ZnO films by Au-catalyzed chemical vapor deposition (CVD) method. The surface of as-synthesized film exhibited hierarchical structure with sub-microstructures with CA about 164.3°, and UV illumination could switch the surface from hydrophobic to hydrophilic. Another way to synthesize a ZnO film is direct oxidation of Zn substrate. Using this method, Hou *et al.* [20] prepared the ZnO film on zinc layer successfully, and the surface was hydrophobically functionalized with n-octadecyl thiol. The modified ZnO film exhibited superhydrophobicity and the water contact angle was $153 \pm 2°$.

The methods used above have their own drawbacks. The hydrothermal method and electrochemical deposition are relatively complex and difficult to apply for mass production. As for the CVD method [19] mentioned above, a catalyst is needed for the growth of ZnO film which means the catalyst layer should be deposited on the substrate in advance. As for direct oxidation method, a Zn seed layer is needed to produce a superhydrophobic surface. Thus the area of as-fabricated surface is limited to the area of seed layer. In this paper, we propose a simple method to fabricate superhydrophobic surfaces by ZnO nanomaterials. The synthesis of ZnO nanomaterials was carried using CVD method. Also in this study, we combined CVD method and surface modification process to fabricate ZnO-based superhydrophobic surfaces, which opens a new potential application for ZnO nanomaterials synthesized by CVD method. Compared to other ways to construct ZnO hydrophobic surfaces, our method is convenient and can be used to fabricate large area surfaces. The contact angle is $155.4 \pm 2°$, and the hysteresis of the superhydrophobic surface is quite low ($3.1 \pm 0.3°$), which means that the surface has self-cleaning property. Besides, the superhydrophobic properties are stable and durable for at least 6 months.

2. Experimental

2.1. Preparation of ZnO Powder

The nanostructured ZnO material was synthesized *via* a simple catalyst-free CVD method [21–23]. The main parameters of CVD method to synthesize ZnO nanomaterials are temperature and gas flow rate. Temperature should be higher than the melting point of Zn (419.6°C) and thus Zn would be in liquid state. So Zn vapor could react with O_2 to form nanostructured ZnO. However, the temperature cannot be too high, otherwise the reaction rate would be too fast and regular ZnO nanostructures would not be found. Gas flow rate is another parameter, as a low O_2 partial pressure could also limit the react rate. In the experiments mentioned in this paper, 640°C and 20 SCCM O_2 flow were found to be proper conditions to synthesize regular ZnO nanotetrapods. The process used was as follows. First, about 2.0 g Zn powder (99.999%) was evenly dispersed on the bottom of an alumina boat which was placed in the middle of a quartz tube. The tube was placed horizontally in a high-temperature tubular furnace. Then, a silicon substrate was placed face-down on the boat. The materials were heated at a moderate rate of about 30°C min^{-1}. When the desired temperature (640°C) was reached, Ar at a flow rate of about 300 SCCM (SCCM denotes standard cubic centimeter per minute at STP) and O_2 at about 20 SCCM were introduced into the system and maintained for 20 min. Finally, the flows of Ar and O_2 were shut down. White ZnO material was collected on the surface of the substrate after the furnace was cooled naturally down to room temperature.

2.2. Surface Modification of ZnO Powder and Fabrication of Superhydrophobic Surface

The micro/nano scale ZnO powder synthesized by CVD method usually agglomerated, thus the material needed to be treated to obtain dispersed particles. First, the as-synthesized ZnO powder was dispersed in ethyl alcohol which was contained in a beaker. Then the suspension was treated with supersonic waves for about 20 min, and the upper suspension was collected in another beaker. This process was repeated four times in order to avoid large particles. To prevent the strong aggregation effect of ZnO nano-particles, aminopropyltriethoxysilane (APS) was added to the suspension [22, 23]. The mass ratio of ZnO, alcohol and APS was 1:100:0.1. Then the suspension was dried in a drying oven and the as-treated ZnO material was ground into a powder. In this process, the APS was used to weaken the agglomeration tendency as well as to enhance the hydrophobicity of the ZnO particles.

The superhydrophobic surface was fabricated on a poly (dimethylsiloxane) (PDMS) film. First, the silicone elastomer base and curing agent were mixed uniformly and the mass ratio was 10:1. PDMS (SYLGARD) and curing agent were procured from Dow Corning Corporation, Midland, Michigan, USA. The mixture of PDMS and curing agent was stirred adequately. The modified ZnO particles were dispersed on the half-dry PDMS film evenly, and then the film was placed in the oven again until solidification was completed. The superhydrophobic ZnO–PDMS surface formed after rinsing the film with deionized water to remove loose ZnO particles.

2.3. Characterization

The morphologies of ZnO powder samples and the ZnO–PDMS surface were characterized with a scanning electron microscope (SEM) (FESEM FEI SIRION) and an optical microscope DZ3 (Union, Japan), respectively. The crystalline structure of the ZnO material obtained was examined by X-ray powder diffraction (XRD) on an X-ray diffractometer D/MAX-2500 (Rigaku, Japan). Furthermore, X-ray photoelectron spectroscopy (XPS) was employed to investigate the surface properties of ZnO particles. Water contact angle characterization of ZnO samples was carried out using an OCA 20 (Dataphysics, Germany) contact angle analyzer.

3. Results and Discussion

3.1. Structure and Morphology Characterization

The SEM micrographs of as-synthesized ZnO material are shown in Fig. 1(f) and 1(g). The ZnO material is found to be tetrapod-like, and the legs are 4–10 μm long. The diameter of each leg is about 200–500 nm. The micrograph with larger magnification in Fig. 1(g) shows that the ZnO particles have submicrometer structures.

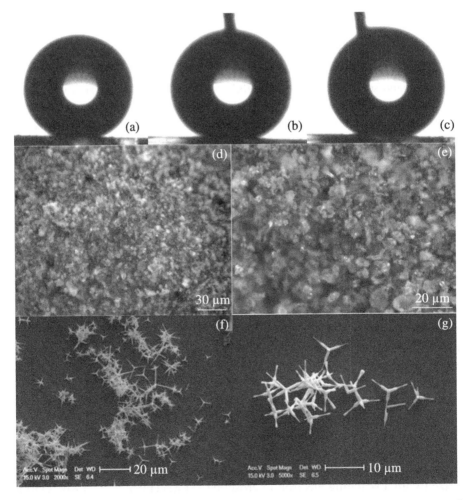

Figure 1. (a)–(c) ZnO–PDMS surface: (a) static contact angle, (b) advancing angle, and (c) receding angle. (d)–(e) Surface morphology of ZnO–PDMS surface constructed by coating PDMS film with ZnO particles. (f)–(g) SEM micrographs of as-synthesized ZnO particles with tetrapod shape.

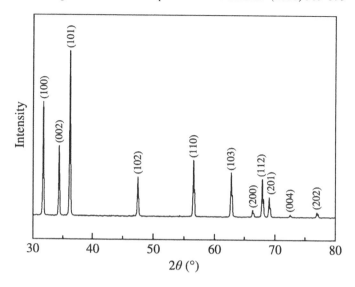

Figure 2. X-ray diffraction pattern of ZnO.

The crystal structure of the ZnO tetrapods was examined with XRD as shown in Fig. 2. The sharp and strong peaks suggested that the products were highly crystalline. All these XRD peaks could be indexed as the hexagonal wurtzite ZnO, with the lattice constants $a = 3.253$ Å and $c = 5.209$ Å. No peaks of zinc or other impurities were found in the pattern, which showed that the tetrapods were rather pure ZnO wurtzite.

The pictures of the superhydrophobic surface taken by an optical microscope are shown in Fig. 1(d) and 1(e). It was found that the tetrapod-like structure of ZnO particles was destroyed during surface modification process, and only a few particles remained as tetrapods in shape. The distribution and outline of the particles are random and irregular, and the diameters of particles range from 3.8 to 10.3 µm. Furthermore, the distance between the particles was also about several micrometers, thus the roughness is considered to be of the same magnitude.

3.2. Floating Test and XPS Examination

In order to study the surface characteristics, non-modified and APS-modified ZnO particles were analyzed by floating test measurements [20]. The floating test was used to measure the ratio of the floated product to the overall weight of the sample after it was mixed in water and stirred vigorously. The ratio above was called the active ratio. Without the addition of silane, the ZnO particles obtained were hydrophilic, and the active ratio was 0.0%. When the ZnO particles were treated with APS, the active ratios of all the samples were above 90.5%. The floating test demonstrates that the hydrophobic organic (APS) molecules had been bonded to the surfaces of the obtained modified ZnO particles, as the hydrophobic property of modified ZnO particles remained even after the particles were washed with hot water and alcohol. Figure 3(a)–(c) shows XPS spectra for the as-treated ZnO mate-

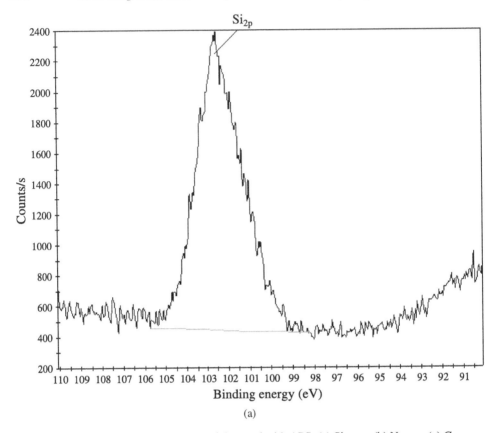

Figure 3. XPS spectra for the ZnO material treated with APS: (a) Si scan. (b) N scan. (c) C scan.

rial. The results indicate that APS is coated on the surface of ZnO particles. Hence, the APS could reduce aggregation and enhance the hydrophobicity of the particles effectively.

3.3. Wettability Study of the Superhydrophobic Surface

The hydrophobic property of the fabricated surface was investigated by contact angle measurements with a precision of ±0.1°, and the smooth PDMS surface was also studied for comparison. Advancing and receding contact angles were measured by adding to and withdrawing liquid from the droplet, respectively. CAs were recorded after they became stable and were checked on several films. The volume of water droplets used for CA measurements was 3.65 µl, and the diameter was about 0.98 mm. Some typical results on ZnO–PDMS surface are shown in Fig. 1(a)–(c). It was found that the contact angle was only 114.5° for smooth PDMS surface, while the contact angle of the superhydrophobic surface was about 155.4 ± 2°. The advancing and receding angles of PDMS surface were 119.6° and 102.2°, respectively. The CAH of the superhydrophobic surface is very low (3.1 ± 0.3°), and the droplet could hardly stay on the surface during the experiment. Compared to smooth

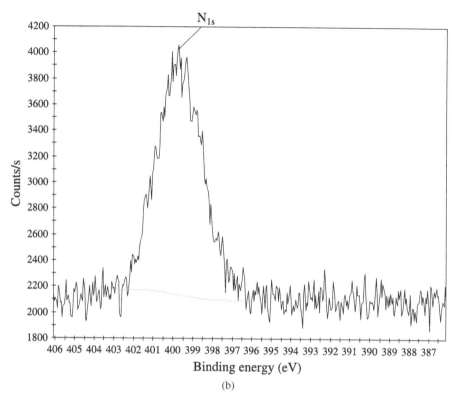

Figure 3. (Continued).

PDMS surface, the superhydrophobic surface constructed by ZnO nanocrystals has much larger CA and smaller CAH. For comparison, the wettability of flat Si substrate modified by APS was studied and CA was found to be 11.9°, which showed that the APS formed a hydrophilic coating. Besides, the superhydrophobicity of ZnO–PDMS surface could be maintained for at least 6 months without any change. Furthermore, we have tested several kinds of silanes including methyltriethoxysilane and ethyltriethoxysilane hoping to find the differences or trends between the three silanes. However, the differences were found on a flat substrate but no significant differences were observed on fabricated superhydrophobic surfaces using the same technique (less than 5°). So only the APS experiments were shown in this paper.

3.4. Dependence of Contact Angle on Droplet Diameter

The static CAs of the droplets with different sizes were measured experimentally and the size dependence of contact angle was observed in our solid–liquid system. As shown in Fig. 4, the CA decreases from 155.9° to 134.2° while the size of droplet decreases due to evaporation. The diameters for the droplet are 1.72 mm, 1.55 mm, 1.42 mm, 1.30 mm, 1.22 mm, 1.09 mm, 0.97 mm, 0.82 mm, 0.66 mm and 0.53 mm, respectively, and the corresponding contact angles are

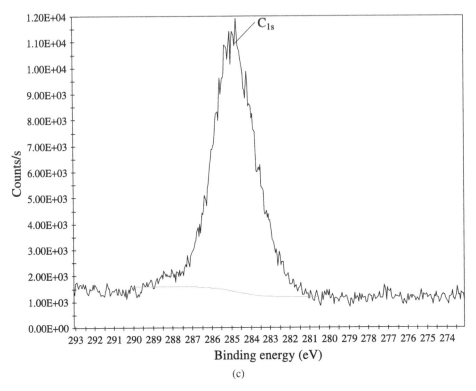

(c)

Figure 3. (Continued).

155.9°, 151.6°, 150.4°, 150.6°, 149.9°, 147.3°, 145.2°, 142.6°, 136.2° and 134.2°. The contact angle varies with the drop size from a few degrees up to 8.6°. This phenomenon has been studied by some researchers, and it was found that droplet follows the evaporation model normally ascribed to hydrophobic surfaces [24]. The size dependence effect could be explained in terms of the so-called line tension, or the tension at the three-phase contact line [25, 26]. The line tension σ, in analogy to surface tension, is defined as the specific free energy of the three-phase contact line or, mechanically, as a force operating at the three-phase line.

Taking line tension into account, the Young's equation can be modified as:

$$\gamma_{LV} \cos\theta = \gamma_{SV} - \gamma_{SL} - \frac{\sigma}{R}, \tag{1}$$

where R denotes the radius of contact area and θ denotes the contact angle.

From equation (1), we learn that the contact angle should increase with increasing drop diameter when the line tension σ is negative, and decrease when σ is positive. For the calculation of line tension, Marmur gave an approximate expression [27]:

$$\sigma \approx 4\delta\sqrt{\gamma_{SV}\gamma_{LV}} \cot\theta, \tag{2}$$

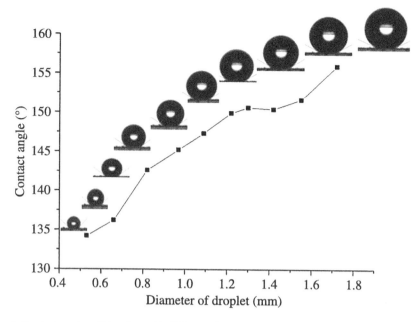

Figure 4. Contact angles of droplets with different diameters: (from right to left) the contact angles are 155.9°, 151.6°, 150.4°, 150.6°, 149.9°, 147.3°, 145.2°, 142.6°, 136.2° and 134.2°, respectively, and the diameters of droplets are 1.72 mm, 1.55 mm, 1.42 mm, 1.30 mm, 1.22 mm, 1.09 mm, 0.97 mm, 0.82 mm, 0.66 mm and 0.53 mm, respectively.

where δ denotes the average distance between liquid and solid molecules. Thus the line tension is negative for an obtuse CA, while σ is positive for an acute CA. As the CA was 134.2°–155.9°, so the line tension was negative in our experiment. Thus the CA should be an increasing function of droplet size, which is consistent with our experimental results.

3.5. Mechanism of Superhydrophobicity

The superhydrophobic properties of the surface are demonstrated by the CAs as shown in Fig. 1(a)–(c). Especially, the hysteresis is quite low, which means that the droplet could move on the surface very easily and the surface is self-cleaning. We attribute the superhydrophobic property to the micro/nano structure of the ZnO particles in the coating. As shown in Fig. 1(f) and 1(g), the as-synthesized ZnO material has micro/submicro structure. And Fig. 1(d) and 1(e) shows that the microparticles are about 1–10 μm in length, thus the roughness of the surface is considered to be of the same magnitude. The micro/nano structures lead to the porous structure of the coating and there is plenty of trapped air in the ZnO coating when the liquid droplet is placed on the surface. In such situation, the hydrophobic behavior relies on Cassie's law. Compared with Wenzel state, a superhydrophobic surface with Cassie state usually has larger contact angles and lower hysteresis. As for the superhydrophobic surface discussed here, the CA and CAH are 155.4 ± 2° and

$3.1 \pm 0.3°$, respectively, which indicates that both adhesion and friction are relatively low. These two properties explain the fact that the droplet could hardly stay on the surface during the experiments.

Furthermore, APS was used to enhance the hydrophobicity and reduce aggregation of ZnO particles. APS has a flexible C4 hydrocarbon chain with one amino group that stretches out in a long zig-zag fashion to form a dense self-assembled layer of packed chains on ZnO as a result of the strong chelating bonds between triethoxy headgroups and Zn atoms on the surface. It is assumed that the layer of packed chains enhanced the hydrophobicity of ZnO particles. However, ASP by itself forms a 'hydrophilic' coating. As mentioned before, there were many trapped air bubbles between the liquid droplet and ZnO particles when the droplet was placed on the surface. Thus, the wettability of ZnO particles has less effect on the apparent CA compared to the porous structure. In a word, both the porous structure and the silane coating lead to the superhydrophobicity of the surface: CA of ZnO surface consisting of micro/nano particles is about $155.4 \pm 2°$ and the CAH is about $3.1 \pm 0.3°$.

However, from energy consideration, the Cassie state is usually metastable and there is energy barrier between Cassie state and Wenzel state. The transition between these two states could occur when fluctuation was introduced in the system. Another parameter, i.e., the energy barrier is associated with the ability of droplet to bounce off the surface [28], has been proposed as the third property of a superhydrophobic surface by some researchers [12]. The impact experiment between droplet and ZnO–PDMS surface and the influence of CAH on droplet dynamic behavior will be discussed separately.

4. Conclusions

A superhydrophobic ZnO–PDMS surface has been fabricated by depositing modified nanostructured ZnO material on the PDMS surface, which is quite convenient compared to other methods. In the experiments, silane was used to avoid the aggregation as well as to enhance the hydrophobicity of ZnO particles. The porous structure of the ZnO layer leads to its superhydrophobicity. Wettability study showed that the CA and CAH were $155.4 \pm 2°$ and $3.1 \pm 0.3°$, respectively. Besides, the superhydrophobicity could be maintained for at least six months without any change. The dependence of contact angle on droplet size was found on as-fabricated superhydrophobic surface, and showed that the static contact angle increased with increasing droplet size. The line tension theory gives a good explanation for this phenomenon. As both ZnO and PDMS have good biocompatibility, the ZnO–PDMS superhydrophobic surface is an interesting subject for further exploration and for possible applications in advanced semiconductor biological devices.

Acknowledgements

This work was jointly supported by the National High-tech R&D Program of China (863 Program, Grant No. 2007AA021803), National Basic Research Program of China (973 Program, Grant No. 2007CB310500), and National Natural Science Foundation of China (NSFC, Grant Nos 10772180, 60936001 and 10721202).

References

1. Y. P. Zhao, L. S. Wang and T. X. Yu, *J. Adhesion Sci. Technol.* **17**, 519 (2003).
2. B. Kakade, R. Mehta, A. Durge, S. Kulkarni and V. Pillai, *Nano Lett.* **8**, 2693 (2008).
3. V. Bahadur and S. V. Garimella, *Langmuir* **23**, 4918 (2007).
4. L. Gao and T. J. McCarthy, *Langmuir* **23**, 3762 (2007).
5. S. L. Ren, S. R. Yang, Y. P. Zhao, T. X. Yu and X. D. Xiao, *Surface Sci.* **546**, 64 (2003).
6. S. L. Ren, S. R. Yang and Y. P. Zhao, *Langmuir* **20**, 3061 (2004).
7. A. Lafuma and D. Quere, *Nature Mater.* **2**, 457 (2003).
8. J. Genzer and K. Efimenko, *Biofouling* **22**, 339 (2006).
9. J. T. Feng, F. C. Wang and Y. P. Zhao, *Biomicrofluidics* **3**, 022406 (2009).
10. B. Balu, V. Breedveld and D. W. Hess, *Langmuir* **24**, 4785 (2008).
11. S. Wang and L. Jiang, *Adv. Mater.* **19**, 3423 (2007).
12. M. Nosonovsky and B. Bhushan, *J. Phys.: Condens. Matter* **20**, 395005 (2008).
13. R. N. Wenzel, *Ind. Eng. Chem.* **28**, 988 (1936).
14. A. B. D. Cassie, *Trans. Faraday Soc.* **44**, 11 (1948).
15. A. Carré and K. L. Mittal (Eds), *Superhydrophobic Surfaces.* VSP/Brill, Leiden (2009).
16. C. Badre, T. Pauporte, M. Turmine, P. Dubot and D. Lincot, *Physica E* **40**, 7 (2008).
17. C. Badre, T. Pauporte, M. Turmine and D. Lincot, *Superlattice Microstructures* **42**, 99 (2007).
18. J. Berger, B. Englert, L. B. Zhu and C. P. Wong, in: *Proc. IEEE International Symposium and Exhibition on Advanced Packaging Materials and Processes*, Atlanta, GA, p. 93 (2006).
19. H. Liu, L. Feng, J. Zhai, L. Jiang and D. B. Zhu, *Langmuir* **22**, 5659 (2006).
20. X. M. Hou, F. Zhou, B. Yu and W. M. Liu, *Mater. Sci. Eng. A* **452**, 732 (2007).
21. F. Q. He and Y. P. Zhao, *Appl. Phys. Lett.* **88**, 193113 (2006).
22. F. Q. He and Y. P. Zhao, *J. Phys. D: Appl. Phys.* **39**, 2105 (2006).
23. B. B. Wang, J. J. Xie, Q. Z. Yuan and Y. P. Zhao, *J. Phys. D: Appl. Phys.* **41**, 102005 (2008).
24. S. A. Kulinich and M. Farzaneh, *Appl. Surface Sci.* **255**, 4056 (2009).
25. D. Q. Li, *Colloids Surfaces A* **116**, 1 (1996).
26. R. Tadmor, *Surface Sci.* **602**, L108 (2008).
27. A. Marmur, *J. Colloid. Interface Sci.* **186**, 462 (1997).
28. P. Brunet, F. Lapierre, V. Thomy, Y. Coffinier and R. Boukherroub, *Langmuir* **24**, 11203 (2008).

Plasma Modification of Polymer Surfaces and Their Utility in Building Biomedical Microdevices

Shantanu Bhattacharya [a,*], **Rajeev Kr. Singh** [a], **Swarnasri Mandal** [a], **Arnab Ghosh** [a], **Sangho Bok** [b], **Venumadhav Korampally** [b], **Keshab Gangopadhyay** [b] and **Shubhra Gangopadhyay** [b]

[a] Department of Mechanical Engineering, Indian Institute of Technology, Kanpur-208016, India
[b] Electrical and Computer Engineering, University of Missouri, Columbia, MO 65211, USA

Abstract
Polymers are widely used in micro-systems for biological detection and sensing and to provide easier alternatives for fabrication of Biomedical Micro-devices (BMMDs). The most widely used polymeric system amenable to micro-fabrication is silicone rubber, particularly poly(dimethylsiloxane) (PDMS). The principal advantage that silicone rubber offers is its ability to get replicated with high aspect ratios by micro-molding. In addition to PDMS, other polymer systems, like resists or epoxies, find extensive use in micro-fabrication providing many aspects such as good interlayer bonding, selective patterning, modified physical properties like variable electrical or optical properties, etc. Most polymer systems are amenable to rapid changes in their surface energies as they are exposed to gas plasmas or UV radiation. Such changes can sometimes be reversible and the exposed surfaces can regain their original configuration with time called hydrophobic recovery. In general, polymer surfaces after such external stimuli become constitutionally highly dynamic and this makes them well suited to prominent applications in fabrication of BMMDs. Our group has extensively worked in the area of polymer surface modification by external stimuli and its characterization and in this paper we have attempted to review some of the group's work.

Keywords
Biomedical microdevices (BMMDs), PDMS, polymers, gas plasma, supercapacitors, bonding, contact angle, FTIR, hydrophobicity

1. Introduction

Biomedical micro-devices (BMMDs) are miniaturized functional systems developed at the microscopic length scale that can be used in surgery, therapeutic management and biomedical diagnostics. These devices have many advantages over their counterparts. They significantly improve the efficiency with smaller samples, faster response time, reduced use of reagents, reduced power requirements, in-

---

* To whom correspondence should be addressed. E-mail: bhattacs@iitk.ac.in

Adhesion Aspects in MEMS/NEMS
© Koninklijke Brill NV, Leiden, 2010

creased speed and accuracy of analysis, increased portability for field usage, and better performance [1].

Conventional semiconductor processing which evolved from the microelectronics industry was primarily adopted for developing the fabrication technology for microdevices and MEMS and the goal of these conventional semiconductor processes was basically miniaturization and densification of 2-dimensional or 2 and a 1/2 dimensional features on planar substrates. The 2 and 1/2 dimensional features are so named because in microsystems realizing feature thicknesses above a certain critical value is challenging. This problem was eliminated by the advent of processes like stereolithography and LIGA (Lithographie Galvanoformun Abformung). MEMS technology has been successfully commercialized in products such as micro-sensors and micro-actuators. However, the extension of silicon based devices to biomedical applications has certain limitations although the majority of the biomedical devices were first fabricated using silicon and glass [2]. The major drawback for silicon based BMMDs is their increased cost and also low levels of biocompatibility. BMMDs for diagnostics and drug delivery are often of "use and throw" kind due to contamination hazards and, therefore, their cost increases tremendously [3]. Their single use and high material cost as well as high processing cost in cleanroom manufacturing makes these devices extremely high-end and unsuitable for mass production.

Thus polymers offer an alternative route for such devices and many disposable biomedical gadgets are made up of polymers. Thus fully polymeric BMMDs or a combination of polymers with silicon/glass platforms (popular as hybrid devices) promises to solve the major problems with silicon such as biocompatibility, total cost, disposability, etc. [3]. In the past few years a number of polymeric BMMDs have been fabricated using materials such as, poly(dimethylsiloxane) (PDMS), polystyrene (PS), polyethylene (PE), poly(ethylene terephthalate) (PET), polycarbonate (PC), SU8, poly(etheretherketone) (PEEK) and poly(methyl methacrylate) (PMMA) [4]. This area is also known as soft lithography and was basically shown for the first time by Whitesides group at Harvard University and involves shaping and working with soft polymeric materials. Several fabrication techniques are used such as replication and molding, micro-contact printing, micro-capillary molding, compression molding, injection molding, nano-imprint lithography, dip pen lithography, etc.

Replica molded silicone rubber, PDMS, has been used in research laboratories all over the world as one of the most favourable alternatives and has been successfully used in diverse applications such as micro-fluidics, genomics, proteomics, metabolomics, etc. One good aspect of this polymer is its ability to change the chemical nature of its surface quickly on exposure to gas plasmas or ultraviolet radiation. This chemical change and also the succeeding dynamic behavior of the polymer surface lead to a variety of novel effects and processes which are extremely useful in fabrication of BMMDs.

This article is intended to provide an insight into the changes that occur on polymer surfaces on plasma exposure and its utility in fabrication with such materials. It begins with the characterization studies of oxygen plasma exposed polymer surfaces like methyl silsesquioxane (spin-on-glass films, SOG) and poly(dimethylsiloxane) (PDMS) which forms a basic model for understanding of the post-exposure events in such polymeric systems. The change in surface chemistry of these polymeric materials leads to high surface energy and a wettable surface. This forms the basis for patterning on such surfaces. This plasma exposure technique also changes important physical properties of the polymeric materials and their surfaces such as their dual layer capacitance and also fluid transport properties. Polymer matrices with embedded graphite nano-particles are excellent demonstrators of the variable dual layer capacitance effects and such capacitance changes drastically on exposure to oxygen plasma. Such materials make excellent sensing electrodes for electrochemical sensing as well as for charge storage and can contribute greatly to the development of super-capacitors. Also, the altered chemical properties of the polymer surface are very important to promote interlayer bonding, especially in multi-level architectures. The bonds so realized are silanol bonds and the bonding mechanism is irreversible. The surface energy change characterized by measuring the contact angle by the sessile drop method is used for determining the plasma parameters which can cause interlayer bonding. We have observed that the parameter window available for good interlayer bonding is small and any change in parameters outside the window results in a bond failure. Another aspect that this article highlights is the behavior of the post-exposed polymer surfaces towards biological entities. One of our works has clearly demonstrated a higher amplification efficiency using polymerase chain reaction starting with only a very few initial ds-DNA template strands using a microchip reactor [1]. Our findings suggest that the post-exposure hydrophobic surface recovery of SOG or PDMS surfaces forming the walls of the reaction chamber leads to a lower non-specific adsorption of ds-DNA molecules. This also endows a reusability aspect to such polymerase chain reaction (PCR) microchips. We also review the work on nano-structured polymeric films prepared by a combination of a highly volatile polymer phase mixed with another polymeric material. The films show very low refractive indices and are highly hydrophobic which makes them good cladding materials for liquid core wave guides.

2. Plasma Generation and Interaction with Surfaces

2.1. Plasma Basics

A plasma is defined as an electrically neutral system composed of positive and negative charge carriers. The first form of plasma observed was the glow discharge, in which equal numbers of positive ions and electrons were present. Plasmas differ greatly in many respects based on several parameters such as pressure, charge particle density, and temperature. Majority of the universe exists in a plasma state

[5], including the stars, which are almost completely ionized because of their high temperatures. In stellar plasmas the ionization is thermally induced and the temperatures of the neutral and charged species are in equilibrium. Laboratory produced plasmas seldom have such an equilibrium state since the production techniques usually involve nonequilibrium processes, which maintains the ionization by raising some of the charged species to a higher temperature than the neutral species. The gas discharge is the most common laboratory phenomenon where a gas or vapor becomes electrically conducting by means of an applied electric potential.

In a glow discharge the electron temperature is significantly higher than the gas temperature surrounding it. Low electron energy plasmas are also called 'cold plasmas'. They include interstellar and interplanetary space, the earth's ionosphere, and some alkali vapor plasmas.

2.2. DC Glow Discharge

The term 'glow discharge' refers to the light given out by a plasma system. The reactor comprises of two parallel plates inside a chamber, which is attached to a vacuum pump and a gas inlet. The plates are attached to a DC supply through a shielding inductor and are also attached to a high voltage source through a switch. The typical pressure for a plasma process is 1 Torr. A high input voltage is provided to generate an arc discharge. At 1 Torr pressure a voltage of the order of 800 V produces an electrical discharge for an electrode spacing of 10 cm [7]. Before the arc is struck, the gas acts as an insulator. When externally supplied voltage is fed the field in the reactor above a certain voltage exceeds the breakdown field of the medium resulting in a conducting path of ions and electrons, and there is an arc through the medium between the two electrodes. This arc causes the following reactions to occur in the medium assuming that the inflow consists of a bimolecular gas XY:

Dissociation	$e^* + XY \rightleftarrows X + Y + e,$
Atomic ionization	$e^* + X \rightleftarrows X^+ + e + e,$
Molecular ionization	$e^* + XY \rightleftarrows XY^+ + e + E,$
Atomic excitation	$e^* + X \rightleftarrows X^* + e,$
Molecular excitation	$e^* + XY \rightleftarrows XY^* + e.$

The 'e' stands for an excess electron and the 'E' stands for energy, e^* indicates an electron with high kinetic energy, X^*, Y^* and XY^* represent high energy monoatomic or diatomic species, respectively.

In general, the superscript '*' represents species whose energy is higher than the ground state. Dissociated atoms and molecules are known as free radicals and are extremely reactive. Ions are the charged atoms or molecules and they essentially have an excess of positive or negative charges. So as a result of the discharge there would be positive and negative charges in the plasma, which are accelerated towards the electrodes on application of a biasing voltage. Due to their small masses electrons travel much more rapidly than the ions. The slowly moving ions travel across the tube and ultimately strike the cathode generating a stream of secondary electrons, which shoot towards the anode. If the biasing voltage is large enough these high-energy electrons collide inelastically with a neutral atom thus creating more ions. This process of creation of secondary electrons and that of more ions sustains the plasma. The charge distribution in the bulk of the plasma is uniform due to an equal number of positive and negative charges. Near the cathode the electron density is lower than the ion density, because of rapid acceleration of electrons towards the anode. Near the edge of this positive region the electrons gain sufficient kinetic energy to create ions. So as one proceeds towards the bulk of the plasma the ion density keeps on rising. However, due to continuous formation of ions there is a sheath of positive charges accumulated near the cathode, which weakens the electric field and thus the ionization rate. Therefore, the ion density rises to a peak and then falls to a constant value in the bulk of the plasma (Fig. 1(a) and (b)) [5]. The glow of a plasma occurs due to moderate energy electrons, whose energy is less than 15 eV [9, 10], exciting the core level electron to a higher state. Beyond this energy level an electron will normally ionize the gas molecule rather than excite it. This state of excitation is of an extremely small duration (10^{-11} s) [11] and is normalized by the electron jumping back to its original orbit thus giving out an optical emission in the process. Electrons very near the cathode are low kinetic energy electrons, as they require certain distance to accelerate to a higher velocity. Similarly,

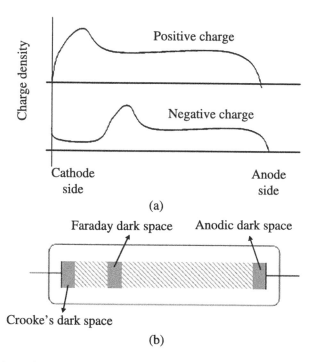

Figure 1. (a) Positive and negative charge densities with variation in position in the plasma. (b) Structure of a DC plasma [10].

near the anode the electron energies are very high and they primarily ionize the gas rather than excite it. Also an anode is an electron sink and the electron density near anode is very low, Fig. 1(a). Thus near both electrodes there are dark regions. The region near the cathode is called the Crooke's dark space (Fig. 1(b)) [5] and is used for most microelectronic fabrication purposes primarily due to the extremely large electric field between the cathode and the sheath of positive ions present close to it. Also there is another secondary dark space called the Faraday dark space near the cathode side where the electrons have gained sufficient kinetic energy to primarily ionize the gas.

Just near the edge of this region, ions that drift and diffuse to the edge are rapidly accelerated towards the cathode and their momentum can be used to drive various physical and chemical processes on the surface of substrates kept on the cathode. The width of this dark space reduces with increase of chamber pressure. This is because with increase in pressure, the mean free paths of the atoms decrease causing an overall compression of the space in the vicinity of the cathode. The opposite behavior is observed on lowering the chamber pressure and the Crooke's dark space width increases. Thus by controlling the chamber pressure one can increase or decrease the momentum with which the ions strike the surface of the cathode. As the Crooke's dark region increases in width with the lowering of chamber pressure the ions attain longer distance to travel before they hit the cathode and thus the momentum transferred to the substrate increases. So, for lowering the impact energy one needs to increase the chamber pressure and *vice versa*. Varying the biasing voltage can also control the impact energy of the ions.

2.3. Radiofrequency Discharge

If the material to be etched on one or more of the electrodes is insulating, then as the ions strike the surface of the substrates that are being plasma modified, they generate secondary electrons and the surface layers become charged in the process. The charge keeps on accumulating on the surface and the electric field is reduced finally leading to the extinguishing of the plasma. One solution could be the use of an AC signal as a biasing voltage. Almost all plasma sources have frequencies in the radiofrequency range (13.56 MHz) [9]. Figure 2(a) [9, 10] shows schematic of a capacitive RF plasma system. Another type of RF discharge called inductively coupled plasma will be described later in this section.

The impedance matching network is used to match the impedance between the chamber and the power source. The capacitor 'C' passes only AC signals thus filtering out the DC component supplied by the power source. When low input frequencies are used, the plasma follows the RF excitation. As the signal frequency exceeds 10 kHz [9, 10], the ions are incapable of following this high frequency due to their excessive mass. The electrons continue to be accelerated. As the polarity of the electrode alternates with every half-cycle the electrons strike both electrodes every half-cycle and this causes the electrodes to develop a negative charge. A plot of the electrical potential along the length of the plasma shows a DC signal for

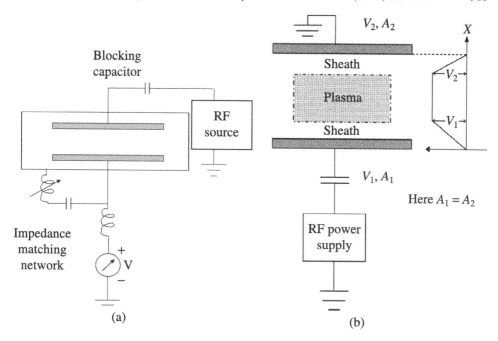

Figure 2. (a) Schematic of an RF plasma system. (b) Plot of DC voltage as a function of position in an RF plasma [9, 10].

most of the bulk of the plasma. This is because most of the bulk of plasma is conducting and does not lead to any substantial potential drop. However, at the edges in the electrode sheath region due to continuous electron depletion there is a large potential drop. This is represented by Fig. 2(b).

The voltage drops V_1 and V_2 can be expressed as follows:

$$V_1 = V_{plasma} - V_{top},$$
$$V_2 = V_{plasma} - V_{bottom}. \tag{1}$$

In case of a parallel plate plasma reactor with unequal electrode areas the relationship between the voltage drops V_1 and V_2 and the surface areas of the electrodes A_1 and A_2 is given by the scaling law

$$\left[\frac{V_1}{V_2}\right] = \left[\frac{A_2}{A_1}\right]^q. \tag{2}$$

Koenig and Maissel [12] found the exponent q as 4, theoretically for low gas pressures and constant current density. Koehler *et al.* [13] found after a series of experiments that $q = 4$ value decreases as the area ratio A_2/A_1 increases.

The objective in most of the micro-fabrication techniques using plasmas is to increase the voltage difference for realizing greater ionic momentum and also bombardment. This objective is achieved by increasing A_2 to obtain a greater value of

V_1. This can be done by connecting the A_2 electrode to the walls of the chamber. This way it is difficult to obtain a voltage ratio greater than $10:1$ [9]. Greater ion bombardment can, however, be obtained by varying the DC bias in Fig. 2(a) by varying the impedance of the LC (Inductor–Capacitor) network.

2.4. Magnetically Enhanced Plasmas

In microelectronic fabrication the ion bombardment or chemical reactions occurring on the substrate surface are a critical aspect. As in a standard plasma, ions and radicals concentrations are only a small fraction of the total gas, therefore increased concentrations of ions and radicals are highly desired. The methods most widely used to produce such kinds of plasmas are magnetically enhanced and electron cyclotron plasmas. In an applied magnetic field the charged particles experience a Lorentz force in a direction perpendicular to both velocity and magnetic field as

$$\mathbf{F} = q\mathbf{V} \times \mathbf{B}, \tag{3}$$

$\mathbf{F}$ is the Lorentz force, q is the total charge on the particle, $\mathbf{V}$ is the velocity of the particle and $\mathbf{B}$ is the strength of the magnetic field.

In such cases, the ambient field induces circular motion on the charge carriers whose radius is directly proportional to the mass and velocity and inversely proportional to the charge and magnetic field. As the ions are heavy they have a large radius of motion and they only pass the plasma with minor deflections and do not move in a complete circle. The electrons, on the other hand, having very small mass have a small radius of motion and, therefore, move through the plasma in a helical path which is many times larger than the normal path in absence of the magnetic field. Thus for a fixed mean free path, the opportunity for impact ionization is much greater in the presence of the magnetic field and thus more ions and radicals are created. Sometimes, the secondary electrons emitted off the cathode are trapped using the magnetic field and returned to the cathode in cycles (Fig. 3) [9].

2.5. Plasma Etching

Etching is the act, art, or practice of engraving by means of an agent, which eats away lines or surfaces left uncovered in metal, glass, etc. An important aspect in

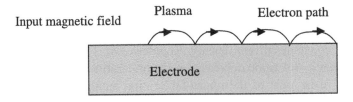

Figure 3. Form of a magnetically enhanced plasma [6].

etching of microstructures is the extent of the undercut or lateral etches. Numerically, this is defined by a parameter called etch anisotropy given as

$$A = 1 - \frac{R_l}{R_v},\qquad(4)$$

A is the etch anisotropy, R_l is the lateral etch rate and R_v is the vertical etch rate.

High-definition structures have a very high value of etch anisotropy. As the etching becomes more and more isotropic R_l becomes comparable to R_v and the etch anisotropy is reduced. Chemically reactive vapors and the reactive species in a glow discharge plasma can be used as etchants effectively. Primarily the dry etching techniques can be categorized into two groups, vapor etching and plasma assisted etching.

2.6. Plasma Assisted Etching

Under this class of etching low-pressure glow discharge plasma having abundance of ionic species and free electrons is directed on the substrate kept at one of the electrodes. Plasma etch processes are easier to start and stop as compared to the other physical or wet etching processes. The etching would stop as soon as the driving voltage which creates plasma is switched off. Also, plasma etch processes are indifferent to small changes in the substrate surface temperature, produce less waste and process byproducts and have anisotropy which can be manipulated to have the desired effect on the substrate. In case of plasma etching the etching process occurs due to physical damage, chemical damage or a combination of both, to the surface. Figure 4 shows a general scheme of classification of the various regimes in a plasma etch process based on chamber pressure.

Any plasma assisted etching process must follow five process steps to successfully etch any substrate. These are: (1) A feed gas introduced into the chamber must be broken down into chemically reactive species by the plasma. (2) These species must diffuse to the surface of the substrate and be adsorbed. (3) Once on the surface they may move around until they react with the surface atoms. (4) The reaction products thus formed must be desorbed. (5) These should then be diffused away from the substrate, and be transported by the gas stream out of the etch chamber. In a typical plasma etch process the surface of the film to be etched is subjected to an incident flux of ions, radicals, electrons and neutral particles.

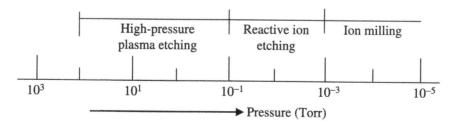

Figure 4. Types of plasma assisted etching in different pressure ranges [10].

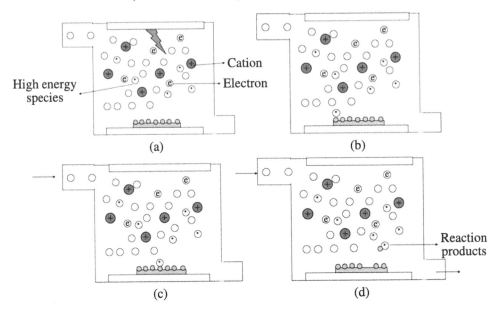

Figure 5. Various steps involved in a plasma etching process. (a) Plasma generation. (b) Free radicals adsorb on the sample surface. (c) Free radicals search for active sites on the sample surface and react with substrate. (d) The product of this reaction is in gas phase and flows out of the plasma chamber [10].

This process is illustrated in Fig. 5. The etch rate is determined by slowest of these steps.

3. Surface Characterization Techniques

An exposure to gas plasmas alters polymer surfaces greatly and the changes thus introduced are physico-chemical in nature. Physical changes may be related to formation of surface cracks and trenches of a few hundreds of nanometers in size. Chemical changes may be related to the surface groups which may drastically alter the surface energy. The post-exposed surfaces are highly dynamic as both the physical or chemical properties keep on changing with time. Thus various real-time surface characterization techniques are often employed to closely study the surface dynamics and as such the techniques need to be well understood. This section will focus on the various available techniques that can give a thorough idea of the polymer surfaces after their interaction with gas plasmas. Some commonly used techniques include time-of-flight secondary ion mass spectrometry (ToF-SIMS), X-ray photoelectron spectroscopy (XPS), Atomic force microscopy (AFM), scanning electron microscopy (SEM), attenuated total internal reflection Fourier transform infrared spectroscopy (ATR–FTIR), etc. Another very commonly used technique which can provide a map of surface energy is contact angle analysis using the sessile drop method. There are numerous techniques which have explored the surface properties such as roughness, polarity, dynamic chemical composition and surface energy of surfaces. Brief descriptions of the main techniques are given below.

3.1. Time-of-Flight Secondary Ion Mass Spectrometry

In this technique a pulsed, highly focused primary ion beam is directed at the surface causing the emission of secondary charged and neutral fragments from surface and near-surface regions. Ga, Bi or C_{60} ion sources are frequently used for a good spectral yield. Positively or negatively charged secondary ions within a narrow energy band are extracted from the surface region and mass-analyzed using a time-of-flight analyzer. The resulting spectrum plots secondary ion intensity as a function of charge/mass ratio. This technique is highly useful for providing the elemental, isotopic and molecular information at extremely high molecular sensitivity [14, 15]. The surface chemistry of plasma exposed polymers is revealed by the molecular information obtained through this technique.

3.2. X-ray Photoelectron Spectroscopy

XPS is a surface measurement technique wherein depths of the order of 5–10 nm can be analyzed. Similar to the ToF-SIMS technique, this also requires the surface to be measured in ultrahigh vacuum to minimize undesirable contamination. X-rays are generally derived from monochromatic X-ray sources or soft X-ray synchrotron beam lines and irradiate the sample surface to cause emission of photoelectrons from the near-surface region. The kinetic energy of these electrons is determined by using a hemispherical sector analyzer, and the electron binding energy is calculated. The spectrum produced show photoelectron intensity as a function of binding energy and essentially maps out the electronic structure of the parent atoms. Characteristic peaks in the spectrum correspond to the electronic core levels in the atoms in the near-surface region and can be used to identify the species present as well as to quantify the relative surface composition. Chemical shifts and curve fitting of peak envelopes with multiple contributions allows the chemical state of surface species to be identified, for example, the oxidation state or bonding environment. XPS also provides quantitative and qualitative information for all elements except hydrogen and helium [15].

3.3. Atomic Force Microscopy (AFM)

AFM finds wide applications for determining the surface topography of a variety of surfaces and its working principle is basically the interrogation of the topography of any surface *via* raster scanning or *via* point-wise measurements of the probe–surface interactions. A very fine pointed probe tip attached to a cantilever film is used for this purpose. The different scan modes that can be executed include probe in direct contact, probe in close proximity but not in contact, or probe vertically stationary or vibrating with high frequency with respect to the scan surface. The technique is primarily used for measuring the surface topography and is able to achieve an atomic resolution in the vertical plane and about 0.01 nm in the horizontal plane

[15]. Post plasma exposed surfaces can be very well scanned with this level of resolution.

3.4. Scanning Electron Microscopy (SEM)

In this technique secondary electrons that are generated from a given surface are used to image the topography of the surface. As the electron beam is rastered on a given surface the secondary electrons generated are collected using an electron detector with a slightly higher surface potential. The images so generated are with a few nanometers spatial resolution [16]. Thus this technique can provide a fairly good idea of the textural change of the plasma treated surface of the substrate material.

3.5. Attenuated Total Reflectance Fourier Transform Infrared Spectroscopy (ATR–FTIR)

The attenuated total reflection Fourier transform infrared spectroscopy (ATR–FTIR) is a commonly used method to characterize the chemical structure of the surface [17]. The ATR technique enables identification of specific molecules and groups located within 100 nm from the surface layer. The process is carried out by passing an infrared radiation into an infrared transmitting crystal to achieve multiple reflections between the ATR crystal and the surface under investigation. ATR–FTIR has been used to detect interactions like chemical bonds between proteins and nitrogen plasma treated polypropylene. This bonding increased the adhesion between the nitrogen treated polypropylene and the hybrid hydrogel [18]. Similarly, the post plasma exposed surface of a spin-on-glass film has been observed to have a sudden increase in surface hydroxyl groups (Si–OH) upon exposure to plasma and then gradual increase in the surface methyl groups (Si–CH$_3$) with time as the post-exposed surface is kept in vacuum [19].

3.6. Contact Angle Analysis

The interfacial tensions at solid–vapor interfaces and solid–liquid interfaces are important parameters in many areas of applied science and technology. These interfacial tensions are responsible for a variety of surface behaviors and properties of different interfaces. However, because of the absence of mobility, a solid phase is very different from a liquid–fluid interface, and hence solid surface tension cannot be measured directly. Therefore, an indirect method by measuring contact angle of a liquid drop on the solid surface is utilized. In this method the angle between the liquid drop and the solid is measured and from this one can determine the surface energy which per unit area is the surface tension force. Various approaches to determine surface energy of solids have been detailed by Etzler [20].

   The contact angle of a drop of liquid on the surface reflects an interplay of the interfacial tensions of three interfaces: the solid–vapor interface, the liquid–vapor interface, and the solid–vapor interface. If we suppose that a drop is in equilibrium with the surface and vapor, then by the definition of equilibrium, an infinitesimal

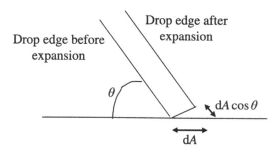

Figure 6. Infinitesimal expansion of a drop on a surface.

change in area, dA should produce a zero change in free energy. Moreover, suppose a water droplet dispensed over a surface expands to its final shape depending on the surface energy (Fig. 6).

Let us also suppose that the area of the solid–liquid interface increases by dA while that of the solid–vapor interface increase by $(\cos\theta)\,dA$ due to the expansion of this water droplet. Each interface has a specific interfacial tension: γ_{lv} represents the interfacial tension at the liquid–vapor interface, γ_{sv} represents the interfacial tension at solid–vapor interface and γ_{sl} represents the interfacial tension at the solid–liquid interface. The sum of the free energy changes due to the infinitesimal change in area must be zero, represented mathematically by equation (5):

$$\gamma_{sl}\,dA - \gamma_{sv}\,dA + \gamma_{lv}\,dA\cos\theta = 0. \tag{5}$$

On rearranging one obtains

$$\cos\theta = \frac{(\gamma_{sv} - \gamma_{sl})}{\gamma_{lv}}. \tag{6}$$

Equation (6) is known as the Young's equation and is useful for calculating contact angles. Unfortunately two of the three interfacial tensions, i.e., γ_{sv} and γ_{sl} are very difficult to determine. The value γ_{lv} is easy to measure and γ_{lv} values for a large number of liquids are available as standard tables. There are three main techniques [10, 17] to determine the contact angles, viz., (a) sessile drop method, (b) captive bubble method, and (c) tilting plate method.

3.6.1. Sessile Drop Technique

This method is based on the equilibrium of an axisymmetric sessile drop on a flat, horizontal, smooth, homogeneous, isotropic, and rigid solid (Fig. 7(a)). Contact angle measurements on polymer surfaces are not only influenced by the interfacial tensions according to Young's equation but also by many other factors, such as surface roughness, chemical heterogeneity, sorption layers, molecular orientation, swelling, or low-molecular weight constituents in the polymer material. These effects have to be considered when contact angle measurements are used to calculate the surface energy of solid polymers. However, if the interest is to compare contact angles of identically prepared surfaces then these factors can be neglected.

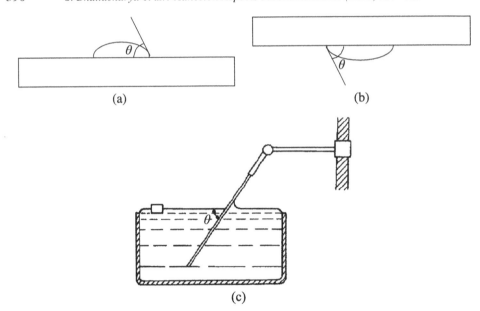

Figure 7. Different techniques for contact angle measurement. (a) Sessile drop technique. (b) Captive bubble technique. (c) Tilting plate technique.

3.6.2. Captive Bubble Method

In order to study highly hydrated polymer layers, captive bubble contact angle technique has been used. As in the case of sessile liquid droplets, the contact angle of the captive bubble is normally measured by the conventional goniometry technique. In this technique a droplet adheres to a surface and hangs off the surface forming a certain angle (Fig. 7(b)).

3.6.3. Titling Plate Method

If a moderately large area of flat solid surface is available, contact angles are usually measured directly from a projection of a sessile drop of the liquid. Alternatively, the tilting plate method shown here can be used; the angle of the plate is adjusted so that the liquid surface remains perfectly flat right up to the solid surface (Fig. 7(c)).

Out of these the most commonly used method for contact angle measurement for determining the surface energy is the sessile drop technique.

4. Plasma Modification of Silicone Rubber (PDMS) and Bonding between Plasma Modified Surfaces

Silicone based rubber PDMS is primarily used in research laboratories all over the world for building chip-based microfluidic devices fabricated using replication and molding processes. The replication technique for device fabrication mostly uses bonding between two or more layers which is mostly achieved by plasma exposure and increase in the surface energy levels of PDMS or glass surfaces. So, it is very critical to study the chemical behavior of the plasma exposed surfaces primarily as the bonding between the surfaces is chemical in nature. As discussed before the changes in the surface texture and chemistry occuring due to plasma exposure of silicone rubber have been widely studied by many research groups using techniques described earlier [6]. Garbassi *et al.* [22] showed that the oxidation of the surface layer increases the concentration of hydroxyl groups [23]. As the silanol groups (Si–OH) are polar in nature, they make the plasma exposed surface highly hydrophilic and this can be observed by a decrease in the advancing contact angle of de-ionized water [16]. These silanol groups then condense with those on the other surface, when two such surfaces are brought into conformal contact [24]. For both PDMS and glass these reactions yield Si–O–Si bonds after loss of a water molecule. These covalent bonds are the basis of the irreversible bonding between the exposed surfaces as they are brought in conformal contact [25]. These irreversible seals have been observed to withstand pressures in the range of 344–413 kPa (50–60 psi) and are practically inseparable. Although the various schemes for formation of these bonds have been very well devised through experiments and hypothesis earlier [26], our group for the first time demonstrated that there exists a common scale based on the contact angle studies of the surfaces of exposed PDMS from which we can predict the right exposure level for successful bonding between two or more layers [27]. Through this work we have demonstrated that the best way to find out quickly the right plasma exposure parameters which would create a thin layer of high density silanols on the exposed PDMS surface (a necessary condition for optimum bond strength between the participating surfaces) is through contact angle analysis. We have determined the bond strength by looking at the separation pressure of a micro-blister of fixed dimension (Fig. 8(a)) which we replicated on a PDMS surface [27]. One of the reports on the blister test was by Allen and Senturia [28]. The micro-blisters have been fabricated in this experiment by using a standardized microfabrication technique wherein a black and white mask is used (Fig. 8(b)). The photopatterned features on these surfaces have been further

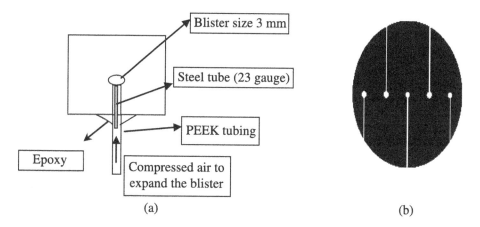

(a)   (b)

Figure 8. (a) Blister assembly. (b) Mask design for microfabricating blisters [18, 27].

replicated on PDMS which is eventually plasma exposed and bonded to another replicated PDMS surface, glass (silicon dioxide) or spin-on-glass (SOG) coated on silicon surface. The chemical name of SOG is also methyl silsesquioxane (MSSQ) and it too is comprised of a highly methylated surface as any other silicone rubber.

We have obtained bond strengths of PDMS–PDMS, PDMS–glass, PDMS–SOG systems. We have found excellent correlation between the contact angles and the bond strengths and have been able to obtain universal curves for these different systems. These curves would provide us with the contact angle values for which the system has maximum bond strength. Table 1 summarizes all the experimental values of bond strengths and contact angles for variations in different plasma parameters for the different surfaces as obtained in this experiment. The bond strength would decrease with rise in contact angle and *vice versa*. Figure 9(a)–(c) shows these correlation trends for the different bonding systems.

Our findings reveal that for PDMS–PDMS and PDMS–glass the bond strength is maximized as the contact angle goes below 10° on the plasma exposed surfaces. For the SOG surface it is maximized below a surface contact angle of 5°. Another important aspect that we have observed is the hydrophobic recovery of the plasma exposed PDMS. We have observed during these experiments that if the exposed PDMS surface is allowed to recover by keeping for a finite length of time without immediately bringing it into conformal contact with another bonding surface then the bond strength is substantially reduced. This is owing to the formation of narrow surface cracks on the PDMS surface which allows the short and long chain oligomers from the bulk of the material to ooze out slowly through these cracks and populate the surface. The micro-cracks can be observed with scanning electron microscopy (SEM). Figure 10(a) and (b) shows SEM images of pre and post-exposed surfaces of spin coated thin films of PDMS. The micro cracks are only observed in the post-exposed surface.

We have also done extensive characterization of the post-exposed SOG and glass (silicon dioxide) surfaces with ATR–FTIR and have found that the areas under the hydroxyl peaks for these surfaces are quite different which is also in agreement with earlier literature where the silanol group density on a surface was found to be the determining factor for bond strength. Figure 11 shows a comparison of the areas under the hydroxyl peaks for exposed SOG and silicon dioxide surfaces. As can be observed the area under the curve for exposed SOG is about 10 times larger than for the silicon dioxide surface.

5. Micro-patterning of Polymeric Surfaces

In this section we review the preparation and plasma assisted surface patterning of highly porous films formulated from polymer-nanoparticle composites. The as-prepared films are highly hydrophobic in nature but a sharp increase in surface energy could be obtained after exposure to low power plasma conditions. The porous films are realized by a simple method wherein a colloidal dispersion of poly(methyl

Table 1.

Contact angle and bond strength for a variety of plasma pressures, powers and exposure times for different surfaces [18, 27]

S.N.	Type of plasma etcher	Chamber pressure (mTorr)	RIE power (W)	Time of exposure (s)	Surface hydrophilicity measured by contact angle		Bond strength measured by separation pressure (kPa)	
					Surface type	Contact angle error ± 2.5°	Bonded system	Pressure error ± 13.8 kPa
1	ICP	20	20	30	PDMS	26	PDMS–PDMS	151
2		40				17		179
3		50				14		193
4		150				14		193
5		250				14		207
6		350				5		276
7		500				5		344
8		600				5		372
9		700				2 (avg. val.)		400
10	PECVD	900	10	30		8		248
11		100				19		158
12	ICP	30	20	30	Glass	19	PDMS–Glass	165
13		150				15		331
14		250				15		358
15		350				12		379
16		500				5		468
17		600				5		455
18		700				5		482
19		1000				2 (avg. val.)		510
20	ICP	700	5	30	PDMS	20	PDMS–PDMS	96
21			10			8		331
22			20			2 (avg. val.)		344
23			30			9		303
24			50			10		262
25			75			15		179
26			100			22		97
27			125			23		83
28			150			28		69
29	PECVD	900	10	30	PDMS	7	PDMS–PDMS	255
30			30			12		207
31	ICP	1000	10	30	Glass	16	PDMS–Glass	344
32			15			15		372
33			20			2 (avg. val.)		468
34			30			7		393
35			50			10		358

Table 1.
(Continued.)

S.N.	Type of plasma etcher	Chamber pressure (mTorr)	RIE power (W)	Time of exposure (s)	Surface hydrophilicity measured by contact angle		Bond strength measured by separation pressure (kPa)	
					Surface type	Contact angle error ± 2.5°	Bonded system	Pressure error ± 13.8 kPa
36			75			14		344
37			100			17		317
38			125			19		289
39			150			28		276
40	ICP	700	20	5	PDMS	28	PDMS–PDMS	179
41				10		27		207
42				15		18		255
43				20		5		351
44				25		2 (avg. val.)		344
45				30		5		296
46				40		8		172
47				45		11		158
48				50		11		138
49				60		17		138
50	PECVD	900	10	35	PDMS	9	PDMS–PDMS	220
51				60		12		241
52	ICP	1000	20	10	Glass	18	PDMS–Glass	331
53				20		5		496
54				30		2 (avg. val.)		482
55				40		10		344
56				45		15		317
57				50		15		303
58				60		16		241
59	PECVD	100	20	35	SOG	34	SOG–PDMS	207
60		400				25		276
61		900				5		538
62		1200				18		310
63		1400				75		34
64	PECVD	900	5	35	SOG	31	SOG–PDMS	276
65			10			5		552
67			20			4		586
68			30			32		358
69	PECVD	900	20	20	SOG	10	SOG–PDMS	414
70				30		4		524
71				35		4		552
72				40		9		448
73				45		8		400

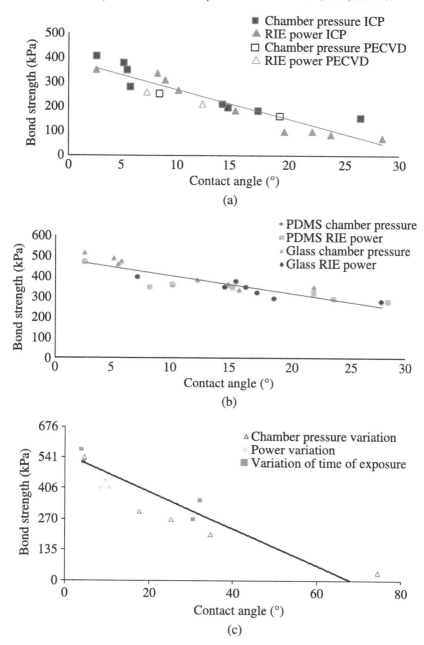

Figure 9. Universal curves of bond strength *vs* contact angle for (a) PDMS–PDMS system, (b) glass–PDMS system, (c) SOG–PDMS system [18, 27].

silsesquioxane) (PMSSQ) nanoparticles in a polymer matrix poly(propylene glycol), PPG are coated on hydrogen passivated silicon substrates and the nanoparticle-polymer composites are decomposed by heating [29]. The increase in entropy of the

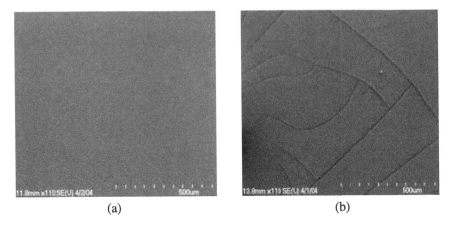

(a)　　　　　　　　　　　　　　　(b)

Figure 10. (a) Undamaged PDMS surface: plasma treated at 10 W power, 0.9 Torr pressure and 30 s exposure, ×110. (b) Cracked PDMS surface after plasma treatment at 10 W power, 0.9 Torr pressure for 70 s exposure, ×110 magnification.

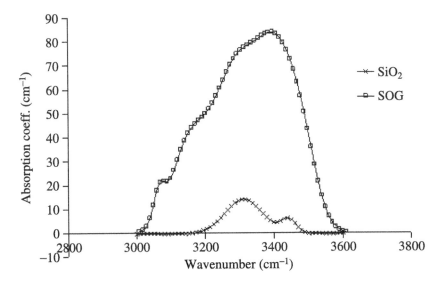

Figure 11. Comparison between SOG and SiO_2 surfaces immediately after plasma exposure (area under the OH peak for SOG = 27 745 (a.u.), area under the OH peak for SiO_2 = 2045 (a.u.)) [1].

system when heated above the decomposition temperature of the polymer provides a degree of mobility to the initially confined nanoparticles. We have observed in this work that when the nanoparticles used in these polymer systems are functionalized with different functional groups helping them to crosslink with each other, and following this the temperature is increased to the decomposition temperature it results in an increased entropy. Figure 12(a) shows schematic of the PMSSQ–PPG system under high temperature curing conditions. The properties of the resulting films, porosity (refractive index) and thickness are highly dependent both on the

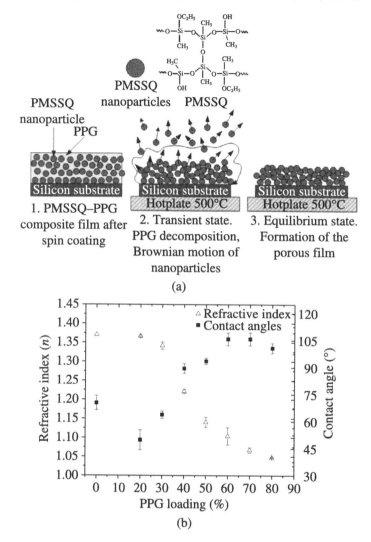

Figure 12. (a) Schematic of PMSSQ–PPG system under high temperature curing condition. (b) Refractive index and contact angles as a function of PPG loading [29].

curing/calcination temperature as well as on the PPG content. Normally all the PMSSQ based films are naturally hydrophobic with a water contact angle of ∼90°. The nanoscale roughness arising from the surface porosity and the nano-particulate nature of the films further increases the contact angle of the films (Fig. 12(b)).

PMSSQ is an organosilicate with the basic empirical formula $(CH_3–SiO_{1.5})_n$, with the Si and O atoms located at the corners of a deformed cagelike structure and one methyl group attached to each Si atom. This highly hydrophobic nature of the films can be attributed to the presence of $Si–CH_3$ and $Si–(CH_3)_2$ in the bonded state as reflected in the FTIR spectra taken at different calcinations temperatures

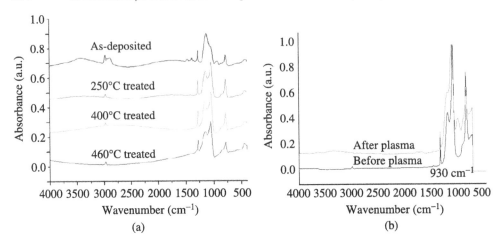

Figure 13. (a) Fourier transform infrared spectra of the PMSSQ–PPG films treated at different temperatures. (b) Fourier transform infrared spectra of the films before and after oxygen plasma treatment [29].

(Fig. 13(a)). The plots show peak at 1275 cm^{-1} indicating stretching vibration of the –Si–CH$_3$ bond, and at 845 cm^{-1} for the –Si–(CH$_3$)$_2$ and 775 cm^{-1} for the –Si–CH$_3$ bending modes. As can be clearly observed the free methyl –CH$_3$ shown by the peak at 2975 cm^{-1} is eliminated due to thermal degradation of PPG within the film when it is calcined above 200°C. We have observed a strong degradation of the bonded methyl on exposure to oxygen plasma of the calcined films and a simultaneous increase in the hydroxyl peak of –Si–OH at 930 cm^{-1}. Figure 13(b) shows the ATR–FTIR spectra of these films before and after oxygen plasma treatment. The broad absorption peak between 3500 cm^{-1} and 3000 cm^{-1} comes from the free moisture adsorption within the highly porous network upon plasma surface modification. The contact angle of the exposed films falls below 5° following plasma exposure and this aspect could be exploited to generate fine micro-patterns which could help in transporting water droplets from the hydrophobic to the hydrophilic region as illustrated in Fig. 14(a)–(d) and thus this patterning can be a basis for fluid transport through surface energy contrast.

6. Plasma Exposed Polymers and Interaction with Biological Entities

The plasma exposed surfaces of polymers such as SOG and PDMS which are quite often used for micro-fluidic architectures recover their original states with time. We have shown with ATR–FTIR that the surface of the SOG coated silicon after heat curing and plasma exposure possesses a dynamic chemical nature. We have found that the post-exposed SOG surface shows an increase in the advancing contact angle with time if it is not brought into immediate conformal contact with another exposed surface. We performed a contact angle study of such a surface. The water contact angle of SOG surface before exposure was 83° which decreased to almost

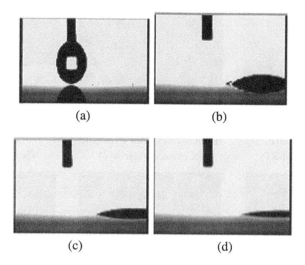

Figure 14. (a)–(d) Droplet dispensed over the hydrophopic region (below the dispenser tip) gradually migrating to the hydrophilic region (towards the right of the image). (These snapshots were successively taken while the droplet moved.)

7° after plasma exposure. Contact angle measurements were taken after 5 min, 1 h, 3 h, 5 h and 2 days. The contact angle increased from 7° after exposure to around 63° after 5 h. No change in the contact angle was observed after 2 days indicating full hydrophobic recovery of the surface and saturation in surface recovery rate after 5 h (Fig. 15). A similar surface recovery was found to occur in PDMS where there is a tendency for the methyl groups to appear on the surface from the bulk of the material due to extensive rearrangement of surface bonds due to surface chain scission reactions [23]. The SOG surface being structurally similar to PDMS should have a similar mechanism of chain scission reactions.

In order to confirm this hypothesis, we recorded the ATR–FTIR spectrum of the SOG surface. A similar study indicating a gradual hydrophobic recovery of PDMS surfaces exposed to corona discharge has been reported earlier [23]. The spectra were taken at 32 cm^{-1} resolution in the 3700–2900 cm^{-1} spectral range covering the –CH and –OH bands. Figure 16(a) shows the Gaussian fits on the absorption data from ATR–FTIR spectra of –CH between 2900 cm^{-1} and 3000 cm^{-1}. A peak has been found at 2970 cm^{-1} which has been assigned to –CH$_3$ (asymmetric) bend stretching vibration [1, 18]. Figure 16(b) shows OH stretching broad band spectra between 3000 and 3600 cm^{-1} for untreated SOG and after exposure to plasma over time [18]. The strong broad absorption band that appears to be at approximately 3500 cm^{-1} (Fig. 16(b)) is attributed to hydroxyl groups, physically adsorbed on the film surface [18, 30, 31]. We found the surface methyl groups to fall to a very low level immediately after exposure. This is followed by a slow recovery of the surface methyl content after 5 h. It has been also observed that the surface hydroxyl groups reach a peak value immediately after exposure and gradually decrease with time.

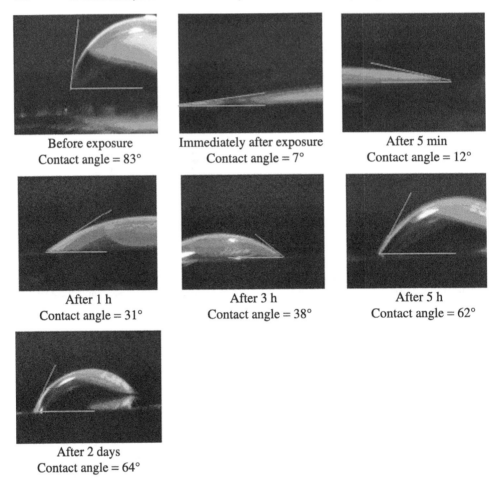

Before exposure
Contact angle = 83°

Immediately after exposure
Contact angle = 7°

After 5 min
Contact angle = 12°

After 1 h
Contact angle = 31°

After 3 h
Contact angle = 38°

After 5 h
Contact angle = 62°

After 2 days
Contact angle = 64°

Figure 15. Contact angle results for post-exposed SOG surface with relaxation time [18].

We have also calculated the total areas under the Gaussian curves fitted over the actual spectral data for both –CH and –OH regions for untreated sample, immediately after treatment, 1 h and 5 h after exposure to plasma (Fig. 17). The area under the CH peak was observed to chang from 4000 arbitrary units (a.u.) to 2000 (a.u.). Simultaneously, the area under the hydroxyl peak was found to change from 0 a.u. to 1400 a.u. The heavily hydroxylated surface corresponds to a contact angle of 7°. The area under the methyl peak rises to 3000 a.u. and for the hydroxyl the peak goes down to 600 a.u. after 1 h of relaxation. The contact angle at this point of time has been found to be 31°. The area under the methyl peak after this does not change much and at the end of 5 h it stabilizes at 3000 a.u. The area under OH peak goes down from 1 h of relaxation to 5 h of relaxation from 600 a.u. to 250 a.u. The contact angle after 5 h of relaxation is 62°, different from the angle of untreated surface (83°). This behavior can be well explained from the area trend. The area under the methyl peak does not get back to its initial 4000 a.u. level which accounts

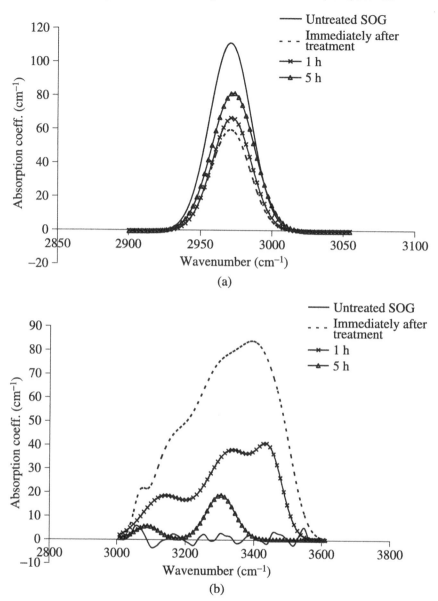

Figure 16. (a) ATR–FTIR spectra of methyl group for untreated SOG, after plasma treatment (immediately after exposure, 1 h and 5 h). (b) ATR–FTIR spectra of OH absorption band for untreated SOG, after plasma treatment (immediately after exposure, 1 h and 5 h) [18].

for a lesser contact angle (62°) after 5 h of exposure. The sudden jump of contact angle from 1 h (31°) to 5 h (62°) can be explained by the reduction of area under the OH peak from 1 h to 5 h. This indicates a direct correlation between the contact angle and the surface hydroxyl groups.

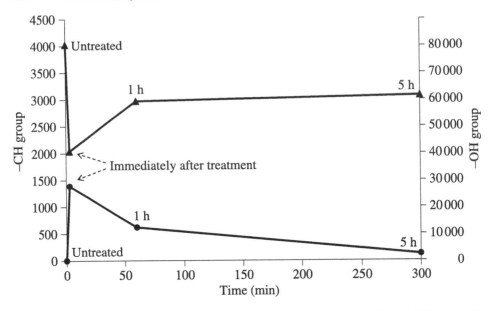

Figure 17. Total areas under the ATR–FTIR spectra in arbitrary units of CH and OH regions for untreated, immediately after treatment, 1 h and 5 h. ▲ is for –CH region and ● is for –OH region [18].

Because of this rapid recovery of the surface, the plasma exposed SOG coating especially on the innerwalls of polymeric microchambers and channels which is left without any conformal contact with the glass or silicon substrate [24] develops a continuous recovery process and becomes increasingly hydrophobic with time. This coating inside such micro-chambers provides a bio-friendly surface owing to its hydrophobic recovery. In an earlier work accomplished by our group we explored such an effect for successfully performing micro-scale polymerase chain reaction in a time domain micro-PCR device [1]. Here we fabricated a hybrid microfluidic device with the base made of silicon and the upper channels and chamber replicated in PDMS with a sandwiched SOG layer. The SOG layer was applied by spin coating on the silicon wafer and was heat cured. This was followed by surface plasma exposure of both the PDMS upper layer and the SOG coating on the silicon substrate after which these were brought into conformal contact. We clearly observed amplifications with initial DNA concentration as small as 0.07 pg/µl which demonstrated the high detection sensitivity of this device. The hypothesis made here to explain the high sensitivity was that due to the negligible non-specific adsorption of the pre-amplified sample DNA on the inner walls of the micro-reactor chamber, all such DNA was freely available in the solution for amplification. Therefore, the micro-chip could amplify even miniscule concentrations of the sample DNA. Figure 18(a) shows schematic of the micro-device and (b) shows a gel image of an amplification trial with 0.07 pg/µl initial DNA concentration and (c) shows the PCR trial on the same chip after washing and without the sample DNA. As there is no band formed in Fig. 18(c) this demonstrates the rigor of the wash cycle on the recov-

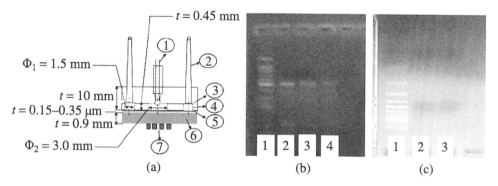

Figure 18. (a) Schematic of the PCR microchip. (Numbers identified in the figure 1: glass housed thermocouple, 2: epoxied inlet/outlet ports, 3: poly(dimethylsiloxane) channels, 4: inlet/outlet reservoirs, 5: SOG layer, 6: thermally oxidized silicon wafer, 7: heaters. (b) Slab gel electrophoresis image of a 0.07 pg/μl of initial sample DNA by amplifying on the on-chip device. (c) Slab gel electrophoresis image of an amplified product without sample DNA on a test chip after washing off the PCR products from the previous run [1].

ered SOG surface which further supports our hypothesis of negligible non-specific adsorption [1].

To further study this non-specific adsorption aspect we incubated FAM labeled RT-PCR products [32] (excitation maximum = 494 nm and emission maximum = 520 nm) inside the PCR chip and then washed it with an elution buffer and found a complete reduction in the fluorescence signal suggesting no non-specific adsorption of ds-DNA on SOG surfaces. In this experiment two PCR devices were fabricated using a 170 μm thick SOG coated glass slide as the base instead of the patterned silicon wafer as was normally the architecture for PCR microchamber [9] (to accommodate imaging modalities of characterization instrument). A 1X50 Olympus inverted fluorescence microscope was used with an emission and an excitation monochromator for characterization of the fluorescence intensity in the device. Ten microliters of FAM labeled RT-PCR products were flown into the microchamber in one of the devices at a rate of 87 μl/min using a syringe pump. The fluorescence level was measured using a photodiode connected to the objective through a monochromator. The response from this photodiode was digitally acquired in the form of voltage data and plotted in time by the computer. Following this, an elution buffer solution [33] (Qiagen Inc., USA) was used to wash off the labeled DNA from the chip in a similar manner as described earlier for 20 min. The corresponding real time change in fluorescence intensity was plotted with time. Figure 19(a) shows plot of the fluorescence intensity change observed in the device during the wash cycle. The intensity changed from 2 V to a constant 0.06 V value after 400 s. At the end of 1200 s, a forced injection of sterile water was provided by utilizing an on-chip plumbing arrangement and then the buffer flow was continued. The residual fluorescence dropped to 0.02 V after the sterile water flow and then stayed constant. The second device in which the labeled DNA had not been flown was injected with the wash liquid in a similar manner as for background measurement purposes. Fig-

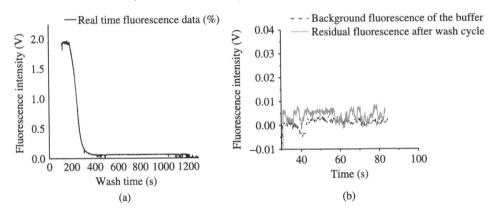

Figure 19. (a) Reduction in fluorescence intensity (V) with wash time. (b) Background fluorescence of the buffer solution and residual fluorescence left over in the micro-chamber [1].

ure 19(b) shows a magnified view of the background fluorescence of the buffer from the second device with the residual fluorescence left over after the wash cycle from the first device. Both these parameters are plotted on the same time scale for an easy comparison. Both values superpose on each other showing that there is no non-specific binding of the labeled DNA inside the chamber or channel.

7. Sol–Gel Entrapped Graphite Nanoparticle Films as Super-capacitors and for Electrochemical Sensing Applications

We have also investigated the effects of plasma exposure on a highly porous film composed of graphite nanoparticles entrapped in a sol–gel matrix. In this work we successfully entrapped nano-sized (average diameter 55 nm with maximum diameter less than 100 nm) graphite particles in a sol–gel matrix and obtained relatively large values of double layer capacitance due to the combination of hydrophobicity of sol–gel with non-hydrophobicity of graphite nanoparticles as well as the enhanced surface area of the films [34, 35]. We also observed that the capacitance increased exponentially when the micro-pores in porous silica dominated the redox reaction resulting in a substantially high double layer capacitance. The capacitance was substantially reduced as the films were exposed to low power oxygen plasma. The graphite/sol–gel electrodes showed a resistivity of 0.052 ohm-cm, suggesting a conductive nature of the carbon nano-structures. The as-prepared graphite/sol–gel thin film electrodes showed a contact angle of 127°. We attributed this to the addition of hydrophobic organo-functional monomers like methyl trimethoxysilane (MTMS), which formed hydrophobic organosilicates as well as micrometer scale roughness arising from the highly porous matrix structure. The pore size distribution of the sol–gel matrix with entrapped graphite nanoparticles was recorded using the BET method and showed two peaks at 1.4 nm (micropores) and around 2.6–3 nm (mesopores). The sol–gel without graphite nanoparticles exhibited peaks at 1.3 nm and 3.3 nm. We further found using SEM that mesopores had a diameter of

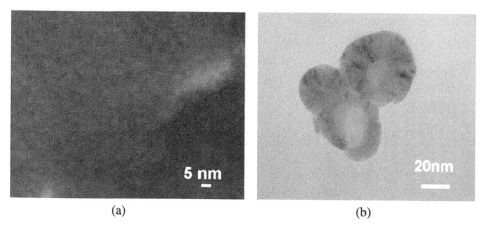

(a)   (b)

Figure 20. (a) Magnified SEM image of sol–gel/nanographite composite film showing mesopores of diameter less than 5 nm. (b) TEM image of spherical graphite nanoparticles of 55 nm average diameter [35].

less than 5 nm (Fig. 20(a)). Spherical nanoparticles of graphite used had an average diameter of 55 nm (Fig. 20(b)).

The double layer capacitance, C, was measured from the current response to a triangular voltage waveform. The electrode area over which the measurements were performed was 0.501 cm^2 and the total graphite mass was around 0.072 mg. Different scan rates (1, 2, 5, 10, 20, 50, 100 and 200 mV/s) for the triangular waveform were used and the change in current was measured from the plateau region of the cyclic voltammogram. The specific capacitance C with units Farads per gram (F/g) was calculated using

$$C = (\Delta i)/(2wm), \tag{7}$$

where w = scan rate, m = mass of graphite nano-particles in the sample, Δi = the maximum difference between the two current readings on the C–V curve. We found out that the specific capacitance of these films was very high, of the order of 36.7 F/g. This high specific capacitance has been attributed to the following reasons: (a) Increased specific surface area because of the use of graphite nano-particles and the micro as well as meso-pores of the sol–gel network. (b) Also the superhydrophobic nature of the sol–gel matrix played a significant role in the high value of the specific capacitance with aqueous electrolytes.

We hypothesized that this large value of capacitance for the graphite nanoparticle/sol–gel composite resulted from non-hydrophobic graphite particles embedded within a hydrophobic porous matrix (Fig. 21(a)). This restricts the solvated ions to tightly confined spaces close to the surface of the graphite electrode, resulting in small separations between the electrolytes and the graphite particle surface and a significant increase in the double layer capacitance (Fig. 21(b)).

The resulting squeezing of the ions within the nanopores brings them closer to the surface causing an increase in the electrochemical capacitance. As low

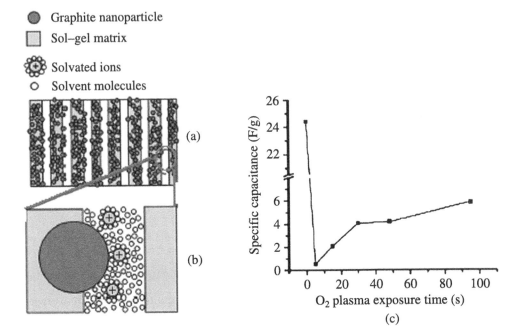

Figure 21. (a) Schematic drawing of the as-prepared films showing the presence of meso/micro-pores with embedded graphite nanoparticles, and (b) enlarged region of the circle in (a) showing the decrease in the distance between solvated ion center and the surface of graphite nanoparticles due to hydrophobicity. (c) Variation in specific capacitance with respect to the plasma exposure time (s) [35].

power oxygen plasma is used for exposing the as-prepared films of the sol–gel/nanographite composites, there is a reduction in the –Si–CH$_3$ bonds, rendering the surface hydrophilic. The films were exposed for different times and the specific capacitance per unit mass was subsequently measured immediately after the plasma exposure. Figure 21(c) shows a plot of the specific capacitance of the films as a function of plasma exposure time. As can be seen from the plot the relatively high value of 24 F/g in the as-prepared sol–gel/nanographite composite films drastically drops to as low as 0.54 F/g at 20 mV/s scan rate after a 5 s exposure to oxygen plasma. This behavior was found to be consistent with earlier hypothesized model. Because of demethylation from the pores the surface suddenly becomes hydrophilic and the squeezing effect of the double layer near this surfaces gone and there is a sudden redistribution of the secondary ions within the nanopores, thereby affecting the double layer around the graphite surface. The charge density of this double layer decreases due to the new charge redistribution resulting in a huge reduction in the capacitance. This is more so as there is a higher density of these nanopores throughout the surface of the film and they certainly outnumber the micropores. As the plasma exposure time of the film is increased there is a slow growth in the capacitance which may be due to the gradual erosion of the film and formation of rough surface on exposure to plasma. To track the demethylation process

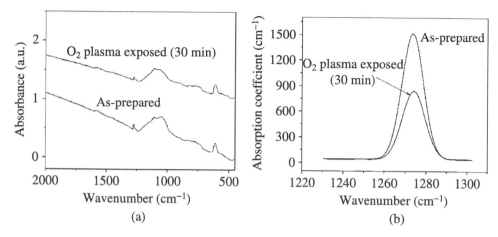

Figure 22. (a) FTIR absorbance spectra of as-prepared and O_2 plasma exposed sol–gel/nanographite film based electrodes. (b) Calculated absorption coefficient in the spectral region of $1280–1250$ cm^{-1} corresponding to Si–CH$_3$ stretching mode [34].

the FTIR analysis was further conducted. Figure 22(a) shows the FTIR spectra of as-prepared sol–gel/nanographite composite films and plasma exposed surfaces and Fig. 22(b) shows the calculated absorption coefficient in the spectral region between 1280 cm^{-1} and 1250 cm^{-1} which corresponds to the Si–CH$_3$ stretching mode.

8. Conclusion

Plasma exposure of polymer surfaces produces considerable chemical changes on their surfaces and this also creates a dynamic environment which can be used for a variety of applications, especially in the fabrication of biomedical micro-devices. We have in this article reviewed some of our group's work in which we have shown the utility of plasma induced changes to realize inter-layer bonding and also the subsequent effects of the hydrophobic recovery of the plasma exposed surfaces.

Acknowledgements

The authors acknowledge the financial support from various funding agencies such as NSF (CRCD), NPB, NIH, Department of Biotechnology (Government of India) and the office of the Dean of Research and Development at the Indian Institute of Technology Kanpur for the work that has been reviewed in this paper.

References

1. S. Bhattacharya, V. Korampally, Y. Gao, M. T. Othman, S. A. Grant, S. B. Kleiboeker, K. Gangopadhyay and S. Gangopadhyay, *J. MEMS* **16**, 1 (2007).
2. N. T. Nguyen, R. Bashir, S. Werely and M. Ferrari, *Biomolecular Sensing, Processing and Analysis*, Vol. 4, p. 93. Springer, Berlin (2006).

3. V. K. Jain (Ed.), *Introduction to Micromachining*, Vol. 1. Narosa Publishing House, New Delhi, India (2009).
4. J. W. Hong, K. Hosokawa, T. Fujii, M. Seki and I. Endo, *Biotechnol. Prog.* **17**, 958 (2001).
5. A. L. Arzimovich, *Elementary Plasma Physics*. Blaisdell Publishing Company, New York (1965).
6. S. A. Campbell, *Science of Microelectronic Fabrication*. Oxford University Press, New York (1996).
7. B. Chapman, *Glow Discharge Processes*. Wiley, New York (1980).
8. L. Spitzer Jr, *Physics of Fully Ionized Gases*. Interscience Publishers, New York (1962).
9. H. K. Yasuda, *J. Appl. Polym Sci., Appl. Polym. Symp.* **42**, 357 (1987).
10. J. S. Judge, in: *Etching for Pattern Definition*, H. G. Hughes and M. J. Rand (Eds). Electrochemical Society, Princeton, NJ (1976).
11. M. Sugawara, *Plasma Etching — Fundamentals and Applications*. Oxford University Press, London (1988).
12. H. R. Koenig and L. I. Maissel, *IBM J. Res. Dev.* **14**, 168 (1972).
13. K. Kohler, D. E. Horne and J. W. Coburn, *J. Appl. Phys.* **58**, 3350 (1985).
14. J. C. Vickerman, *Surface Analysis: the Principal Techniques*. John Wiley & Sons, West Sussex, UK (2003).
15. R. J. B. Reed, *Electron Microprobe Analysis and Scanning Electron Microscopy in Geology*. Cambridge University Press, Cambridge, UK (1996).
16. H. Hillborg and U. W. Gedde, *IEEE Trans. Dielectrics Electrical Insulation* **6**, 5 (1999).
17. R. Snyders, O. Zabeida, C. Roberges, K. I. Shingel, M. Faure, L. Martinu and J. E. Klemberg-Sapieha, *Surface Sci.* **601**, 112 (2007).
18. S. Bhattacharya, Y. Gao, V. Korampally, M. T. Othaman, S. A. Grant, K. Gangopadhyay and S. Gangopadhyay, *Appl. Surface Sci.* **253**, 4220 (2007).
19. H. J. Hettlich, F. Ottenbach, C. H. Mittermayer, R. Kaufmann and D. Klee, *Biomaterials* **12**, 521 (1991).
20. F. M. Etzler, in: *Contact Angle, Wettability and Adhesion*, K. L. Mittal (Ed.), Vol. 3, pp. 219–264. VSP, Utrecht (2003).
21. L. D. Eske and D. W. Galipeau, *Colloids Surfaces A* **154**, 33 (1999).
22. F. Garbassi, M. Morra, L. Barino and E. Occhiello, *Polymer Surfaces. From Physics to Technology*. Wiley, New York, NY (1994).
23. H. Hillborg and U. W. Gedde, *Polymer* **39**, 1991 (1998).
24. M. L. Chabinyc, D. T. Chiu, J. C. Mcdonald, A. D. Stroock, J. F. Christian, A. F. Karger and G. M. Whitesides, *Anal. Chem.* **73**, 4491 (2001).
25. J. C. Mcdonald, D. C. Duffy, J. R. Anderson, D. T. Chiu, H. Wu, O. J. A. Schueller and G. M. Whitesides, *Electrophoresis* **21**, 27 (2000).
26. K. L. Mittal, *Polym. Eng. Sci.* **17**, 467 (1977).
27. S. Bhattacharya, A. Datta, J. M. Berg and S. Gangopadhyay, *J. MEMS* **14**, 590 (2005).
28. M. G. Allen and S. D. Senturia, *J. Adhesion* **25**, 303 (1988).
29. V. Korampally, M. Yun, T. Rajagopalan, P. K. Dasgupta, K. Gangopadhyay and S. Gangopadhyay, *Nanotechnology* **20**, 425602 e-pub (2009).
30. H. A. Benesi and A. C. Jones, *J. Phys. Chem.* **63**, 179 (1959).
31. M. L. Hair, *Infrared Spectroscopy in Surface Chemistry*. Marcel Dekker, New York (1967).
32. J. E. Rice, J. A. Sanchez, K. E. Pierce and L. J. Waugh, *Prenatal Diagnosis* **22**, 1130 (2002).
33. Hints from optimum elution of DNA from spin columns, Issue 4, 1998, *available online at* www.qiagen.com.

34. B. Lahlouh, J. A. Lubguban, G. Sivaraman, R. Gale and S. Gangopadhyay, *Electrochemical and Solid-State Letters* **7**, 338 (2004).

35. S. Bok, A. A. Lubguban, Y. Gao, S. Bhattacharya, V. Korampally, M. Hossain, R. Thiruvengadathan, K. D. Gillis and S. Gangopadhyay, *J. Electrochem. Soc.* **155**, K91 (2008).

Printed and bound by CPI Group (UK) Ltd, Croydon, CR0 4YY

23/10/2024

01778246-0002